Mathematics Today
Twelve Informal Essays

Mathematics Today
Twelve Informal
Essays

Edited by
Lynn Arthur Steen

VINTAGE BOOKS
A Division of Random House
New York

First Vintage Books Edition, September 1980

This material was prepared with the support of National Science Foundation Grant OSS-7526057. However, any opinions, findings, conclusions, or recommendations expressed therein are those of the authors and do not necessarily reflect the views of the NSF.

Library of Congress Cataloging in Publication Data

Main entry under title:
Mathematics today.

Reprint of the 1978 ed. published by Springer-Verlag, New York.
Bibliography: p.
Includes index.
CONTENTS: Steen, L. A.   Mathematics today.—
Hammond, A. L.   Mathematics, our invisible culture.
—Richards, I.   Number theory. [etc.]
1. Mathematics—Addresses, essays, lectures.
I. Steen, Lynn Arthur, 1941-
[QA7.M3447 1980]   510   80-10888
ISBN 0-394-74503-5
Manufactured in the United States of America

Preface

The objective of the present book of essays is to convey to the intelligent nonmathematician something of the nature, development, and use of mathematical concepts, particularly those that have found application in current scientific research. The idea of assembling such a volume goes back at least to 1974, when it was discussed by the then-newly-formed Joint Projects Committee for Mathematics (JPCM) of the American Mathematical Society, the Mathematical Association of America, and the Society for Industrial and Applied Mathematics. Currently, the nine members of the JPCM are Saunders Mac Lane (Chairman) of the University of Chicago, Frederick J. Almgren, Jr. of Princeton University, Richard D. Anderson of Louisiana State University, George E, Carrier of Harvard University, Hirsh G. Cohen of the International Business Machines Corporation, Richard C. DiPrima of Rensselaer Polytechnic Institute, Robion C. Kirby of the University of California at Berkeley, William H. Kruskal of the University of Chicago, and George D. Mostow of Yale University.

The JPCM decided to make production of this volume its first major project and requested the Conference Board of the Mathematical Sciences (CBMS), of which its three sponsoring societies are all member organizations, to approach the National Science Foundation on its behalf for support of the undertaking. A proposal submitted by the CBMS in December 1974 and in revised form in July 1975 was granted by the Foundation in May 1976, and work on assembling the volume got under way. Direction of the project has been carried out by a five-member Steering Committee formed for this purpose by the JPCM. The Steering Committee has consisted of Paul R. Halmos of the University of California at Santa Barbara, Allen L. Hammond of *Science* magazine, Jack C. Kiefer of Cornell University, Harry Schwartz of the New York Times, and Jacob T. Schwartz of New

York University (Chairman). Throughout the course of operations, Lynn A. Steen of St. Olaf College has served as Editor and coordinator, while CBMS headquarters has provided administrative management.

The Steering Committee would like to acknowledge the contributions of all those who have been involved in this essay project. First of all, it is to the individual essay authors that prime credit will be due for whatever success the volume achieves. All who served on the JPCM while the project was in progress deserve thanks for their helpful advice and support. Lynn Steen's able and energetic work throughout as Editor and coordinator was invaluable, as was the role of the CBMS in securing the grant for the project and managing operations under it. The publisher, Springer-Verlag, specifically its production and editorial staff, has cooperated in admirable fashion in the production of the volume. Finally, for itself and on behalf of the JPCM and the CBMS, the Steering Committee would like to express deep appreciation to the National Science Foundation and its Office of Public Understanding of Science for support of this project.

August, 1978

Jacob T. Schwartz
Chairman, Steering Committee

Contents

Part Four

Mathematics Today

Lynn Arthur Steen

━━━━━━━━━━━━━━━━

> *It saddens me that educated people*
> *don't even know*
> *that my subject exists.*
> —Paul R. Halmos

Mathematics does exist. It is inherent in the rational power of man, as much a part of his nature and history as language, art, or religion. Today it is having an enormous (but often unnoticed) impact on science and society. Abstract mathematical ideas, some more than a century old, helped make possible, for instance, the revolution in electronics that transformed the way we communicate and the way we think. Neither radio, television, telephone, satellites, calculators, nor computers would be possible were it not for numerous results of "pure" mathematics. More recent advances in the mathematical sciences have helped improve our ability to predict the weather, to measure the effects of environmental hazards, to study the origin of the universe, and to project the outcomes of elections. Mathematical methods have become indispensible to the proper functioning of our technological society.

As technology requires the techniques of applied mathematics, so applied mathematics requires the theories of the core of pure mathematics. From mathematical logic to algebraic topology, from number theory to harmonic analysis, abstract structures of pure mathematics are used extensively by the contemporary applied mathematician. Few specialties of pure mathematics are immune to the demands of application. Problems and conjectures rooted in science and technology create fast-growing thickets of theory that soon become nearly impenetrable. Across the landscape of contemporary mathematics new species blend with old growth, creating a diverse and vigorous living resource for solving problems and understanding structure.

Yet many educated people are oblivious to the existence or significance of this resource. What concerns they have center on "why Johnny can't add" or on "mathophobia," the modern jargon for the traditional feeling that "I never was any good at math." Visions of draconian teachers demanding insane memorization of meaningless mumbo-jumbo prevent a large number of people from reacting normally to the opportunities offered by contemporary mathematics.

Jeremy Bernstein, writing recently in *The New Yorker* about the difficulty scientists have in communicating with nonscientists, observes that the *sine qua non* for effective popular science writing is that the writer know what he is talking about. Biologists, astronomers, chemists, and physicists have all succeeded in communicating the excitement and essence of their fields to interested laymen willing to read a book that requires careful attention. Mathematicians, reports Bernstein, are in last place in communicating with the general public.

Certainly there is no shortage of mathematicians who know what they are talking about. Although one would never know it from observing the popular scientific press, there are nearly as many practicing mathematicians as there are economists or physicists, and many times as many mathematicians as astronomers. One problem mathematicians face in communicating with the general public is the abstract, otherworldly vocabulary of their subject. Molecules, DNA, and even black holes refer to things with a material sense, providing the chemist, biologist, and physicist with an effective communication link based on physical reality. In contrast, not even analogy and metaphor are capable of bringing the remote vocabulary of mathematics into range of normal human experience.

A lesser problem is a complacent isolation, subscribed to by both the general public and the mathematical community. The mathophobic public generally prefers to avoid constant reminders of its mathematical handicap, while many mathematicians, in their turn, welcome the lack of public interest since it saves them the difficult and possibly embarrassing task of justifying the sometimes extreme abstractions of contemporary mathematics.

Not all mathematicians have taken this position. First-rate mathematicians have occasionally spoken to the general public about the nature and significance of their work. (See, for example, the bibliography at the end of this volume.) But, while these classics of popular mathematics are still useful, they do not reflect the stunning mathematical progress of recent decades. Mathematics in the 1970's is not the same as mathematics in the 1950's or the 1930's. It is important for each generation to articulate where it is, to set forth examples and evidence of current work as additional milestones in the long journey of mathematics. That is the purpose of this present volume.

The twelve essays in this volume reflect well the state of the mathematical sciences as we enter the final quarter of the twentieth century. Rather than attempt to cover systematically the vast mathematical landscape, we have selected for detailed examination certain regions that illustrate the diversity of pure and applied mathematics. Although many important areas have necessarily been omitted from this selection, the topics treated do convey a sense of the methods, accomplishments, and challenges of contemporary mathematics.

History

Part of the difficulty laymen have in comprehending contemporary mathematics is that the subject has such an incredibly long and sophisticated history. The physical sciences build on concepts and theories at most a few hundred years old; much of scientific biology, medicine, statistics, and psychology is of even more recent vintage. Yet mathematics measures its history in millenia. Major advances occurred in ancient Greece: the sophistication of Euclidean logic, although now viewed as rather informal by professional logicians, is a hurdle sufficiently high to eliminate nearly half of the educated population from effective knowledge of mathematics more recent than 2,000 years old. Those who complete a strong high school mathematics program reach, roughly, the middle of the seventeenth century while the first year of college calculus carries some students only as far as eighteenth-century mathematics.

Here, in the time of Leonhard Euler, Pierre-Simon de Laplace, and the younger Bernoulli brothers Daniel and Nicholas, modern physics has only recently been born, chemistry is still an embryo, and biology, as we now know it, has not yet been conceived. Mathematics, however, is in its prime: the calculus of Gottfried Leibniz and Isaac Newton has been developed and applied for a hundred years, and has solved an impressive variety of significant mathematical problems. Less than one percent of the population of the United States now learns anything of mathematics beyond this golden age of the late eighteenth century. Nothing that the majority has learned does anything to dispel their image of mathematics—archetypically Euclidean geometry and Newtonian calculus—as a completed discipline that has resolved once and for all every problem of measurement and motion.

But in the nineteenth and early twentieth centuries both science and mathematics were transformed by the discovery of basic principles from which general theories swiftly grew. The accomplishments of such giants as Darwin in biology, Maxwell in physics, and Freud in psychology were matched by those of Gauss and Riemann in mathematics; these accomplishments revolutionized sciences and mathematics. Current undergraduate students of science begin their education with this revolution; most undergraduate students of mathematics finish theirs at this point. This gap, which now spans nearly two centuries, makes public communication of mathematical ideas enormously difficult. Not only the vocabulary, but the momentum and entire rationale of current mathematics depends on the revolution of the nineteenth century. We can hardly expect to appreciate the achievements of the present without at least a rudimentary knowledge of the challenges posed by the past.

Nineteenth-Century Mathematics

Mathematics in the nineteenth century grew swiftly in two apparently opposite directions. It refined the tools of calculus into a rigorous system of analysis that made possible powerful theories of mathematical physics; these theories led ultimately to quantum mechanics, relativity theory, and consequently to a deeper understanding of the fundamental nature of matter and of space. At the same time nineteenth-century mathematics, by vigorously questioning the logic of calculus and geometry, discovered whole new worlds of mathematics in the theories of infinite sets and non-Euclidean geometries; these theories ultimately let twentieth-century mathematicians to more profound understanding of the foundations of their own subject.

The remarkable thing about these two themes—one applied, concrete, and externally influenced, the other theoretical, abstract, and introspective—is that they grew from common roots. One of these roots is Joseph Fourier's insight that every mathematical function, no matter how complex, can be expressed as the sum of certain basic simple functions, namely those representing the periodic vibrations that form the pure tones in music or the pure colors in light. Fourier's suggestion, motivated by his analysis of the distribution of heat in unevenly heated bodies, is one of the boldest and most significant ideas in the history of mathematics. This idea was explored by several mathematicians before Fourier, but the effectiveness of Fourier's applications helped history attach his name to the concept.

The investigation of Fourier's idea lasted throughout most of the nineteenth century and involved many of the greatest mathematicians of that time, including Peter Gustav Lejeune-Dirichlet, Georg Bernhard Riemann, Karl Weierstrass, and Georg Cantor. These successors of Fourier discovered what made his methods work (carefully controlled "convergence" of infinite sums) and what might cause them to fail. Theoretical physicists, following the path of what worked, transformed classical physics with these new tools; and mathematicians, exploring the myriad paths along which Fourier's methods did not work, discovered the vast and hitherto unknown world of infinite sets.

Another great achievement of nineteenth-century mathematics was the discovery that Euclidean geometry was not the only possible kind of geometry. This disconcerting discovery was apparently first made in the early nineteenth century by Carl Friedrich Gauss—the greatest mathematician since Newton. Gauss never published his discoveries, perhaps for fear of ridicule. The credit for public disclosure belongs to Nikolai Lobachevsky and Janos Bólyai. The key to these "non-Euclidean" geometries is the discovery that the crucial fifth (or "parallel") postulate in Euclid's system need not be true in every geometry: it might as well be assumed to be false because there are some examples in which it is false. This discovery led to the pluralization of mathematics (itself already a strangely plural noun): where we once had geometry, we now have geometries and, ultimately, algebras rather than algebra, and number systems rather than a number system. This pluralization continued unchecked well into the twentieth century, an abundant but sometimes bewildering legacy from nineteenth century exploration.

The major achievements of nineteenth-century mathematics illustrate well the delicate symbiosis between pure and applied mathematics. Fourier's analysis of the heat problem led, ultimately, to Georg Cantor's highly abstract theory of infinite sets, while purely abstract considerations of non-Euclidean geometry led to models essential for relativity theory. Other major contributions from the nineteenth century reveal the same pattern. Group theory, Boolean algebra, and the algebra of matrices arose out of purely intellectual considerations in response to the mathematician's need to clarify certain outstanding theoretical issues. Yet they have since been applied extensively in, for example, atomic physics, electrical engineering, and economics. On the other hand, complex analysis (the study of functions of the "complex" numbers which incorporate the "imaginary" number $i = \sqrt{-1}$) arose simultaneously with Fourier analysis in intimate relation with mathematical physics. Yet today complex analysis plays a key role in the solution of intricate problems in the very pure theory of prime numbers.

The message of the nineteenth century concerning the role of mathematics is unequivocal: although mathematics is molded both by requirements of pure form and by determination of scientific fact, both molds produce similar structures. Whether mathematics is pursued for its own sake or for the sake of its applications to science, the problems it generates and the structures required to solve them share a common logical basis; they differ primarily in the form in which they are expressed. No wonder Paul Halmos is saddened by the realization that most educated people have not had sufficient experience in mathematics (or in science) to appreciate this fundamental paradigm of mathematical and scientific research.

> *The paradox is now fully established*
> *that the utmost abstractions are the true weapons*
> *with which to control our thoughts of concrete fact.*
> —Alfred North Whitehead

Twentieth-Century Mathematics

The great, versatile German mathematician David Hilbert ushered mathematics into the twentieth century with a forceful challenge to the 1900 International Congress of Mathematicians that included a list of 23 major unsolved problems facing mathematics. These problems exercised special influence on the direction of mathematics in the first half of the twentieth century; those who solved them achieved instant acclaim. But perhaps more important than the specific problems was Hilbert's proclamation of faith that in mathematics there can be no *ignorabimus*. Hilbert's argument—indeed, the example of his entire life—was that it is in the nature of mathematics to pose and to solve problems; there was no possibility, in Hilbert's view, of never knowing. The tools of pure thought, in the minds of creative mathematicians, should be sufficient to solve any specific mathematical problem.

To document his case, Hilbert (and his students throughout the world) embarked on a program to codify and formalize the process of mathematical proof. They had good reason for believing that formalization would introduce into mathematics the same kind of certainty that Newton's laws had introduced into mechanics two centuries earlier. But, just as quantum mechanics devastated Newtonian determinism, so the "undecidability" result of Kurt Gödel in the 1930s devastated Hilbertian certainty. Gödel proved what could well be one of the most profound results in the history of thought: no significant formal system can ever be strong enough to prove or to refute every statement it can formulate. Gödel proved, in Hilbert's words, that in mathematics, there is *always* an *ignorabimus*.

Another major influence on twentieth-century mathematics was Cantor's introduction of infinite sets into the vocabulary of mathematics at the end of the nineteenth century. Set theory provided a rich idiom in which many long-standing problems could be rephrased and resolved. For example, questions concerning the stability of solutions of differential equations (where solutions represent paths of moving objects) were translated into problems concerning the geometry of certain sets of points called surfaces, thereby helping to establish the emerging field of topology. Questions concerning the common structure of matrices, groups, and sets led in the twentieth century to the subject now widely known as "modern" or "abstract" algebra. And similar methods when applied to nineteenth-century analysis led to "abstract" analysis in which the integrals and derivatives of calculus are applied in spaces of infinite dimension.

These three disciplines—algebra, analysis, and topology—represent the common culture of the modern mathematician. The theories, definitions, and methods of these fields form the classics of mathematics education: no one can consider himself truly literate in mathematics who cannot read mathematics as written in the language of algebra, analysis, and topology. From these three early twentieth-century fields has emerged the incredible variety of contemporary mathematics.

There are now close to a hundred recognized subdisciplines within the core of the mathematical sciences; if we add to this the applied mathematical sciences of statistics, computer science, operations research, and theoretical physics, the number of distinct subdisciplines would easily number several hundred. Research in these fields is scattered among roughly 1,500 journals published in nearly 100 different languages. In 1975 alone, *Mathematical*

Reviews indexed over 24,000 research articles. It is entirely understandable that nonmathematicians would find all this quite bewildering; so too do many mathematicians. Mathematics today is an enormous, powerful, complex enterprise largely beyond the language or intuition of the nonspecialist.

> *I find both a special pleasure and constraint*
> *in describing the progress of mathematics,*
> *because it has been part of so much speculation:*
> *a ladder for mystical as well as rational thought*
> *in the intellectual ascent of man.*
> —Jacob Bronowski

The Mathematical Sciences

The world of mathematics may be visualized as many concentric layers built on the core of pure mathematics. This core is still red-hot with new ideas, new structures, and new theories. Ideas from the core percolate through the outer layers of the mathematical sciences, providing a constant supply of intellectual fuel for some of the incredibly complex problems of the more applied fields. And, in return, problems arising in the outer layers—in the diffuse boundary where pure mathematics blends with applied science —provide the central core with new structures, new methods, and new concepts.

Core mathematics is the science of significant form. It is nourished both by internal energy, like a self-sustaining atomic reaction, and by new fuel supplied by the outer layers that are in closer contact with the surface of human problems. Layers near the core employ sophisticated techniques in the service of external objectives. Theories in these layers are directed more towards solving problems than towards discovering basic form. Layers remote from the core employ mathematics more as metaphor than as theory: applications blend with technique so thoroughly that a totally different discipline emerges. Theory and problems diffuse through the ill-defined boundaries between these layers, each enriching the other and nourishing both mathematics and science.

In the past quarter century the world of mathematics has evolved from a single discipline to a cluster of intertwined subjects now usually termed "mathematical sciences." These sciences include, in addition to core fields such as number theory, mathematical logic, differential topology, and algebraic geometry, the applied areas of operations research, statistics, computer science, combinatorics, and mathematical programming. The use of mathematical models in science has proliferated to such an extent that there now are distinct disciplines (with their own professional journals) in mathematical economics, mathematical biology, mathematical psychology, and mathematical linguistics. Even fields such as political science and history have not been immune to the expanding influence of mathematics: for example, the recent (and frequently controversial) introduction of statistical techniques into historical research has given rise to a new specialty termed "cliometrics" (from Clio, the muse of history). There remain few areas of human intellectual activity that have not been shaped significantly by the mathematical sciences.

The essays in this volume were selected to represent the entire sphere of mathematical science. Five of the essays examine recent developments in classical fields: mathematical problems with roots more than a hundred years old are still being solved. Ian Richards ("Number Theory") describes some very recent contributions to centuries-old problems about the pattern of prime numbers and the solution of equations in whole numbers Jonathan Alperin ("Groups and Symmetry") shows how symmetry patterns in geometry and algebra—first studied in the nineteenth century—have led mathematicians to the threshold of a complete analysis of finite groups, one of the "elementary particles" of mathematics.

New developments based on nineteenth-century ideas undergird the three other articles in this part as well. Roger Penrose ("The Geometry of the Universe") shows how paradoxical models of non-Euclidean geometry—models that cannot even be faithfully represented in our three-dimensional space—may be "true" models of the universe. Philip Thompson ("The Mathematics of Meteorology") describes the long history of mathematical techniques in meteorology, and shows how recent improvement in approximate methods for solving the hydrodynamic equations of the atmosphere has revealed the possible presence of an uncertainty principle of prediction in which increased accuracy of computation may not necessarily lead to any further increase in accuracy of weather forecasts. Concluding this part, Kenneth Appel and Wolfgang Haken ("The Four-Color Problem") describe the extensive man-machine intellectual symbiosis that led to their recent computer-based solution of the century-old puzzle concerning the number of colors required to distinguish different regions on a map.

The five essays in the next part introduce areas of the mathematical sciences that have emerged primarily in recent years. Each shows not only how new mathematics can be used to advance some aspect of science, but also how mathematics frequently leads to surprising conclusions that could not have been anticipated by even well-informed common sense. For example, Ronald Graham ("Combinatorial Scheduling Theory") uses mathematical reasoning to demonstrate that increasing the number of personnel in a production process may increase (rather than decrease) the time required to complete the process and goes on to discuss new ideas from combinatorial mathematics that provide mathematical models for similar scheduling problems. David Moore ("Statistical Analysis of Experimental Data") discusses various problems related to the design of medical experiments and shows how extensively mathematics is used in this important human enterprise. Moore shows how careful statistical design can minimize the risks to patients in clinical studies and how it is possible to design an experiment so as to reduce the chance of certain types of erroneous conclusions to exactly zero.

The final three essays in this part examine widely different examples of contemporary applied mathematics. Martin Davis ("What is a Computation?") demonstrates how some very abstract ideas related to Gödel's undecidability result are now bearing fruit in the analysis of computer algorithms by making it possible to distinguish between solvable and unsolvable problems. Jacob Schwartz ("Mathematics as a Tool for Economic Understanding") uses mathematical analyses of game-like economic models to show that there can exist states of the economy in which appropriate centralized intervention can improve everyone's economic welfare simultaneously (rather than merely transferring wealth from one group to another) by moving the economy away from "suboptimal equilibria." Finally, Frank Hoppensteadt ("Mathematical Aspects of Population Biology") describes some mathematical techniques used to study ecological phenomena, and shows, for instance, how economic circumstances can conspire with biological facts to make immediate extermination of a species the most profitable long-range decision for a particular industry.

Each of these essays makes contact both with core mathematics and with some aspect of science, citing numerous uses of mathematics never dreamed of by the originators of the mathematical theories. These examples illustrate well the adage that it is impossible to predict which mathematical theories will eventually prove to be useful, or in what way. The corollary of this adage is a paradox: problems requiring new mathematics for their solution cannot always be solved by straightforward attack. The process by which ideas move between the outer and inner layers of the mathematical sphere is not that direct. Ideas from the central core of mathematics not originally connected to the original problem may turn out to be required for its solution. Conversely, applied problems generate mathematical ideas which take on life of their own, frequently providing the means for solving problems other than those that generated them.

This paradox is an important point frequently lost in discussions of the continual tension between pure and applied mathematics, and between basic and applied research. Even those impatient to get on with results usually realize, after some reflection, the importance of basic research. But what they often fail to realize is that while fundamental ideas may be comandeered for some urgent application, they cannot be created upon command. Several million dollars combined with years of experimental research can convert $E = mc^2$ into an atomic bomb, but no money in the world can buy or create the mind of an Einstein. Constraining the creative spirit can destroy it, as much in mathematics as in art. Those who wish to ensure a continuing flow of basic ideas for future application have little alternative but to encourage the development of all basic ideas, even ideas that may at first seem remote from any possible application.

This does not mean that one idea is as good as another, nor that there is no way to judge among alternative avenues for the development of mathematics. Quite the contrary, despite minor arguments among proponents of different specialties there is remarkable agreement among experts as to which mathematics is really deep and essential, and which is merely routine and dull. Yet despite an objectivity about mathematical results that has no parallel in the world of art, the motivation and standards of creative mathematics are more like those of art than like those of science. Aesthetic judgments transcend both logic and applicability in the ranking of mathematical theorems: beauty and elegance have more to do with the value of a mathematical idea than does either strict truth or possible utility. Nevertheless, through some totally mysterious common element of the human mind, creative mathematicians share a strikingly similar set of aesthetic standards. Even though one mathematician may have little interest in another's specialty and little expertise in its details he can usually perceive the beauty in the other's results.

The Nature of Mathematics

We conclude by reflecting a bit on three important mysteries concerning the nature of mathematics: how is it created, why is it relevant, and from where does it draw its objectivity? These are the themes of the opening and closing essays of this volume. In "Mathematics—Our Invisible Culture," Allen Hammond interviews practicing mathematicians and reviews statements from great mathematicians of the past in order to illustrate how the creation of mathematics is akin to a process which distills pure thought from the complex sensory input of our brain. There is little doubt that the major advances in mathematics have occurred only when a mathematically talented mind has employed intense and continual concentration to extract significant form from the clutter of data provided by specific examples. We know practically nothing about the circumstances that make this type of creative genius possible, apart from the very evident fact that not everyone can create mathematics by squeezing mathematical thoughts from his brain.

Felix Browder and Saunders Mac Lane ("The Relevance of Mathematics") explore the diverse ways in which mathematics comes to be relevant to mankind, noting that the ultimate relevance of a particular topic has no relationship to the motivation of its creator, to the degree of its abstractness, or to its apparent relevance at the time it is created. Despite experiences of applied mathematics spanning several thousand years, we apparently cannot predict the utility of particular mathematical innovations any more than we can predict the source of mathematical creativity. We do know a little about the conditions that foster significant mathematical activity, and about the conditions that repress it. (For example, mathematics, both pure and applied, flourished in ancient Greece and in nineteenth-century Europe, but it was virtually nonexistent in ancient Rome or in America before 1870.) And we know from innumerable examples that we cannot rely only on the concepts that have been applicable in the past unless we wish to rule out all possibility of significant innovation.

So where, in conclusion, does mathematics stand today? Its health is excellent: the energy and vigor displayed in current research is about as strong as it has ever been. What weaknesses exist seem to be due to competition from other fields for the service of the really gifted minds of each generation. There is no doubt that the revolutions in biology and in computers have opened new inviting alternatives for the young student wishing to leave his mark on future generations.

The usefulness of mathematics to society is greater now than ever before in history. The experience of several decades of interaction with an increasingly scientific society has helped mature the relation between mathematics and its scientific clients More scientists have training in mathematics than ever before, so the time required for ideas to percolate from the theoretical core to the applications-studded surface of the mathematical sphere should continue to decline. Since the scientific revolution, the typical lag between an idea and its utilization has declined from centuries to decades and, recently, to years.

Yet despite these outwards signs of good health, there is today a professional malaise in many parts of the mathematical community, often tinged with Halmos's sense of sadness. Mathematics is progressing along the edge of a fault-line in society's intellectual landscape: on one side is the nonscientific majority, almost refusing to acknowledge that mathematics exists, while on the other is a tiny, youthful minority who achieve important results that even their teachers cannot always comprehend. This schism suggests a serious danger, that the gulf between mathematics and society may now be so great that public support for the objectives and values of mathematics may be inadequate to sustain the type of research program that our technology requires. A society which demands a technology that it cannot comprehend may lose its ability to sustain that technology.

Faced with this mildly hostile environment, many mathematicians retreat into the security of their own research, yielding to the disinterest of outsiders and too easily maintaining their self-imposed isolation. But others, including the authors of this volume and the members of the committees that undertook this project, believe that certain steps must be taken to improve communication between the general public and the mathematical community. This volume contains one part of a potential dialogue; its readers will, we hope, supply another part.

Mathematics is, in Allen Hammond's phrase, our invisible culture. It is an odyssey into a man-made universe, one that few people ever take. It represents one of the supreme achievements of the human mind, a triumph of mind (logic) over matter (science). Whitehead argued that only music rivals mathematics as the most original creation of the human spirit. It gives "special pleasure" to the few who, like Jacob Bronowski, have sampled its riches. No wonder mathematicians are sad that educated people do not even know that their subject exists. But mathematics does exist, in profusion, as a necessary product of the mind of man: *cogito, ergo sum.*

Part One

Mathematics —
Our Invisible Culture

Mathematics —
Our Invisible Culture

Allen L. Hammond

An inquiry into mathematics and mathematicians might begin with certain curious facts. One is that mathematics is no longer an especially uncommon pursuit. Never mind that a multitude of mathematicians seems a contradiction in terms. The universities are simply teeming with them. The latest figures compiled by the National Science Foundation show that there are as many mathematicians in the United States as there are physicists or economists. Mathematicians are not a rare breed, simply an invisible one. It is a multitude singularly accomplished at keeping out of the public eye. Who has ever seen a mathematician on television, or read of their exploits in the newspapers?

A second fact about this reticent profession is even more startling. All those people are busy doing something, including some very remarkable somethings. In all its long history extending back 25 centuries, mathematics has never been more vigorous, more active than now. Within this century mathematicians have experienced philosophical upheavals and intellectual advances as profound as those that have catapulted physicists into fame or transformed economists into the indispensible advisors of governments. The foundations of mathematics itself have been challenged and rewritten, whole new branches have budded and flourished, seemingly arcane bits of theory have become the dicta for giant industries. Yet this drama has been played out in near obscurity. The physical concepts of relativity and subatomic particles have entered the language, the gross national product is reported to millions of living rooms, but it is as if the very texture of mathematics is antithetical to broad exposure. What is it in the nature of this unique field of knowledge, this unique human activity that renders it so remote and its practitioners so isolated from popular culture?

In searching for a foothold to grapple with this elusive subject, an Inquirer is struck by the contradictions that abound. For example, mathematics is nearly always described as a branch of science, the essence of pure reason. Beyond doubt mathematics has proved to be profoundly useful, perhaps even essential, to the modern edifice of science and its technological harvest. But mathematicians persist in talking about their field in terms of an art—beauty, elegance, simplicity—and draw analogies to painting, music. And many mathematicians would heatedly deny that their work is intended to be useful, that it is in any sense motivated by the prospect of practical application. A curious usefulness, an aesthetic principle of action; it is a dichotomy that will bear no little scrutiny in what is to come.

A further contradiction arises from the stuff of mathematics itself. It is in principle not foreign to our experience, since the root concepts are those of number and of space, intuitively familiar even to the child who asks "how many" or "how large." But the axiomatization and elaboration of these concepts has gone quite far from these simple origins. The abstraction of number to quantitative relationships of all kinds, the generalization of distance and area first to idealized geometrical figures and then to pure spatial forms of diverse types are large steps. Somewhere along the lengthy chains of logic that link modern mathematics to more primitive notions, a transmutation has occurred—or so it often seems to outsiders—and we can no longer recognize the newest branches on the tree of mathematics as genetically related to the roots. The connection is obscured, the terminology baffling. Is any of it for real? Do these abstractions and elaborations genuinely expand our understanding of number and of space, or do they amount to an empty house of theorems?

Mathematicians bristle at such questions. But it is not surprising that there is a popular tendency to dismiss much of this unfamiliar stuff as the subtle inventions of clever minds and having no important relationship to reality. What is surprising is that mathematicians do not agree among themselves whether mathematics is invented or discovered, whether such a thing as mathematical reality exists or is illusory. Is the tree of mathematics unique? Would any intelligence (even a nonhuman one) build similar structures of logic? How arbitrary is the whole of mathematical knowledge? These too are points worth additional inquiry.

We might also learn something of the end result, however incomprehensible, if we could see the process by which it is made and know more of the makers. Should we pity the poor mathematician, condemned to serve his or her days bound to a heavy chain of cold logic? How does that image jibe with the white-hot flashes of insight, the creative "highs," so often reported, or the intensely human character of mathematicians in the flesh? Clearly, a suitable subject for this inquiry is the nature of mathematicians themselves, their motivations, their trials, their rewards, and how they spend their days.

A final question might be directed toward the place of mathematics in our culture. There are those, including Plato, who have identified mathematics with the highest ideal of civilization—a lofty claim indeed. A claim more often made and subscribed to by mathematicians is that mathematics is one of the finest flowerings of the human spirit, a cathedral of enduring knowledge built piece by piece over the ages. But if so it is a cathedral with few worshipers, unknown to most of humankind. Mathematics plays no role in mass culture, it cannot claim to evoke the sensibilities and inspire the awe that music and sculpture do, it is not a significant companion in the lives of more than a very few. And yet it is worth asking whether mathematics is essentially remote, or merely poorly communicated. Perhaps it is a remediable ignorance, not an inability, that now limits appreciation and enjoyment of mathematical intuitions by a wider audience; perhaps our culture is only reaching the stage at which mathematics can begin to penetrate a larger consciousness.

Three Mathematicians

This Inquirer was himself originally trained in a mathematical subject. But he does not—as mathematicians say—*do* mathematics and he approaches this inquiry from the great camp of outsiders. The principal sources of information that he consulted and here reports on are of two kinds: what mathematicians of note have written about themselves and their craft, unfortunately a sparse literature; and three live data points, talking mathematicians, captured in a joint conversation before an inquiring microphone. Let us introduce them. In order of descending age, we have: Lipman Bers, a professor of mathematics at Columbia University; Dennis Sullivan, a professor of mathematics at the Institute des Hautes Etudes Scientifiques in Bures-sur-Yvette, France; and Miller Puckette, a mathematics special student at the Massachusetts Institute of Technology.

"Papa" Bers, as he is known to his students, was born in Latvia. He evokes the filmmaker's European emigree with his accent and continental manners, an image that is enhanced by his small mustache, greying hair, and glasses. He is a kindly man, active in the cause of scientists imprisoned abroad, and enthusiastic when he speaks of mathematics. His research is in analysis, a branch of mathematics that began in Newton's time with the calculus but now spans a range extending from numerical and algebraic subjects at one extreme to geometric and spatial subjects at the other.

Sullivan is one of the world's experts on things spatial; he is a topologist, a field sometimes described as an essay on the word "continuous" because it deals with the properties of objects undergoing transformations in their shape that do not involve cutting or tearing. Sullivan is a vigorous man who gestures and moves his body when he speaks. He fairly radiates intensity; with his long untamed locks, full beard, and glasses, he is a fierce, almost primal figure on first meeting. His laughter, however, turns out to be rich and infectious and frequently punctuates the discussion.

Puckette is trim and clean-shaven, bright-eyed in his youth. He is intent on the discussion but not seemingly fazed by his elders. Puckette is not yet a practicing mathematician but is well launched in that direction; in high school he was twice a member of the U.S. team entered in the International Mathematics Olympiad. He is studying combinatorics—one of the newer branches of mathematics on the numerical side of the family which deals with how objects are distributed into classes. For the sake of the completeness that mathematicians hold dear, he is to represent the class of things numerical.

The diversity of backgrounds and personal styles, deliberately contrived for this inquiry, emphasize what is all too easily forgotten. Mathematics is universal and unified, but those who do it are varied indeed. For all its abstractness, it is a very human endeavor. As the four of us sit down in Bers' office at Columbia to begin our discussion*, trucks whine by on the highway below the window and the heating system hisses continually. The high-ceilinged old room is a bit bleak, but the atmosphere is warm, almost jovial.

"I sometimes work like I used to do when I was young, in long stretches, thinking only about the same problem. Sometimes I have to force myself to sit down to write down something or to read something. In general, one waits—I don't know how long for other people—one waits for the few pleasant moments when suddenly something that was obscure and mysterious becomes clear."

Bers is talking about how he spends his time, what he actually does when he does mathematical research. His description hints at the mental self-discipline and intensity of thought required. Listen to Sullivan:

"In one period I would work like this. I would wake up in the morning thinking about something, and think about it until I ran into some other talking mathematician, then I would start talking about it. I would talk about it all day long, go through all kinds of deformations and interactions about what I was doing. I would go home in the evening to my family, have supper, do what I had to do to lead a normal life, then sit on the couch and start thinking about it again. I would think about it until I fell asleep, then start over again the next day. It's this very funny process, slowly the topic is revolving and changing, but there is never any rhyme or reason about it. You must follow different trains of thought; they don't always lead to something. It's sort of a recycling process. I keep thinking things over and over again for years, maybe not always the same things, but mixed in with other things."

It is a picture of single-minded pursuit. The question arises as to whether the thinking process is consciously organized, directed by some plan of attack.

"You have to have an overall prejudice or goal," Sullivan says. "But I see

---

*In what follows, individual comments have been lightly edited and sometimes excerpted to focus the discussion.

the process as sort of eating [mathematics], seeing new examples; I sort of like the examples, the phenomena of mathematics, like different types of entertainment."

BERS (interjecting): Do you invent or do you discover? What is your gut feeling? [See box on p. 20.]

SULLIVAN: Sometimes you come upon something sort of natural, it's like you're discovering it. But sometimes you just make something up out of thin air, so to speak; maybe force it a little bit.

BERS: What gives you more pleasure, inventing or discovering?

SULLIVAN: I haven't invented very often, so that was very, very pleasureable. I would say that inventing and discovering give different kinds of pleasure. Most mathematicians are basically driven in their desire to simplify, to understand. You need a balance. The ideal situation is where you have a rich set of phenomena and you vaguely feel that you are about to understand it, but if you understand it too well, then it gets a little low on entropy. This sort of paradoxical situation is the best. By the way, would you prefer to find an example or prove a theorem?

BERS: I want to know the truth.

SULLIVAN: But suppose I gave you a choice, to find an example or to prove that there is no such example, which would you prefer?

BERS: I cannot imagine myself having this power. Only one thing is true.

INQUIRER: Does the understanding that results from a new mathematical example or theorem reflect an aspect of reality, an innate truth about the universe, or is it an aspect of an arbitrary, invented system?

BERS: I'm not sure we can answer this question. I was educated in a philosophical tradition, logical positivism, that leads me to believe that evolution developed the human mind to understand nature, and that nature is written in the language of mathematics, as Galileo said. But when I actually do mathematics, I have the subjective feeling that there is a real world to discover, the world of mathematics, which is much more eternal, unchangeable, and real than are the accidents of physical reality. This is the feeling that I have. I actually feel a little embarrassed: here I am, a grown up man, worrying about whether the limit set of a Kleinian group has positive measure and willing to invest a great deal of effort to find the answer.

SULLIVAN: I prefer to talk from the point of view that there are two kinds of textures in mathematics. One is the context of structure of mathematics: if you see the structure unfolding, you feel, "that's something that is there." Then the idea of examples, of phenomena—that's like something you invented. I mean you just write it down; it's pretty inventive.

BERS: The human mind is made so that when we do mathematics, we have the feeling that we are dealing with some reality. There are some parts of mathematics that are more real than others. To me the value of a mathematical theory depends on how much it deals with things made, in some sense, by God and not by man.

Discovery or Invention?

The argument among mathematicians over whether mathematical truths are invented or discovered has been going on a long time. It is not an argument that is easily resolved but it is revealing of how mathematicians think about their work.

The two points of view are at first glance quite distinct. One holds that mathematicians discover a piece of reality no less firm than physical reality, a truth not of their own making but rather an inherent part of the universe. "God made the integers," as one nineteenth century mathematician put it—they did not arise out of some Greek geometer's fertile imagination but rather from human experience. Hence the properties of the integers encompassed in simple arithmetic and in the more sophisticated theorems of number theory are viewed by most mathematicians in much the same way as astronomers think of the planets—discovered elements of the heavens. This absolutist or Platonist viewpoint—mathematics as reality revealed—extends to other areas of mathematics as well and is in fact the dominant dogma in the mathematical community. But even among the integers there lurk some telltale signs of human inventiveness. The Arabic number system we use today, including the concept of the number zero, was invented by mid-eastern mathematicians during the Middle Ages and has had a substantial impact on mathematics ever since, making possible such things as the binary arithmetic of computers.

The second point of view emphasizes the role of human creativity in inventing mathematical structures. Clearly there is an element of human creativity involved, but how much, where to draw the line? Any system of mathematics rests ultimately on a series of axioms, for example, and there is in many instances an element of choice as to which axioms to use. Euclidean geometry was based on five supposedly basic and self-evident axioms about the nature of space, but just how arbitrary such choices can be was shown by the discovery (in physics, not in mathematics) that space is not Euclidean after all but rather Riemannian—that an alternate set of axioms due to Riemann provided a geometry that corresponds more to physical reality than that of Euclid's. Other mathematicians have since invented and explored the properties of still other geometries, all good mathematics, but with an element of choice. Nor is this element of human inventiveness confined to geometry; there exist alternate, "nonstandard" models of the real number system too. Even the laws of logic on which all of mathematical reasoning depends are not universally regarded as absolute. Some mathematicians have argued that there too there is an element of choice and convention. In fact many mathematicians will admit in private that they think they create something, a good example or a good idea. In its most extreme form, this humanist or constructivist outlook rejects the idea of mathematicians as passive discoverers of a remote reality and instead asserts that mathematical phenomena are created by humans alone and do not otherwise exist.

Distinguishing between the two points of view is a little like peeling an onion—there are layers and layers of examples and counterexamples. Calculus as a technique of calculation was invented but the related mathematical phenomena of slopes and areas under a curve correspond to real things. A theorem may be discovered but its proof is usually invented. Group theory may be concerned with the properties of abstract, invented concepts but it also seems to be the language in which some important realities about the universe are revealed to mathematical physicists. The distinction between invented and discovered mathematics verges on being a subjective one, or at least a psychological one. Einstein is reported to have felt that he invented the concept of relativistic spacetime, after the fact, but that he felt he was discovering an aspect of reality while he was doing the work. It is a feeling familiar to many mathematicians: "we are all Platonists in the trenches," as one put it. But the debate continues wherever mathematicians gather. "Do you invent or do you discover?"

SULLIVAN: I have a more down-to-earth feeling. First of all, the whole of mathematics is built up from essentially two concepts. One is number, and one is space. Both of these things are part of reality. I think that mathematics is very much tied to physical reality by these two concepts. In fact, mathematics always had these two sides. The more sophisticated side is often attached to number, the more intuitive side is often attached to spatial ideas. Of course these two sides keep crossing and interacting.

Science of Art?

The notion that mathematics is closely related to physical reality has had many champions. John von Neumann, father to both the theory of games and the modern computer, has written that "mathematics is not an empirical science ... and yet its development is very closely linked with the natural sciences." It is undeniable, he says, "that some of the best inspirations in mathematics—in those parts of it which are as pure mathematics as one can imagine—have come from the natural sciences." Soviet mathematicians have argued along similar lines. A.D. Alexandrov, for example, says that it is the high level of abstraction peculiar to mathematics that gave birth to notions of its independence from the material world, but that "the vitality of mathematics arises from the fact that its concepts and results, for all their abstractness, originate ... in the actual world."

Not all mathematicians accept the idea that their creations are intimately dependent on physical reality, of course. Some prefer to think in terms of man-made intellectual structures built upon axioms that are essentially arbitrary—which just happens, in some cases, to have enormous applicability to phenomena in the real world. Others, steeped in modern philosophy, point out that physical reality as mirrored in the intricacies of quantum theory is none too firm a hitching post in any case—let alone mathematical "things made by God." But the real thrust of those whose vision of the nature of mathematics de-emphasizes the importance of the physical connection is not epistemological but rather aesthetic, concerned with criteria internal to mathematics and bound up with the process by which new mathematics is created.

The great French mathematician Henri Poincaré wrote about "the feeling of mathematical beauty, of the harmony of numbers and forms, of geometric elegance. This is a true aesthetic feeling that all real mathematicians know." He went on to describe the creative process in mathematics in terms astoundingly like those now used by psychologists studying the specialization of the two halves of the brain. Poincaré distinguished "two mechanisms or, if you wish, the working methods of the two egos:" the one logical, capable of the intensive work upon a problem that usually precedes an intuitive breakthrough and the calculations that follow it, which he identified with the conscious mind and which psychologists now describe as left-brain functions; the other closely linked to an aesthetic sense and capable of recognizing that pattern, among all the possibilities that present themselves, that is both beautiful and important. The recognition of the pattern that solves the problem, Poincaré says, cannot be willed but comes of its own accord in what seems like a sudden flash of intuition from the unconscious mind; psychologists now attribute pattern-recognition to right-brain activity. Much remains to be ascertained about the creative process, but clearly there begins to be a physiological rationale for the insistence of mathematicians that aesthetics, a sense of the beautiful and the elegant, is an important element in mathematical success.

One who has championed the aesthetic aspects of mathematics forcefully is the English number theorist Godfrey Hardy, whose proudest boast was that he had never done anything useful in the sense of practical applications. Hardy described mathematicians as makers of patterns of ideas; he asserted that for them, as for other artists, "beauty and seriousness [are] the criteria by which [their] patterns should be judged." Beauty, he says, "is the first test: there is no permanent place in the world for ugly mathematics." And he summed up his life's work in this way:

"The case for my life, then, or for that of any one else who has been a mathematician in the same sense in which I have been one, is this: that I have added something to knowledge, and helped others to add more; and that these somethings have a value which differs in degree only, and not in kind, from that of the creations of the great mathematicians, or of any of the other artists, great or small, who have left some kind of memorial behind them."

Another who has described and defended mathematics as a creative art is the American algebraist Paul Halmos. "It is, I think, undeniable that a great part of mathematics was born, and lives in respect and admiration, for no other reason than that it is interesting—it is interesting in itself . . . I like the idea of things being done for their own sake," he says. "Is there really something wrong with saying that mathematics is a glorious creation of the human spirit and deserves to live even in the absense of any practical application?" Halmos compares mathematics to music, to literature, and especially to painting:

"The origin of painting is physical reality, and so is the origin of mathematics—but the painter is not a camera and the mathematician is not an engineer . . . In painting and in mathematics there are some objective standards of good—the painter speaks of structure, line, shape, and texture, where the mathematician speaks of truth, validity, novelty, generality—but they are relatively the easiest to satisfy. Both painters and mathematicians debate among themselves whether these objective standards should even be told to the young—the beginner may misunderstand and overemphasize them and at the same time lose sight of the more important subjective standards of goodness." Mathematics, he argues, "is a creative art because mathematicians create beautiful new concepts; it is a creative art because mathematicians live, act, and think like artists; and it is a creative art because mathematicians regard it so."

If mathematics is the most intellectual of the arts, however, it is also strikingly like a science, particularly in its insistence that there is only one version of truth. So useful have the logical structures and ideas of mathematics been as a language for physics, so intimately interwoven has been their evolution, that many scientists have remarked on what physicist Eugene Wigner calls "the unreasonable power of mathematics." Let us rejoin the discussion as it considers this dual character of mathematics.

BERS: Objectively, I certainly feel that mathematics is part of science. There is no doubt.

INQUIRER: Yet some of the words you were using earlier about elegance, and emotional reaction, are not necessarily objective.

BERS: Evolution has made our minds such that what is elegant and beautiful is what is useful and powerful. And the history of science seems to confirm this.

SULLIVAN: I feel mathematics is like this tree that starts from very basic things like space and number, but it has these wonderful branches and flowers that go beyond. I mean, it's too fantastic—the same way that music is too fantastic. It's sort of unbelievable: you start concentrating on one theorem, some really fantastic theorem, and you can't really hold it. It's so amazing.

There is this aspect I mentioned earlier, of lying on a couch thinking, of pure thought which is kind of denser in mathematics than in art. Somehow it is sort of distilled in mathematics: [pure thought] is almost all there is.

BERS: There is also the absolute standard of being right. The requirement of being correct, and also that anyone can check whether you are right, exists only in mathematics. It does not exist in—I almost said in the other arts. This is unique. In mathematics you have complete freedom with a complete lack of arbitrariness.

SULLIVAN: Two physicists can argue about some point, [and they] argue and argue and argue. You can have two mathematicians doing exactly the same thing, and then suddenly one of them will say, "You're right." One of them will completely see the point and it will be right, because these points can be decided. We could sit here and argue about whether there is a counterexample to some theorem or other, but if you change the statement and have an argument, it can be decided.

Beauty and Power

INQUIRER: How does one distinguish good mathematics? What are the criteria?

PUCKETTE: I would go back to [an earlier] description of sudden joy to say what, to a pure mathematician at least, is important and unimportant. [I think] others feel the way I do that a theorem or a result is only as useful as it is beautiful. Or else perhaps sometimes beauty and importance coincide.

BERS: But this begs the question.

SULLIVAN: No it doesn't beg the question to my mind. Because I sort of appreciate [certain difficult and complex theorems]. But I actually like or feel more satisfied by results that are very simple, that may even be trivial to prove, if they explain something or have a lot of applications.

INQUIRER: Can you give an example?

SULLIVAN: Look at two planes. Generally they would intersect in a line, but they might coincide. You consider a plane and a line—the line might cut a plane or lie in the plane. You observe this pattern and there's hardly anything to prove—it's obvious. But as you pursue this pattern in the study of spaces and manifolds and so on, it has great universality, [it is] very profound, has many ramifications, yet comes from a simple thing in linear algebra. I like that idea and I like to see all the things that rush out of it. . . . This is a very down to earth, naive apprehension of something, different from writing down a complicated function and observing that it satisfies certain equations and existence theorems. And it almost gives me more joy, is almost more important, a simple thing like this. You watch this mathematician Thom [René Thom, a French geometer] work; he uses ideas like this over and over again until he just overcomes complicated problems. It's very powerful.

BERS: Simplicity. I think you've hit here on a very important point, simplicity. I think mathematicians generally agree that simplicity and beauty are important, and there is no trouble in recognizing them when you see them.

SULLIVAN: I agree with what Bers has said, that what is striking and beautiful in mathematics is pretty universally agreed upon. There are other words, such as importance, that we could be discussing here.

INQUIRER: Has your taste for what is beautiful changed over your career?

SULLIVAN: Well, I now like things to be simpler, more geometric, to be more explained by examples; I have been slowly sliding over from the idea that one can understand structure to wanting to see lots of examples passing in review.

BERS: Certainly historically the conception of what is beautiful has changed. And in general when an older man said something was not beautiful, he was wrong. Poincaré did not like Lebesgue integrals and space-filling curves, things like that.

SULLIVAN: I don't like them either.

BERS: Well, you're wrong.

SULLIVAN: You know, that only proves that there is a subtle balance between richness and complicatedness. You like a subject to be rich, but not too complicated. On that particular day, Poincaré may have felt that the phenomena in question were simply too complicated. Later it turned out to fall out of the general theory naturally.

PUCKETTE: So it was just the way he [Poincaré] was looking at it that made it seem disconnected.

INQUIRER: Do you find it surprising that mathematics is so useful in the physical sciences?

SULLIVAN: I hate to keep repeating this, but mathematics is built on the concepts of number and space—what could be more useful? I don't find it surprising at all that mathematics is to the point for physical reality.

BERS: The fact is that in the days before calculus and for a long time thereafter much of mathematics was created to facilitate applications in physics. And this may be the reason why before mathematics can become really successful in biology or psychology, mathematics itself will have to be changed. New mathematics will be created. At least it is not inconceivable.

SULLIVAN: It is also true that a lot of people working in biology, even in physics, don't really know that much mathematics.

BERS: But when they say "to know something," it is different than when I say to know something. I remember a lecture here at Columbia on algebraic topology and its applications to physics. T.D. Lee [a Columbia physicist] was in the audience and he said, "I don't believe that algebraic topology will be useful in physics, but if it should turn out that I am wrong, then every theoretical physicist will have to take out two months and learn it." And they would do it. They would learn a few examples and the rules for how to compute. That's all they need.

SULLIVAN: No, what I was saying is that many scientists don't even know mathematics in the sense of taking two months out. There are lots of nice examples around, many theories are explained by one or two nice examples, and if you have the culture of knowing these examples, you might be able to draw on this stuff and use it. I have the feeling that a lot of mathematics can be applied that exists now.

Isolation and Frustration

The nearly universal recognition that mathematics is a powerful scientific tool might be thought very gratifying to mathematicians. In fact, the identification with science is more often a source of frustration, because it is essentially the only thing most people know about their subject. Listen to the anguish with which Halmos says, "It saddens me that educated people don't even know that my subject exists. There is something that they call mathematics, but they neither know how the professionals use that word, nor can they even conceive why anybody should do it."

It is useless to debate whether this gap between mathematicians and the rest of the world is due to some deep-seated fear of mathematics or to the failure of more than a very few mathematicians to make the effort to communicate something of what is going on in their subject to a broader audience. The formidable abstractness of many of the ideas of mathematics and the fussy preciseness that makes many mathematicians insist on using their technical jargon even with nonspecialists certainly add to the problem. It is further complicated by pedagogy: until the university, mathematics is usually taught by people who do not really understand it themselves on more than a very superficial level. Certainly intellectual capability is not really the main barrier to a wider apprehension of at least many mathematical ideas. In experimental programs in which practicing mathematicians have taught very young ghetto children advanced algebra and similar subjects, the rapidity with which the students absorbed concepts far advanced beyond those normally taught at their grade level, the enthusiasm for their subject, and the rapport between students and mathematicians have astounded professional educators. Nonetheless, such efforts and successes in bridging the barrier are the exception, not the rule. Mathematics is one of the few subjects that a student can study through high school and even a few years into college without coming into contact with any results invented since 1800. Those of us who do not major in mathematics or a few areas of physics can pass entirely through the educational system and never encounter any of the revolutionary ideas of twentieth century mathematics. And many mathematicians see nothing wrong with this state of affairs.

The gap extends beyond a general poverty of knowledge of mathematical ideas to a lack of familiarity with even the trappings of mathematical culture. Most educated people have heard of physicists such as Einstein, Heisenberg, and Fermi and have some rough idea of their work; more would make it a point of pride to be familiar with at least the names of the major composers, painters, and writers not only of the past but of the present. Yet how many can name a couple of first-rank mathematicians in this century or identify even a single concept associated with such great geniuses of the past as Gauss, Cauchy, or Riemann?

All this tends to leave mathematicians isolated, dependent only upon themselves and their colleagues for appreciation and recognition. Isolation is not the only stress placed upon would-be practitioners. As a mathematician pointed out in a plea for understanding addressed to a widely read scientific journal, there is also the loneliness of total involvement in a problem over days or weeks to the exclusion of nearly everything else; there is the risk of frustration when such enormous efforts end, as they most often do, in failure; there is the likelihood of a lifetime of insignificance, since nearly all the really major innovations in mathematics are the work of the exceptional genius, not the average contributor; and there is the near certainty of early obsolescence. With such a catalogue of discouragements, it may seem remarkable that anyone would want to become a mathematician. What is the nature of the experience, and what are its rewards?

The Joy of Understanding

BERS: I think the thing which makes mathematics a pleasant occupation are those few minutes when suddenly something falls into place and you understand. Now a great mathematician may have such moments very often. Gauss, as his diaries show, had days when he had two or three important insights in the same day. Ordinary mortals have it very seldom. Some people have it only once or twice in their lifetime. But the quality of this experience—those who have it know it—is really joy comparable to no other joy. The first time you learn something, you also have this joy. So—this has been observed by many people—the work consists in preparing yourself for this moment of understanding, which comes as a result of an unconscious process.

PUCKETTE: Most of my mathematical experience has been rigged, so far. I have been taught but not gone off on any research efforts where the outcome was uncertain. Most of the experience given to me has been moments of joy, simply because it is in fact possible to bring them to me. You stare at a problem on a competition or exam for 30 minutes or so and suddenly the whole thing flies together. The attraction [of mathematics] is that it's great fun.

SULLIVAN: There are always rare moments of understanding that are very nice . . . I remember one experience. I was already a graduate student and I was trying to learn some of Milnor's work [John Milnor, a mathematician at the Institute for Advanced Study in Princeton]. I remember going back and trying to think it through one more time. Suddenly I got the picture that he was actually trying to present in his whole book, a sort of geometrical picture that involved this idea of transversality which I've described before. From that, the whole book just sort of fell away. That was the point from which I reproduced the whole book, even though before it was all in my head as a big complicated thing. The whole thing just fell away and for the first time I realized that I really could understand some nontrivial mathematics. Before that I never really felt the master of something; I could do it forwards and backwards, but this was sort of a geometrical idea that was very strong. It was a very vivid experience.

This is part of the [problem of] lack of confidence about being able to understand mathematics. It's not so easy, mathematics. The idea that it is possible to really understand something very well and in a very simple way was a kind of thought process that I just didn't know. Before everything had been like history; I mean, how do you understand history? You know it, talk about it, go on and on and on; but you can't just suddenly see it coming out of a point. That to me was a big step. And the idea that you can have ideas like this, new ideas, not just understanding other people's ideas—that was really getting somewhere.

BERS: I remember why I decided to study mathematics. When I was in high school, I liked mathematics, but I did very little reading because there were few books about mathematics to read where I lived in Latvia. But I once constructed a one-to-one correspondence between an infinite set and a subset. I thought something must be wrong. I showed it to my geometry teacher who said, "This is the stupidest question I ever heard." [Today such examples are taught in beginning courses in set theory.] So I went away and felt very bad. I also tried to read one book which we had in the library which said the basic concept of mathematics is sets, but didn't explain what they were. I was lost. I applied to an engineering school. Then I saw in a bookstore—I was a little over 17—a small book on set theory by Kamke, and I said "Aha, I will learn what a set is." So I read it and it really hit me, and I felt, this is what I want to study.

The second such experience was when I first discovered something that I considered nontrivial. It was when I was working on gas dynamics, solving partial differential equations. It's too technical to explain; however, I remember how it was—a fantastic feeling, which one is lucky to have five or six times during a lifetime.

PUCKETTE: I suppose the first thing of a mathematical nature that I was really interested in was in kindergarten—no, this is true—where I learned the first 5 or 6 powers of 2. Just by adding 2 and 2 to get 4, 4 and 4 to get 8, and so on. I don't know why I decided to do this.

I got to where I couldn't add very well after awhile, but there were always second and third graders who I could ask for the next number. I don't remember what it was about it that fascinated or excited me. I couldn't have added 2 and 3 if you had asked me, so I certainly wasn't hit by any sense of structure (or at least none that I can remember now). But there was something about it that just fascinated me, and continued to do so for several months.

SULLIVAN: Although there is a point about what Professor Bers was saying about the joy of your own great moments of lucidity and discovery, mixed very sparsely amid years of confusion, there's also a more daily relationship with mathematics when you understand what someone else is doing. I think of it as a learning process. Oftentimes you are learning something new, but often it is something old, so it is not your own discovery. But there's a lot of joy in just daily thinking about mathematics. I mean, it's very rich. It's actually easier than you make it sound.

This is why there is often very little scholarship and sense of the past. A good advisor can sort of plop you down on that square there and you can start thinking about it; and because mathematics is very rich and current, you can *find* things. It's not that hard to do mathematics. I was amazed [to discover that] it's harder to understand mathematics than to do it. Of course it's hard to find something really good.

This is one of the great arguments: You see people around like this Thurston [William Thurston, a Princeton mathematician] who is sort of a geometrical magician and you say, "Well gee, why should I be trying to work on this? If he spends ten minutes on these problems he'll go so much farther than I." But this doesn't disturb one because there is such a vast range of problems that you can start almost anywhere, and there are nice things to see. It's like having a fantastic landscape and countryside [on which] you can go anywhere you like.

It's a pleasant *and* painful process. There's this dull pain all the time you have to think and concentrate and try to understand. But there are more frequent levels of understanding—a sort of superficial understanding when you can just say something empirically, then a little better understanding when you start seeing relationships, then [the best level] when you find a totally new area, which happens very rarely.

For me the attraction of turning to mathematics is . . . that in mathematics it is possible to actually make progress in a train of thought. The levels of understanding are conservative; I mean, every year I feel I change.

PUCKETTE: I would say my attraction to mathematics has been rather manipulated, so far. I've never come yet to the point where I can sit down and actually see a frontier that I can push back a little bit. I unconsciously spend a moment thinking about the Poincaré conjecture, the Fermat conjecture, and so on. I'll probably never even hope to see any of these solved. But it's great fun.

It's not obvious to me now why I'm interested in studying mathematics. But I know that I *am* very interested in studying mathematics.

BERS: Mathematics is growing so rapidly now that it is an effort just to understand some things that have come along. At some point one has to reconcile oneself that one will never understand too many things. There are very few mathematicians that have a real overview of all of mathematics. And at some point one has to understand that one will die without knowing many things one would like to know. I envy the people who are getting their education now because they will learn much more than I did.

Most of us have to accustom ourselves to the fact that there is only a certain level that we can reach. There always will be people, and one can name them, who will be definitely better than you, or go faster. It is somehow like a big cathedral, mathematics. You don't expect that you will do the overall design. But you can still be a good worker.

SULLIVAN: One of the things I like about mathematics is that you should be able to enjoy a discovery that has just been made as much as if you made it yourself.

BERS: And also the enjoyment of others. Most professional mathematicians spend a large part of their time teaching. Teaching calculus, for example, I find a very enjoyable occupation, because I get to communicate very great ideas. And mathematicians are like a small village: they gossip shamelessly; they know each other. There are too many nowadays really, but they are still a small village, partly because they are all alone. A chemist works with plastics or an electrical engineer with transistors; people have some feel for what these are, but what does a mathematician do? Nobody understands. So there is a certain amount of loneliness. The advantage is that you are doing something that in a way is very real and imperishable. And impersonal. And—I don't claim that mathematicians are more honest than other people—but in your work you have to be honest, because if you fool yourself you are lost. And this means something.

Some of the mathematical writings of Archimedes are written in the form of letters. Obviously there was communication with other mathematicians from the very beginning. At first often letters. And now, especially in a center like where we are, one can communicate with many people. Social intercourse was always necessary for mathematics. After all, the idea of a proof must have developed from the idea of discourse. To prove something was to convince somebody.

This [discourse] is one of the secondary attractions of mathematics. Of course when you study the works of a great old man, you really are in a sense in a very intimate intellectual intercourse. Secondly, on a more mundane level, you have friends all over the world. I mean, we know people, and some of them quite closely at least on the intellectual level, in Japan, in Red China, in Russia, and all over the world. When you meet them, you are able to talk about mathematics very easily, to communicate across these boundaries, and also across the age barriers. In what other profession can you see a 20-year-old kid lecturing at the blackboard and old venerable people sitting there and taking notes. This is a wonderful thing.

SULLIVAN: There are only risks and hazards if you think that the objective is to make great discoveries in your own name. If you just want to wander out in mathematics and manipulate it, eat it, feel it, try to understand it, communicate it and so on, there's a tremendous amount to do that is enjoyable. Painful too, because it's frustrating when you don't understand things. But there is a lot to do that is enjoyable. This process can go on as long as you can stay awake.

Pattern and Form

However natural eating and breathing mathematics may be for a practitioner, it is not part of the normal diet of most of us. But we nonetheless live in a world in which mathematical concepts and generalizations based on them increasingly determine our outlook. Plato, in a lecture that puzzled his contemporaries in ancient Greece, expressed a vision of a universe organized on mathematical principles which he identified as the Good. A modern interpretation of that idea, according to the functional analyst Felix Browder, might begin with a vision of mathematics as "the science of significant form. . .the ultimate and transparent form of all human knowledge." He argues that the eighteenth century vision of a rational, mechanistic universe was influenced in no little way by the Newtonian synthesis of mathematics and a Copernican cosmos. Who would argue that the Einsteinian synthesis expressed in the theory of general relativity has not altered the prevailing world view in our own century?

A similar point of view was expressed by the mathematician turned philosopher Alfred North Whitehead: "The notion of the importance of pattern is as old as civilization. Every art is founded on the study of pattern. The cohesion of social systems depends on the maintainance of patterns of behavior, and advances in civilization depend on the fortunate modification of such behavior patterns. Thus the infusion of patterns into natural occurrences and the stability of such patterns, and the modification of such patterns is the necessary condition for the realization of the Good. Mathematics is the most powerful technique for the understanding of pattern, and for the analysis of the relation of patterns.

"Here, we reach the fundamental justification for the topic of Plato's lecture. Having regard to the immensity of its subject matter, mathematics, even modern mathematics, is a science in its babyhood. If civilization continues to advance, in the next two thousand years the overwhelming novelty in human thought will be the dominance of mathematical understanding."

Let us explore the idea of mathematics as culture and as a formative influence on culture further in the thoughts of our captive mathematicians.

SULLIVAN: There is a famous remark about the texture of mathematics that the things we work on today could be explained very simply to Euclid or somebody like that. There is a great permanence of ideas; really great ideas are somehow very permanent. This is a kind of impact that doesn't really affect how we get our bread every day, but human knowledge is a big piece of culture, and good ideas just increase. There's not much in existence that's really so good it will, like mathematics, last thousands of years.

BERS: It can suddenly become exceedingly important to humanity the moment we make contact with an extraterrestrial intelligence, if there is any. Mathematics is the only common language we will have. If we get a signal from somewhere, 2, 3, 5, 7, 11, 13... [the sequence of prime numbers], then we will know, "This is it." We cannot conceive of an intelligence with a different kind of mathematics. This one thing would be unmistakable.

There is also a certain indirect influence between general ideas of mathematics and the way people think about other problems. I used to illustrate this with an analogy between Euclid's mathematics and the Declaration of Independence, which of course was written consciously as a mathematical proof, and the Lobachevsky or Hilbert point of view represented in the Gettysburg address; there is a self-evident axiom in the Declaration of Independence, ("All men are created equal") and there is in the Gettysburg address a proposition to be tested by consistency proof.

SULLIVAN: So what you're saying is like what I was trying to say, that there is some increase or change in human thought, and the achievements of mathematics are a part of that. You don't have to postulate a hypothetical communication with outer space to get at the idea of an absolute, proven thought achievement.

BERS: Other great achievements, like music and painting, are easily enjoyed by a large number of people. We can hope that as education and leisure develop more people will enjoy them. But the number of people who really enjoy mathematics is very small. Thus far our attempts to communicate the beauty of our art to a wide audience have failed miserably.

SULLIVAN: We could have on television a curve tracing back and forth and sort of converging to a space-filling curve [see box on p. 33]. Just in 5 or 10 seconds you could give everybody who saw it an intense feeling of geometrical joy, but without understanding the definitions or anything. And it would do this, I'm sure. I've seen an audience watch such a curve form. It's fantastic! There's a pattern happening there, an infinite process, and suddenly the whole screen is filled with a curve. It's an artistic thing.

INQUIRER: You're saying that there's something in the recognition of that pattern, something inherent in the way we look at space such that everybody has reactions to it whether one knows mathematics or not.

A Space-filling Curve

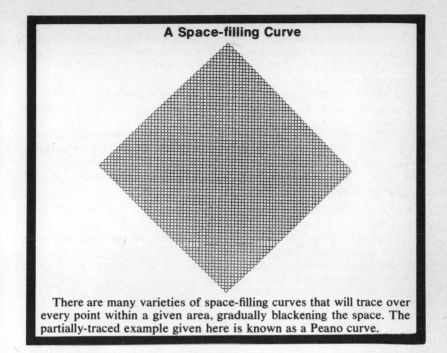

There are many varieties of space-filling curves that will trace over every point within a given area, gradually blackening the space. The partially-traced example given here is known as a Peano curve.

SULLIVAN: Sure. With a computer and television now you can do all kinds of things. It would be a dynamic piece of art.

I think that mathematics could be part of the flowering of the human spirit in some fundamental way. I mean the fact that children can be very amused, like with what Puckette was doing with 2s, making little squares, doing various patterns. These are the same kind of things a mathematician enjoys, except often they are clothed in highly technical language.

There are all kinds of games which illustrate logic and deduction, and mystery stories are enjoyed by many people, so there's not anything special about mathematics that can't be enjoyed by the common man. It's part of everyone's brains. Actually I don't particularly feel that it's important that mathematics be enjoyed by a lot of people; that would be very nice, but it's more important that mathematicians work on good problems and pursue mathematics.

Clearly mathematicians take their subject seriously and believe it an important part of the human enterprise. This is underscored by a light-hearted exchange that occurred in the discussion when at one point Sullivan observed that mathematics is, for many mathematicians, a substitute passion or religion. To which Bers replied, "I didn't know it was a substitute." It is of course possible that mathematicians deceive themselves about their role in the scheme of things. But it is also possible that, in our ignorance of their world, we miss a fundamental and increasingly important aspect of our own.

Bibliography and Acknowledgments

The author would like to acknowledge the generosity with which Lipman Bers, Dennis Sullivan, and Miller Puckette gave of their time and themselves in serving as resources for this inquiry. Remarks by other mathematicians quoted above were obtained from published works included in the following list of additional reading.

Alfred Adler, Reflections—mathematics and creativity, *New Yorker*, **47** (February 19, 1972) pp. 39–45.

A. D. Alexandrov, et al., *Mathematics—Its Content, Methods, and Meanings*, Vol. 1, MIT Press, Cambridge, 1963.

H. J. M. Bos, and H. Mehrtens The interaction of mathematics and society. *Historia Mathematica* 4 (February 1977) 7–30.

Felix E. Browder, Is mathematics relevant? And if so, to what?, *University of Chicago Magazine*, **67:3** (Spring, 1975) pp. 11–16.

Paul R. Halmos, Mathematics as a creative art, *American Scientist*, **56** (April, 1968) pp. 375–389.

Jacques Hadamard, *Psychology of Invention in the Mathematical Field*. Dover, New York, 1945.

G. H. Hardy, *A Mathematician's Apology*, Cambridge University Press, New York, 1969.

John von Neumann, The mathematician, in R.B. Heywood, *The Works of the Mind*, University of Chicago Press, Chicago, 1947, pp. 180–196.

Henri Poincaré, Mathematical creation, *Scientific American*, **179** (August, 1948) pp. 54–57.

Donald R. Weidman, Emotional perils of mathematics, *Science*, **149** (1965) p. 1048.

Alfred North Whitehead, Mathematics as an element in the history of thought, in J.R. Newman, *The World of Mathematics*, Vol. 1, Simon and Schuster, New York 1956, pp. 402–416.

Part Two

Number Theory
Groups and Symmetry
The Geometry of the Universe
The Mathematics of Meteorology
The Four-Color Problem

Number Theory

Ian Richards

Number theory is concerned with the properties of whole numbers or "integers." It is one of the two oldest branches of mathematics, geometry being the other. But whereas the classical "circle and triangle" geometry of Euclid is essentially dead as a research science, number theory still contains a large body of unsolved problems. Indeed, some of its most intriguing problems go back to Euclid's time.

Let us start with a few of the problems which have been solved. Consider any prime number p. (A "prime" number is one which is not divisible by any smaller number except 1.) Say we could take the thirty-nine digit number

$$p = 170, 141, 183, 460, 469, 231, 731, 687, 303, 715, 884, 105, 727,$$

which was the largest prime known until the advent of modern computers. Suppose we multiply together all of the numbers less than $p - 1$ (where p is the large number given above): that is, we take 1 times 2 times 3 and so on up to $p - 2$. The result is called "$p - 2$ factorial" [and written $(p - 2)!$]. I have not computed this number, and neither has anyone else, and neither has a computer. Even to write it down would take more paper than there is in all the books in all the libraries in the world; even if a computer could compute it (which it couldn't), there would not be enough paper in the world on which to print out the answer. Yet the number must exist, for we can define it logically, and we can think about it.

Well now, suppose we take this number "$p - 2$ factorial" and divide it by p; there will be a certain quotient and then there will be a remainder (just as when dividing 72 by 7 we get a quotient of 10 and a remainder of 2). Of course, for the reasons mentioned, no one has ever carried out this division. But we know what the remainder would be. It would be 1. We know this because of a theoretical result called "Wilson's theorem."

Wilson was an eighteenth century English scholar, and few people have ever become immortal for less reason. He didn't prove his "theorem," and someone else, in a forgotten article somewhere, had actually stated *and* proved the theorem earlier. (I've forgotten the original author; history can be cruel!) Wilson's "discovery" was published by a sycophant friend, who predicted that the theorem would never be proved—indeed could never be proved—because the human race has no good *notation* for dealing with prime numbers. This remarkable statement was communicated to Gauss, a man regarded by many as the greatest mathematician of all time. (For more about Gauss, see the box on page 39.) Gauss proceeded to prove "Wilson's theorem" in five minutes, standing on his feet. Gauss remarked about Wilson that "he doesn't need *notations*, he needs *notions*."

There is a myth that a mathematician must be "very fast with figures." This is false—mathematics mainly involves intuition and the ability to grasp theoretical ideas. If a mathematician has these talents, he need have very little else. But Gauss was an exception. He was good at everything! He was a master of languages, he had memorized the whole table of logarithms—and he was a mathematician. Like many of the great mathematicians, he was also precocious. Before he was seventeen he had settled the age old problem of "trisecting the angle," by proving that it was impossible. When he was twenty-four (in 1801), he published the first systematic treatise on the theory of numbers, a book called *Disquisitiones Arithmeticae*. In the opening paragraph of this book he lists those men who he considers his predecessors in number theory; there were exactly four of them—Fermat, Euler, Lagrange, and Legendre. Furthermore, Gauss leaves the reader in no doubt that he considers himself the first among equals. For instance, in discussing a certain result (called today the "law of quadratic reciprocity") which others before him had discovered but which he, Gauss, had been the first to prove, he writes:

> The fundamental theorem [i.e., the reciprocity law] must certainly be regarded as one of the most elegant of its type. No one has thus far presented it in as simple a form as we have done ... What is more surprising is that Euler already knew other propositions which depend on it and which should have led to its discovery ... After Euler, the renowned Legendre worked zealously on the same problem in his excellent tract, 'Recherches d' analyse indétérminée', ... Legendre also attempted a demonstration and, since it is extremely ingenious, we will speak of it at some length in the following section. But since he presupposed many things without demonstration (as he himself confesses ...) ... the road he has entered upon seems to lead to an impasse, and so our demonstration must be regarded as the first. Below we shall give *two other demonstrations* of this most important theorem, which are totally different from the preceding and from each other.

In other words, what no one else could prove, Gauss proved once and then two more times for good measure.

Carl Friedrich Gauss

One day in 1785, a German schoolteacher, in order to keep his students busy, told them to add all the numbers from 1 to 100. There is a formula for this, which the teacher knew. But of course the students (they were "second graders") did not know it, and so the teacher could be assured of an hour's peace. (They had discipline in the schools in those days!) However one of the students, Carl Gauss, immediately walked up to the front of the room and presented the correct answer: 5050. Gauss knew the formula too!

Luckily for the history of mathematics, the teacher was big enough to realize that something extraordinary had happened. For there was no way that Gauss, the son of uneducated parents, could have been taught this formula: he must have discovered it himself. Thus began the career of a man generally regarded as the greatest mathematician of all time.

There is no space here to describe Gauss's work, or to attempt to analyze his complicated and frequently contradictory personality. He was a strange man, sometimes cold and aloof, but also a man of the most exacting intellectual standards. Today his name appears everywhere in mathematics and its applications: just to give one example, there is a unit of electromagnetism called the "Gauss."

A modern successor to *Disquisitiones Arithmeticae* is a book published in 1975 by the young English mathematician, Alan Baker. (In fact, Baker begins his book with a quotation from *Disquisitiones*.) Baker's work, which is in the fields of "transcendental numbers" and "Diophantine equations," has caused a revolution in number theory. We shall say more about these developments later.

Whole Numbers

Diophantine equations, named after a mathematician of ancient Greece called Diophantus, are concerned with the solution of equations in whole numbers. The restriction to whole numbers is what makes this a "number theoretic" as opposed to a purely computational problem. For, while an equation may have many solutions, only in special cases will these solutions be whole numbers. Thus, consider the following example:

$$x^2 = 2y^4 - 1.$$

This equation has solutions in which x and y are positive whole numbers. To be precise, it has exactly two of them. One is easy to see: $x = 1$ and $y = 1$. But probably no one would quickly find the second. It is $x = 239$, $y = 13$, and the Norwegian mathematician W. Ljunggren proved in 1942 that there are no others. (To be fair, I should add that this result is so special and so bizarre that it is not considered particularly central in number theory. I have stated it here just for fun.)

Another Diophantine problem, with a much richer history, is that of writing an arbitrary whole number as a sum of squares. The perfect squares are 1, 4, 9, 16, . . . , something which every schoolchild knows. Other numbers, like 6, are not squares, but they can be written as *sums* of squares, for example $6 = 4 + 1 + 1$. So far, so good, but the next question is harder. Suppose we want to write *any number* as a sum of squares (for instance, we have just seen that 6 is the sum of the three terms, 4, 1, and 1). How many terms will we need in general? Is three enough? Do we need four? Or five? Or, for very large numbers picked at random, say

5, 149, 176, 235, 882, 197, 318, 266, 512,

is it possible that we would need to put together very many perfect squares in order to generate the number? The answer is that four squares are always enough, whereas sometimes three squares are not. This fact, like so many others, was discovered by the ancient Greeks. But in number theory, unlike geometry, the Greeks were rarely able to *prove* their discoveries. The sharp contrast between the ease of stating such theorems and the difficulty of proving them is one of the things which attracts mathematicians to number theory. We have a simple statement like:

Every positive whole number is the sum of four or fewer perfect squares.

All right, you may say that you don't care. That is perfectly reasonable; but you do understand the statement. And yet, in all probability, neither you nor I could prove it in a thousand years.

That was a touch bombastic, but the point must be made—that just because we understand what something *means*, it doesn't necessarily follow that we can *do* it. With physical things, say climbing Mt. Everest, everyone understands this point intuitively. And most of us have at least some begrudging respect for the people who climb mountains. Yet to my taste, the "four-squares theorem" is far more interesting. Here is a theorem which was true before the pyramids were built, before there was any man who could understand or appreciate it. Furthermore, when we first discover it (as the ancient Greeks did), it comes as a speculation: something which *seems* to be true, but which we cannot prove. It seems to be true because it works in every case that we try, perhaps thousands of cases, or if we are moderns and have a computer, perhaps millions. Still, since this is a mathematical problem, no matter how many cases we test, there are always infinitely many lying beyond them—and one of these might contain the exception which destroys the rule. Only a logical proof can encompass infinitely many cases at the same time, and bridge the gap between speculation and established truth.

For a long time (no one can predict how long), the "discovery" remains a mere conjecture. Then finally someone proves it—or disproves it. At least we hope that this will happen. In the case of the four-squares theorem, it did happen, after about two thousand years: the conjecture was proved to be true. Euler, one of the two greatest mathematicians of the eighteenth century, worked at the problem on and off for much of his life. But it fell to his rival Lagrange to find the proof. (For more information on Euler and Lagrange, see the boxes on pages 42 and 43.) Since then other proofs have been given (curiously, the first proof is still the simplest), and some of these proofs involve unexpected tie-ins with mathematical physics. (These tie-ins come via the so-called "elliptic functions.") It would be fair to say that there is a substantial segment of modern mathematics which has been influenced by the four-squares theorem.

It is a curious fact about the human mind that people will work harder to do something which captures their imagination than they will for any practical purpose. A recent project carried on by J.B. Rosser, L. Schoenfeld, and J.M. Yohe at the University of Wisconsin's Mathematics Research Center illustrates this point. The three mathematicians were using a computer to gain evidence in favor of a certain number-theoretic conjecture. The conjecture is called the "Riemann Hypothesis," and it is somewhat technical to state. What we say here may only give the flavor of it. However, for our present purposes, that is sufficient; we will discuss this hypothesis later in more detail.

Leonhard Euler

It is hard to imagine how one man could write a hundred books. Still, some men have. But only one man in history has ever composed that much mathematics. Leonhard Euler (1707–1783) was one of the five or six greatest mathematicians of all time. Born near Basel, Switzerland, he spent most of his working life in Berlin and St. Petersburg. His "Euler characteristic," the first discovery ever made in topology, is sometimes taught to bright students in high school; and every calculus book written today is the direct descendant of a treatise which Euler wrote. But other parts of his work are so advanced that even now they are just barely understood. The man who did this work was a devoted family man and a devout Christian. He was genuinely humble and sometimes withdrew his papers to allow younger men to publish first. And it is interesting to note that Euler, like some other great figures in the history of culture, continued working after he went blind.

Joseph Louis Lagrange

The chief rival of Euler, and one of the two greatest mathematicians of the eighteenth century, Lagrange was a man of very different temperment. Where Euler was devout, Lagrange was a skeptic, and where Euler was incredibly prolific, Lagrange published only his most brilliant ideas. Yet the two men achieved an almost equal impact on mathematics. As a man Lagrange was generous, even to the point of forgiving and hoping for improvement in a younger mathematician who continually stole ideas from him. Yet he was also prone to melancholy, and when he felt his mathematical powers slipping late in life, he settled into despair. (Photo by permission of "Germanisches National Museum," Nuremberg.)

The theory of prime numbers leads to a certain "function of a complex variable" called the "Riemann zeta-function." The problem concerns the "zeros" of this function (that is, points where the function equals zero). It is known that, with certain trivial exceptions which we can classify, all of these zeros lie inside a certain infinite strip in a 2-dimensional plane. Also it is known that there are infinitely many zeros. The Riemann Hypothesis asserts that all of the zeros lie on the central line of the strip. This hypothesis was made by the German mathematician Bernhard Riemann in 1859, and it remains unproved to this day. It is one of the most famous unsolved problems in mathematics, for a large part of the so-far undecided theory of prime numbers hangs on it. (For more on Riemann, see the box on page 45.)

Rosser, Schoenfeld, and Yohe, in their research on the Riemann Hypothesis, used a computer to show that the first three million zeros of the zeta-function do lie on the central line of the strip. Of course, such information does not constitute a proof—computer data rarely does—for there are infinitely many zeros, and we have considered only the first three million of them. Still, three million cases without a single exception might seem strong evidence. However, the history of number theory provides examples showing that such finite reasoning is quite unreliable. There is, for instance, a discovery made by the English mathematician J.E. Littlewood in 1914. He found that a certain equation is true for all numbers up to a very large value (called today "Littlewood's constant"), but that this equation is false for infinitely many numbers lying beyond this value. The best estimates to date suggest that Littlewood's constant exceeds 10^{100}—that is, a 1 followed by 100 zeros—a magnitude easily greater than the number of atoms in the visible universe.

There is another amusing sidelight to the Rosser, Schoenfeld, Yohe research. Pure mathematicians are somewhat suspicious of computers. This is because the computer is a machine, and machines can make mistakes. But if the machine did make a mistake, how would you know? The results coming out of a computer generally have to be accepted on faith. Recognizing this, and thus wary, Rosser, Schoenfeld, and Yohe approached their work with great caution; and they checked and rechecked their programs with monastic zeal. In the course of this checking, they discovered that the computer itself contained several errors in its internal logic. The computer had been used for years without anyone's discovering this fact. To be sure, the errors were subtle and not likely to affect routine computations. Still, three people working on an impractical problem in one of the most ethereal domains of mathematics were the first ones to ferret out these mistakes.

Prime Numbers

A number greater than one which is not divisible by any smaller number except one is called "prime." Prime numbers are important because they form

Bernhard Riemann

There is an old cliché that "an artist must suffer." Bernhard Riemann did, for much of his life. This was not because he was unknown. He was in fact recognized as a genius. He just had no job! And to make matters worse, several members of his family depended on him for support. So Riemann shared his meager income—starved—and worked! His work in mathematics has influenced almost every branch. Just to give one example, Einstein's general theory of relativity is based on mathematics that Riemann developed. The eight-page paper which contains the Riemann Hypothesis (mentioned prominently in this essay) is the most important paper ever published in the theory of numbers.

Finally in 1859, Riemann's luck changed. He obtained a professorship in Göttingen, and three years later he married and had a daughter. But the years of neglect caught up with him, and in 1866, at the age of thirty nine, he died of tuberculosis.

the "atoms" from which arbitrary numbers are built up by multiplication. For example, $15 = 3 \cdot 5$, and $8316 = 2 \cdot 2 \cdot 3 \cdot 3 \cdot 3 \cdot 7 \cdot 11$. (The dot indicates multiplication; thus for "$3 \cdot 5$" read "3 times 5.") For more examples involving prime numbers and their relation to other numbers, see the box on page 47.

Despite the simplicity of their definition, surprisingly little is known about prime numbers. We know that they are basic to modern arithmetic and algebra. But of the laws governing their distribution, we know very little. This is not due to neglect: prime number theory has attracted the attention of some of the best mathematicians of modern times. Perhaps no field of mathematics contains so many unproved conjectures.

What do we know about primes? Well, Euclid proved that there are infinitely many of them, and for the next two thousand years, this was the only *theorem* in the subject, although many conjectures were made. For instance, it was noted in antiquity that there are numerous cases of "twin primes," i.e., pairs of consecutive odd numbers both of which are prime, such as

$$17, 19 \quad \text{or} \quad 29, 31 \quad \text{or} \quad 41, 43.$$

The ancients conjectured that there are infinitely many of these twin primes. This conjecture is still unproved. However, recently the Chinese mathematician Jing-run Chen demonstrated a somewhat weaker form of it: he showed that there are infinitely many pairs of consecutive odd numbers, p and $p + 2$, in which the first number p is prime, and the second one $p + 2$ has at most two prime factors. (In a "twin prime" we would require *both* p and p + 2 to be prime; here we could say that p is prime but p + 2 is merely "almost prime.")

Despite the somewhat fragmentary nature of this result, Chen's theorem has rightly been hailed as a major achievement. For one thing, no one knows how long it will take before anyone is able to do better. Also, Chen's proof draws on almost everything that has ever been done in prime number theory: to write it down, in a form that would be comprehensible to a mathematician who had not specialized in this problem, would require two books of about three hundred pages. By this I mean that the entire contents of both books—six hundred pages of equations and arguments, written in a style whereby the Pythagorean theorem of Greek geometry could be proved in about four lines—would have to be mastered before Chen's theorem could be understood. (Incidentally, I have in mind books which actually exist. The first might be any one of several standard treatises on prime numbers, and the second is a book titled *Sieve Methods* by H. Halberstam and H.-E. Richert. Chen's theorem is given as the culmination of the second book.)

To shift to a different quiestion: how do we tell whether a given number is prime? We know in principle how to do this. Take a number n. If n is not prime, then n has some smaller divisor, say 2 or 3 or 4 . . . . So we could test whether n is prime by dividing it by all smaller numbers. However, there are

Prime and Composite Numbers

Mathematicians divide the positive whole numbers into three classes:

The single number 1, which is called the *unit*. Multiplying by 1 produces no effect, e.g., $3 = 3 \cdot 1 = 3 \cdot 1 \cdot 1 = 3 \cdot 1 \cdot 1 \cdot 1$ etc.

The *primes*, that is those numbers which have no divisors except themselves and one. Thus 17 is a prime, whereas $15 = 3 \cdot 5$ is not.

The *composite* numbers, that is numbers greater than 1 which are not prime. For example, $4 = 2 \cdot 2$, $6 = 2 \cdot 3$, and $9 = 3 \cdot 3$ are composite.

There are twenty-five primes less than one hundred, namely:

2, 3, 5, 7, 11, 13, 17, 19, 23, 29, 31, 37, 41,
43, 47, 53, 59, 61, 67, 71, 73, 79, 83, 89, 97 .

(No even number except 2 is prime since every even number has 2 as a divisor.) The prime numbers form the "atoms" from which arbitrary numbers are built up by multiplication. For example, $12 = 2 \cdot 2 \cdot 3$, and $210 = 2 \cdot 3 \cdot 5 \cdot 7$, and there is no other way to write 12 or 210 as a product of primes. (Of course, we ignore trivial variations in the order of the terms, e.g., writing $12 = 3 \cdot 2 \cdot 2$ instead of $2 \cdot 2 \cdot 3$.) The same statement applies to any number, and this fact, the unique factorization of integers into primes, forms the foundation of theoretical arithmetic.

faster procedures, the most famous of which is the "Sieve of Eratosthenes" (see Figure 1).

A related problem is to generate all of the primes up to a given limit. The Sieve of Eratosthenes also works well for this purpose: thus in Figure 1 we have generated all of the primes less than twenty. Much work has been done in this area. For example, the American mathematician D.N. Lehmer in 1914, before the days of computers, calculated all of the primes up to ten million.

It will be useful for our discussion to consider how Lehmer did this. First he started with the numbers from one to ten million. Then immediately he eliminated from his list all numbers ending in the digits 2, 4, 6, 8, or 0, since these numbers are even, and hence not prime. Similarly he could eliminate all numbers ending in a 5, since they are divisible by five. (Of course the numbers ending in a 0 are also divisible by five, but they have already been eliminated.) From here on, the work gets more difficult. For it is not so easy to tell whether a number is divisible by three or seven or eleven, etc. Lehmer's method was the same as that shown in Figure 1: to eliminate every third number, then every seventh number, every eleventh number, and so on. To aid him in this work he invented various sorts of mechanical

Erathosthenes, one of the Greek mathematicians of antiquity, invented a process to "screen" numbers to sort out primes from composites. His process is called a "sieve" because of the way it depends on successive eliminations. We show how the sieve would be used to generate all of the prime numbers less than twenty. At the same time, of course, it tells for each particular number less than twenty, whether it is prime or not:

$$1 \; ② \; ③ \; 4 \; ⑤ \; 6 \; ⑦ \; 8 \; 9 \; 10 \; ⑪ \; 12 \; ⑬ \; 14 \; 15 \; 16 \; ⑰ \; 18 \; ⑲ \; 20$$

| | 2 | 2 | 2 | 2 | | 2 | | 2 | | 2 | 2 |
| | | 3 | 3 | | 3 | | 3 | | 3 | | |

What we have done is to eliminate all multiples of 2 or 3 (or both); the circled numbers which remain are prime. Exception: the special number 1 is called the "unit" and does not count, while 2 and 3 themselves are prime and are not eliminated.

The justification for checking only for multiples of 2 and 3 is as follows: If a number n has any factors (i.e., if n is *not* prime), then some factor is less than or equal to $\sqrt{n}$; for if $n = a \cdot b$, then either a or b must be less than (or equal to) $\sqrt{n}$. Also, any factors of n must contain prime subfactors, so only prime factors need be considered. For $n = 20$ or below, $\sqrt{n} < 5$, and the only primes less than five are 2 and 3: just the ones we used!

Figure 1. The Sieve of Erathosthenes

aids—essentially mini-computers; and these may have been among the first mechanical calculators ever put to effective use. (The idea of computers is much older, but many of the early models never got off the drawing board.)

The Sieve of Eratosthenes, while valuable in doing computations, was for centuries considered theoretically useless. The trouble was that the successive eliminations seemed to be too random and impossible to predict. The situation changed when in 1920 the Norwegian mathematician Viggo Brun introduced a new method based on the "principle of inclusion-exclusion." This principle is charmingly simple, and the reader who wishes to pursue it may turn to the box on page 50-51. Before we imagine that it is *so* simple, however, I would like to ask you to put yourself in Brun's place. We have a problem: to find a *theoretical* method, applicable in advance and not involving exhaustive calculations, for saying something about the prime numbers lying in an arbitrary interval. This interval might be too large to handle computationally. For example, we might ask: How many primes are there less than a billion billion billion? If we cannot calculate this number exactly (and no one could), can we at least *approximate* it? The answer turns out to be "yes," and Brun's sieve furnishes one of the main methods for attacking such problems. Now, if you will turn to the box on page 50-51 and see what Brun did, I think you will find some of the initial steps quite surprising. Starting with the problem of prime numbers, Brun developed a new approach, involving areas of mathematics more commonly associated with information theory and linguistics. After a long detour, his method eventually comes back and solves the original problem. (Actually, it only partially solves the problem, but Brun's theory and its extensions are still the best methods that we know.) This kind of originality is what separates the master from the routine practitioner.

Let us return to the question: how do you tell whether a given number is prime? We have seen that, in principle, the Sieve of Eratosthenes gives an answer. I say, "in principle," because if the number is too large the Sieve may be quite impractical. It turns out that for certain *special* numbers there are faster methods. The book *Mathematical Snapshots* by the great Polish mathematician Hugo Steinhaus contains the provocative statement:

The seventy-eight digit number:
$2^{257} - 1 = 231, 584, 178, 474, 632, 390,$
$847, 141, 970, 017, 375, 815,$
$706, 539, 969, 331, 281, 128,$
$078, 915, 168, 015, 826, 259, 279, 871 ,$
is composite; it can be proved to have divisors but these are not known.

That book was once a Christmas present for me. My father, a slightly more practical-minded type (a biologist) snorted that Steinhaus's statement seemed ridiculous—how could you know that a number has divisors without knowing what they are? I didn't understand it either, but I thought it was pretty neat! In fact this "nonconstructive" theorem (in which someone had proved that a number existed without being able to find it) is the main thing I remember from Steinhaus's book. It was years before I had any idea how such things were done. The method is roughly this: There is an obvious principle which says that if you put seven leaflets into six letter boxes, then at least one box must contain a duplication of leaflets. In another version, if there are 100,000 seats in a football stadium and every seat is filled, but only 99,999 tickets were sold, then somebody has sneaked in without paying. (Try to find him.) So it appears that by counting you can often prove the *existence* of something, whereas a much more exhaustive search would be necessary to *find* it. In fact, if the numbers involved are large enough (say seventy-eight digits), then the "exhaustive search" becomes a "preposterous search"—there are not that many atoms in the solar system! Yet a clever mathematician may be able to do the count; not one step at a time, of course, but by using "arithmetic." That, in essence, is how mathematicians prove the existence of numbers which they cannot actually find.

The Prime Number Theorem

Up to now we have considered mainly unsolved, or partially solved, or occasionally solved problems. The reader may be wondering whether anything gets done in this subject. One of the clear victories is a result now called the "Prime Number Theorem." It is interesting to trace the history of this problem. Its theme is the average density of primes in the interval starting

Brun's Sieve and the Law of Inclusion-Exclusion

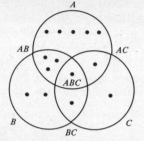

Suppose each of the regions above contain a finite number of elements:

A = number of points in top circle = 10
B = number of points in left circle = 7
C = number of points in right circle = 4
AB = number of points in left "football" = 4
AC = number of points in right "football" = 2
BC = number of points in bottom "football" = 2
ABC = number of points in center "triangle" = 1

(The "footballs" show where two regions intersect; the "triangle" is the intersection of three regions.) Suppose we want to count the total number of elements in the three circles combined (= 14). We could take

$$A + B + C = 21, \quad \text{but this is too big} \tag{1}$$

(for the points in AB, AC, and BC are counted more than once). We could take

$$A + B + C - AB - AC - BC = 13, \quad \text{but this is too small} \tag{2}$$

(for the points in the triple intersection ABC have been subtracted too often). Finally, if we take

$$A + B + C - AB - AC - BC + ABC = 14, \tag{3}$$

we have the right answer.

The reason is simple; at the end each point has been included and excluded just the right number of times. An obvious principle says: to count a set correctly, count everything once. In this case, each point has finally been counted exactly once. For example, consider the point in the center ABC. Look again at

$$A + B + C - AB - AC - BC + ABC.$$

The point gets counted three times with $A + B + C$, then excluded three times with $- AB - AC - BC$ (running total so far = 0); then finally the point is put back in by adding in ABC. So, similarly, it goes for all the other points.

The relevance of this to the sieve problem becomes clear if we use the law of inclusion-exclusion to count the primes in some interval, say the interval from 11 to 30. Actually we shall count the composite (i.e., nonprime) numbers, and then deduce the number of primes by subtraction. First, we observe that there are altogether twenty numbers in the interval 11, 12, 13, ... , 30. Every composite number in this range must be divisible by at least one of the primes 2, 3, 5 (possibly, of course, by several): this holds since the next prime is 7, and the smallest composite number with *only* 7's or larger primes as factors is $49 = 7^2$. Now we have a pattern which is strikingly similar to the previous three-circle figure:

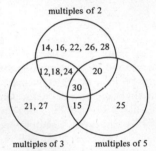

(Of course 21, say, is also a multiple of 7; but we have no space for that category.) There are fourteen composite numbers between 11 and 30, and twenty numbers altogether, so there must be six primes. And indeed there are: the primes 11, 13, 17, 19, 23, 29.

While this procedure may seem no better than merely counting the primes, the logical principle behind it is applicable to large intervals where counting would be impractical, say the interval from nine billion to ten billion. To give the idea, note that the "multiples of 2" circle contains altogether ten points, and *that this is no accident:* there are twenty numbers from 11 to 30, and half of them are even! Similarly the "multiples of 5" circle contains $^{20}/_5 = 4$ points. The "multiples of 3" circle contains seven points, whereas $^{20}/_3 = 6.666 \ldots$. (What causes the bonus?)

Now we come to the most surprising part of Brun's method. Let us consider once again the inclusion-exclusion formulas (1), (2), and (3) which we developed for counting the points in overlapping circles. Recall that the formula (3) is exact, while (1) and (2) are respectively too big and too small. Brun's trick was to avoid (3), and use (1) and (2) instead. His reasoning was that, even though the formulas (1) and (2) are not exact, they are *simpler* than (3), and they give reasonable approximations. Actually, to keep this in perspective, we have to assume that instead of the three sets A, B, and C (as in the pictures drawn above), there are thousands of sets. Then the number of combinations grows beyond all sensible bounds (more than the number of atoms in the solar system—that sort of thing), and cutting this number down was Brun's primary concern.

with zero and going up to a given magnitude x. For instance, as we saw above there are 25 primes less than 100. Similarly, we could count the number of primes less than 1000, or less than 1,000,000, and so on. If we do this, we find that the average density of primes decreases. Thus the numbers of primes less than one hundred, one thousand, and one million are:

	primes < 100	primes < 1000	primes < 1,000,000
Number of primes	25	168	78,498
percentage of primes	25%	16.8%	7.8%

By looking at just such tables as this, Gauss and Legendre around the year 1800 came to a remarkable conjecture. They noticed that the number of primes in various regions seemed to be related to the function log x, a relationship which at that time seemed completely inexplicable.

The function log x, sometimes called the "natural logarithm," arises in differential calculus in connection with problems of growth and decay. For example, consider a loan at 100% interest, compounded in tiny bits, so that it grows continuously. (The fact that 100% interest is usury need not detain us.) Then log x represents the number of years which must pass before 1 dollar turns into x dollars. Mathematicians call this process, whereby money increases under compound interest, the "law of exponential growth." By changing the description of the process slightly, it can be made applicable to such varied problems as population growth, radioactive decay, cooling bodies (and even compound interest charged at reasonable rates).

What Legendre and Gauss observed was that the *the number of primes less than x is approximately equal to x/log x*. The approximation, in percentage terms, grows better and better as x gets larger (see Figure 2). We notice that the percentage errors are still rather large, even when x is a billion. However, Gauss was not just a number theorist. He also founded the field of mathematical statistics and applied it to problems as diverse as astronomy and prime counting. Gauss analyzed the behavior of the errors in the prime formula, using what statisticians today call "the method of least squares." He came to the conclusion that, as x approaches infinity, the errors will eventually approach zero.

This "Prime Number Conjecture" (it was a conjecture in 1800) struck mathematicians as most remarkable. On the one side of the equation we

x	Number of primes < x	$x/\log x$	Percentage error
one thousand	168	144.8	16.0%
one million	78,498	72,382	8.4%
one billion	50,847,478	48,254,942	5.4%

Figure 2. Data Supporting the Prime Number Theorem

have the prime numbers; on the other side, the function log x coming from calculus and related to population growth. A marriage of the discrete and the continuous!

It took fifty years before anyone made any progress towards proving the Gauss–Legendre conjecture. The first person to do so was the Russian mathematician Pafnutii Chebychev around 1850. He got a partial result, and his ideas were then imitated by others. But eventually it turned out that his methods would not go any further, and they were abandoned. In 1859 Bernhard Riemann, a newly admitted member of the Germany Academy of Sciences, availed himself of one of the privileges of membership—publishing papers in the Academy's journal. (Although he was barely over thirty, he was to die a few years later.) His eight-page paper had the title "On the Number of Primes Less Than a Given Magnitude." The paper was terribly sketchy, and the arguments contained large gaps. Very little was definitely proven. But nearly everything that has been done in the theory of prime numbers since then has been influenced by that paper.

For about thirty years, other mathematicians tried to prove the main results enunciated in Riemann's paper—but to no avail. Then finally, in 1894 (thirty-five years after Riemann's paper had appeared), the French mathematician Jacques Hadamard scored an important breakthrough. Still the original conjecture of Gauss and Legendre remained open. But it was not to last much longer. Two years later, in 1896, Hadamard and de la Vallée Poussin working independently proved the Prime Number Theorem.

Their proofs (both of which were based on the work of Riemann) used very indirect methods. These methods came from the theory of functions of a complex variable (a branch of advanced calculus which uses "imaginary" numbers). In fact Hadamard, in making the breakthrough mentioned above, developed some new techniques in complex variables, and these techniques later found application in the theory of radio waves. They can be used to show whether a filter, which is a device used to remove the static from radio signals, is likely to blot out the information contained in the signal.

All of this was very nice from the viewpoint of a growing science. But some mathematicians felt dissatisfied. The proof of the Prime Number Theorem was so roundabout that no one really understood it. They could follow it step by step—it was correct—but what it had to do with *prime numbers* seemed very obscure. For years people sought an "elementary" proof, that is, one using only the basic properties of prime numbers, and avoiding such esoteric methods as complex variables and wave analysis. Again this development took a long time, but a proof of the desired type was finally found by Atle Selberg and Paul Erdös in 1948. However, the new proof turned out to be just as hard to understand as the old one. It is technically "elementary" in the sense that each step is elementary. But there are so many steps, and the way they fit together is so complicated, that no simple picture emerges. Probably this is unavoidable. There seems to be a principle of "conservation of difficulties" which says that a hard theorem is hard, no matter how you approach it.

The Riemann Hypothesis

Our next example is more technical than the previous ones, and to discuss it we shall have to use a few technical terms. Still, because of its importance, it would be a shame to omit it. Riemann, in the paper of 1859 mentioned above, made a conjecture which has never been solved. This conjecture is now called the Riemann Hypothesis, and it has become a cliché. It is taken as the standard model of a terribly hard or terribly desirable mathematical result. Thus, for example, when I was an undergraduate I once held a small research contract. I expressed doubts about my ability to do anything significant. To reassure me, my advisor said, "Well, they don't expect a Riemann Hypothesis out of you!"

At first glance, this hypothesis looks disarmingly simple. It involves the solutions to a certain equation. Now while that equation may seem complicated to a nonspecialist, everything in it is familar to students of university calculus in their second year. Yet, while we can easily write the equation down, nobody knows how to solve it. Here is one form of it:

$$1 - \frac{1}{2^s} + \frac{1}{3^s} - \frac{1}{4^s} + - \cdots = 0 .$$

The unknown s in this equation is presumed to be a complex variable, that is, $s = a + ib$, where $i = \sqrt{-1}$.

The Riemann Hypothesis asserts that (with certain specific exceptions which are well known), every solution to this equation lies on the line in the complex plane where $a = \frac{1}{2}$ (see Figure 3). A proof of it would automatically imply sharp improvements in many of the known results about prime numbers. Edmund Landau, an outstanding German number theorist, considered the matter so important that he had a whole chapter of his famous 1927 treatise *Vorlesungen über Zahlentheorie* entitled "Under the Assumption of the Riemann Hypothesis." He did not prove any of the "theorems" in this chapter, but merely showed that they would follow from the Riemann Hypothesis.

Riemann's equation has an infinite number of solutions—that much is known—and the important ones all lie inside a certain infinite strip in the complex plane. The Riemann Hypothesis asserts that the solutions all lie on the central line of this strip. As we noted earlier, three mathematicians at the University of Wisconsin, using a computer, verified that this holds for the first three million solutions. However, as we also noted before, such computer results can never prove anything definite about the infinitely many solutions which lie beyond the range of the computer's search.

There is another viewpoint, the purely theoretical one. Since the solutions are confined to an infinite strip, we can think of them as coming in a definite order, like cars on a highway. Looking at the problem in this way, we can ask whether (since we cannot prove the Riemann Hypothesis) we can at least

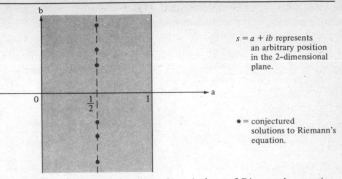

$s = a + ib$ represents an arbitrary position in the 2-dimensional plane.

● = conjectured solutions to Riemann's equation.

The Riemann Hypothesis involves the solutions of Riemann's equation

$$1 - \frac{1}{2^s} + \frac{1}{3^s} - \frac{1}{4^s} + \cdots = 0.$$

It is known that (with certain trivial exceptions) all of these solutions lie somewhere inside the shaded "critical strip," and it is conjectured that they all lie on the central line of this strip. A proof of this conjecture would have far-reaching consequences for the theory of prime numbers. However, the conjecture has remained open since Riemann made it in 1859.

Figure 3. The Riemann Hypothesis

show that a *positive percentage* of the solutions lie on the critical line. This was done by Selberg in 1942. However the percentage that Selberg was able to establish was extremely small—a tiny fraction of one percent. In 1974, Norman Levinson of MIT succeeded in proving that *at least one third* of all the solutions lie on the critical line.

Levinson's early work was not in number theory, but rather in ordinary differential equations, a field related to mathematical physics. His work on the Riemann Hypothesis came shortly before his recent death from cancer. This achievement could stand on its own, but—coming at the end of a long and distinguished career—it should help to dispel the notion that mathematics is a young man's game.

The Solution of Equations in Integers

Remember the equation $x^2 = 2y^4 - 1$ which we considered earlier? It has exactly two solutions in positive integers (i.e., whole numbers), namely $x = 1$, $y = 1$, and—rather more interesting—the pair of values $x = 239$, $y = 13$.

The theory of solving equations using only integers is called "Diophantine equations" after the ancient Greek, Diophantus, who originated it. Because of the inherent simplicity of its statements, this theory has remained at the center of number theory from antiquity up to the present time. Mathematicians feel that, since the equations look so simple, one *should* be able to solve them! However, as the reader will have guessed by now, frequently one cannot. Again we encounter the contrast between simplicity of statement and almost insuperable difficulty of proof that keeps number theory alive as a research science.

Our first equation had only two solutions (meaning solutions in integers, of course), and the question naturally arises: Do all such Diophantine equations have only a finite number of solutions? And if so, is there some way to find them?

The answer to the first question is no. For example, the equation

$$x^2 - 2y^2 = 1$$

has infinitely many solutions in positive integers. The first few are $x = 3$, $y = 2$; then $x = 17$, $y = 12$; then $x = 99$, $y = 70$; . . . . The reader may be wondering how these solutions are found. Actually, there is a formula for them, discussed in the box on p. 57.

So some Diophantine equations have only a finite number of solutions, while others have infinitely many. What makes the difference? People wondered about this for a long time before any definite answer emerged. Pierre de Fermat, in the 1600's, was the first person to prove anything significant about Diophantine equations.

Here is an aside: Fermat had the annoying habit of not writing down his proofs. Consequently his successors, like Euler, had to do everything over. The question arises: since he never wrote them down, how do we know Fermat really *had* proofs for his propositions? Well, he said that he did, and he was a very clever fellow. Also, he did give some hints as to his methods. And, finally, almost everything (with one important exception) which he stated has since been proved to be true. The one exception is the famous "Fermat's Last Theorem." It asserts that for integers x, y, z, none of which are zero, and an integer $n > 2$, there are no solutions to the equation

$$x^n + y^n = z^n.$$

If the reader has ever read any other expository account of number theory, he has probably seen this equation—for it is one of the standard topics. Therefore, in this article, we will let Fermat's Last Theorem rest in peace. No one has ever proved it (unless Fermat himself did).

For a long time, there was really no such thing as a *theory* of Diophantine equations. Each single equation, if it could be solved at all, required a special theory that pertained only to that equation. One equation—one theory. Eventually this grew tiresome, and by the end of the nineteenth century the subject of Diophantine equations had fallen into disfavor.

Then in 1909, the Norwegian mathematician Axel Thue announced a result which opened up the subject in an entirely new way. Recall our earlier

Solving a Diophantine Equation

We give a general formula for solving the equation:

$$x^2 - 2y^2 = 1.$$

It goes like this: If x, y is one solution, then to get from the numbers x, y to a new solution x', y', use the formulas:

$$x' = 3x + 4y,$$
$$y' = 2x + 3y.$$

For instance, putting in $x = 3$, $y = 2$ (which satisfy the original equation since $3^2 - 2 \cdot 2^2 = 1$) gives $x' = 17$, $y' = 12$; these also work since $17^2 - 2 \cdot 12^2 = 289 - 288 = 1$. To verify that this will always work, simply compute $x'^2 - 2y'^2$, substituting from the above formulas.

question: Why do some Diophantine equations have only finitely many solutions, whereas others have infinitely many? Thue's result essentially answered this question. The first equation

$$x^2 = 2y^4 - 1,$$

has only finitely many solutions because it is of "degree 4"—meaning that there is a term $2y^4$ with exponent 4. Thue proved (with certain technical restrictions which we omit) that equations in two variables x, y of degree three or more have only a finite number of solutions. (On the other hand, equations of degree two—like our friend $x^2 - 2y^2 = 1$—can very well have infinitely many solutions.)

Thue's theorem was the first truly general result in the field of Diophantine equations—for it applied to a wide class of equations rather than to one particular equation. However, it had one major flaw. Its proof was not logically "effective." What does this mean? To give the idea, I will begin by mentioning another famous problem which was solved recently.

Gauss, in his famous book *Disquisitiones Arithmeticae*, introduced a certain class of numbers. Never mind the definition. Gauss thought they were important, and we will call them here "Gauss numbers." Gauss succeeded in finding nine of these numbers; they were

$$1, \ 2, \ 3, \ 7, \ 11, \ 19, \ 43, \ 67, \ 163.$$

He conjectured that these nine were all that existed, but this he could not prove. Nor could anyone else, from 1800 until 1934. Then in 1934, H. Heilbronn and E.H. Linfoot proved that there are *at most ten* Gauss numbers! Gauss had found nine, and Heilbronn and Linfoot proved there are at most ten. But as to the tenth: the proof of Heilbronn and Linfoot gave no clue as to whether or not it existed. They showed no way to find it if it did exist; and they showed no way ever to be sure, if you hadn't found it, that it wasn't there somewhere, just outside the range of your calculations. The "tenth Gauss number" might exist or it might be a ghost.

This is what we mean by a logically "noneffective" proof.

Most of the recent advances in Diophantine equations have involved turning previously noneffective arguments into effective ones. The Gauss problem was solved in 1966 by the American mathematician Harold Stark. He showed that there are only nine Gauss numbers; the tenth one doesn't really exist—which is what everyone had believed all along. At about the same time, the English mathematician Alan Baker put Thue's theory on an "effective" foundation. This has had wide-ranging repercussions, and the end is not in sight. We conclude by mentioning just one of the applications of Baker's theory. Using it, R. Tijdeman in Holland recently made substantial progress in settling a hundred-year-old conjecture of Catalan. He showed that, except possibly for a finite number of exceptional cases, there are only two powers of integers—squares, or cubes, or fourth powers, and so on—which differ by exactly one. These, of course, are 3^2 and 2^3, i.e., 9 and 8. (The box on p. 59 explores Tijdeman's theorem in a bit more detail.)

The Old and the New

In this final section, we will describe how one of the oldest results in number theory has recently found an application in computer programming. The old result is called the "Chinese Remainder Theorem"; it was used by the ancient Chinese to predict the common period of several astronomical cycles. The application to programming involves the desire to achieve "multiple precision"—that is, to make the machine give more accuracy than it is normally designed for.

Begin with something commonplace: Everyone knows the idea of "odd and even"—children play marbles using it, and in Austria they will settle an argument by saying "let five be even" (meaning "have it your way, I don't want to fight about it"). Of course the idea of "odd and even" has to do with the number 2. Less well known is the fact that there is an idea similar to "odd and even" which applies to 3 and larger numbers. Take 3. Imagine a group of schoolchildren being divided into three teams; typically the teacher will line them up and ask them to count off: "one, two, three, one, two, three, one . . ." The children are divided into three groups, and the process is cyclic—"one" followed by "two" followed by "three" and then back to "one." (To see the connection with "odd and even," notice that if, instead, the children had counted "one, two, one, two, one, two . . .", then the "ones" would be in the odd positions—the 1st, 3rd, 5th, etc.—and the "twos" would be the evens.)

The extension of this notion to period three is called "congruence modulo 3," and, of course, longer periods are possible too. Well, periodic events occur all the time: weeks, years, phases of the moon, etc. The ancient

Tijdeman's Near Solution of Catalan's Problem

Catalan's problem concerns the powers of integers, that is

the squares: 1, 4, 9, 16, 25, ...
the cubes: 1, 8, 27, ...
the fourth powers: 1, 16, ...
etc.

(We do not include the first powers, because every integer is a first power, and then there would be nothing to say.)

More than a hundred years ago, Catalan conjectured that the only two powers of integers which differ by exactly one are 3^2 and 2^3, i.e., 9 and 8. Recently R. Tijdeman in Holland, using ideas of the English mathematician A. Baker, proved this conjecture except for finitely many cases.

[The number of "exceptional cases," while finite, is far too large to check by computer, and another idea will be needed before Catalan's problem is completely solved. Several examples of the inordinately large numbers which arise in number theory are given elsewhere in this article.]

It is interesting to take a closer look at Catalan's conjecture. We recall that it involves all the powers of integers (squares, cubes, etc.). Upon examination, we find that there are powers of integers differing by 1:

$$9 \text{ and } 8 \quad (\text{i.e., } 3^2 - 2^3 = 1),$$

and by 2:

$$27 \text{ and } 25 \quad (\text{i.e., } 3^3 - 5^2 = 2),$$

and by 3:

$$128 \text{ and } 125 \quad (\text{i.e., } 2^7 - 5^3 = 3),$$

and by 4, in two different ways:

$$125 \text{ and } 121 \quad (\text{i.e., } 5^3 - 11^2 = 4),$$
$$8 \text{ and } 4 \quad (\text{i.e., } 2^3 - 2^2 = 4).$$

Where does this stop? And in which cases are there several solutions? Catalan's conjecture says that, for a difference of one, 9 and 8 is the only solution.

Chinese, in studying astronomical cycles, were led to the question: what happens when you put two periodic events together? The problem is mainly interesting when the two events have different frequencies: weeks and years, say. Then the patterns which emerge are quite remarkable. Several examples, showing the different possibilities, are given in the box on pages 60–62.

The Chinese Remainder Theorem

We will illustrate the theorem for the case of two periodic cycles with periods 3 and 5. Note that $3 \cdot 5 = 15$. The Chinese Remainder Theorem says that the numbers 1, 2, 3, . . . , 15, if arranged in a certain way, will just fill up a 3×5 box:

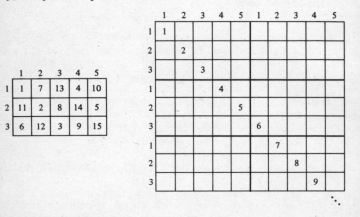

The right-hand figure, with many 3×5 boxes, gives rise to the left-hand one if we superimpose the corresponding patterns. It is rather like the method a trick photographer uses to make multiple images. In both figures, we count down and to the right. The significance of this is that both the vertical and horizontal positions ("modulo 3" and "modulo 5") advance by one step at each stage. Notice how in the "3 to 4" transition, the sequence on the left disappears off the bottom of the picture and reappears on the top. On the right, of course, the 4 goes into a different box.

The Chinese Remainder Theorem simply says that all of the spaces in the small rectangle get filled up. Is it obvious that this must happen? Let's take another case, with "periods" 4 and 6:

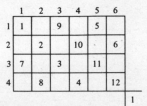

The space where 13 would go is already filled by 1, so only half of the squares got filled. What is wrong? The trouble is that 4 and 6 are both multiples of a common divisor, namely 2. When this happens, there will be trouble; whereas if the two sides of the rectangle have *no common divisor*, then things go well, as they did in the first figure above. (One way to ensure that there is no common divisor is to make both sides *prime* ; for prime numbers have *no* divisors except themselves and one.)

Now for computers: Imagine a "baby computer" which can only handle the numbers from 1 to 5 (this would be a pretty silly computer, of course). Suppose that a programmer, using this machine, needed to have the numbers from 1 to 15. The machine likes little numbers and balks at big ones. So the programmer uses *two little numbers* to represent (or "code") one big one. Thus consider the 3 × 5 rectangle we had above (and which is repeated below as "code A"). Each number from 1 to 15 has a definite place (for example, 10 is in the 1ˢᵗ row, 5ᵗʰ column). So two numbers no bigger than 5 suffice to code any number from 1 to 15.

	1	2	3	4	5
1	1	7	13	4	10
2	11	2	8	14	5
3	6	12	3	9	15

Code A The "CRT" code.

	1	2	3	4	5
1	1	4	7	10	13
2	2	5	8	11	14
3	3	6	9	12	15

Code B A less clever code.

Still, the skeptical reader might ask: why go to all this trouble? Why not use an easier code, like the one shown in "code B." An answer to this question is suggested by the figure below.

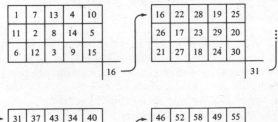

The reason the CRT is useful in programming is that it behaves "consistently" with respect to the basic operations of addition and multiplication. To give the idea: Suppose we fix our attention on two columns, say the 2ⁿᵈ and the 3ʳᵈ. Pick at random *any* numbers from those two columns, say 2 and 13. Then the product 2 · 13 = 26 will lie in the 1ˢᵗ column (in its own box), and this will be true no matter which numbers we pick, so long as they come one each from the 2ⁿᵈ and 3ʳᵈ columns.

Try it again, still using the 2nd and 3rd columns: take 12 and 3. Then $12 \cdot 3 = 36$, again in the 1st column (although in a different box).

Try one more time: take $12 \cdot 13 = 156$; this is off the end of the scale, but if the above pattern were continued, 156 would fall in the 1st column.

Similar things work with other columns. It also works the same way with the rows. For example, the product of any two numbers in the 2nd row will lie in the 1st row in its appropriate box. (Example: $2 \cdot 14 = 28$.)

Here is how the computer would use these properties in doing arithmetic. Say the computer has to multiply 2 times 8. The computer reads these numbers as

$$2: \quad 2^{nd} \text{ row, } 2^{nd} \text{ column},$$
$$8: \quad 2^{nd} \text{ row, } 3^{rd} \text{ column}.$$

We have seen that, so to speak

$$(2^{nd} \text{ row}) \cdot (2^{nd} \text{ row}) = (1^{st} \text{ row}),$$
$$(2^{nd} \text{ column}) \cdot (3^{rd} \text{ column}) = (1^{st} \text{ column}).$$

So $2 \cdot 8$ should lie in the 1st row, 1st column. And in fact it does: $2 \cdot 8 = 16$, and 16 lies in the 1st row, 1st column of its box.

The "pedestrian" code B does not have these properties.

We conclude this section by saying a little bit about how this "ancient wisdom" has come to have an application in modern computer programming. Computers are built to handle "words" of a certain size. Of course, the people who design computers try to pick a reasonable "word length" (this is computer jargon for the number of digits the computer can handle: e.g., 21 has two digits and 4,172,316 has seven). If the number of digits were too small, the machine would be useless, whereas allowing too many is wasteful. Normally about fifteen digits are allowed. But sometimes one needs more. Since the computer is programmed internally to handle words of a certain size, it is apt to throw a tantrum if you try to change its word length. When a programmer needs longer words, he frequently finds it easier to leave the machine alone and—in a manner of speaking—play a trick on it.

To give the idea, imagine a computer which can handle numbers of up to fifteen digits. Suppose now that we need thirty digits. The machine likes "little" (fifteen digit) numbers, and balks at big ones. So the programmer uses *two little numbers* to represent (or "code") one big one. There are various ways to do this. The simplest is just to break the big number in half, producing two smaller ones (like breaking 2143 down into 21 and 43). But this is not the best. The trouble is that breaking a number in half in this crude way leads to a rather confusing kind of "arithmetic," which furthermore the machine is not naturally built to handle. (Of course the programmer could figure it out; the problem is that, if he *has* to figure it out, then he is wasting time.) It is curious that for many purposes, the thousand-year-old Chinese

Remainder Theorem gives the most efficient way of handling this problem.

To conclude this essay, I will make a few observations about the place of number theory within mathematics and about its relation to the applied sciences. Number theory, called by Gauss "the queen of mathematics," has also been called, by people who do not like it, "a useless subject." In this connection, we may note that Gauss also called mathematics "the queen of the sciences," since all the other sciences spring from it. Obviously the concept of number lies at the core of mathematics; the question is how much the systematic study of numbers benefits the applied sciences—and in what way.

Several applications of number theory to other subjects have been mentioned in this article. Moreover, the current development of computers, information theory, and related fields, suggests that number theory will be applied more extensively in the future. (A similar thing happened with a subject called "group theory" fifty years ago: this formerly "pure" field, which had been studied by mathematicians since around 1825, turned out to be one of the crucial tools in the new quantum mechanics.)

Still, it would be ingenuous to suppose that these applications are the main reasons mathematicians do number theory. This becomes particularly clear when one compares the relatively few applications with the enormous effort which has gone into building up the subject. Why has anyone made this effort? Why did a great mathematician like Euler, a man who made contributions to virtually every branch of mathematics and whose collected works fill more pages than a complete set of Encyclopaedia Britannica, work part-time for many years in an unsuccessful attempt to prove the four-squares theorem? The answer is that great mathematicians are professionals, and professionals love their craft.

The French mathematician, Henri Poincaré, in trying to isolate the distinction between first-rate and second-rate mathematics, said "there are problems that one poses, and there are problems that pose themselves." Clearly the number theory questions which we have considered in this chapter "pose themselves." Their extreme simplicity of statement, combined with difficulty of proof, stamp them as genuine problems. Moreover, the resolution of such problems had led to the development of whole theories.

A good theorem will almost always have a wide-ranging influence on later mathematics, simply by virtue of the fact that it is *true*. Since it is true, it must be true for some reason; and if that reason lies deep, then the uncovering of it will usually require a deeper understanding of neighboring facts and principles. In this way number theory, "the Queen of Mathematics," has served as a touchstone against which many of the tools in other branches of mathematics have been tested. This, in fact, is the real way that number theory influences pure and applied mathematics.

Suggestions for Further Reading

General

Philip J. Davis, *The Lore of Large Numbers*. New Math. Libr., No. 6. Random House, Westminster, Maryland, 1963; M.A.A., 1975

Ivan M. Niven, *Numbers : Rational and Irrational*. New Math. Libr., No. 1. Random House, Westminister, Maryland, 1961; M.A.A., 1975.

Oystein Ore, *Invitation to Number Theory*. New Math. Libr., No. 20. Random House, Westminster, Maryland, 1967; M.A.A., 1975.

Technical

H. Davenport, *The Higher Arithmetic*. Hutchinson House, New York, 1952.
A charming introduction to number theory. Do not be misled by the title—this book is for serious students only.

Godfrey H. Hardy, and E. M. Wright, *An Introduction to the Theory of Numbers*. Oxford, 1938.
Except for the first two chapters, this book is rather difficult. However, there are no really easy books on number theory. This is one of the classics. Each chapter treats a different topic, and the chapters are to some degree independent.

A. E. Ingham, *The Distribution of Prime Numbers*. Cambridge University Pr, Cambridge, 1932.
This is the best introduction to the subject written in the English language. The style is crystal clear, and the sections in small print are superb; they show how a curious mind will find further questions after the "main questions" have supposedly been answered. Requires a knowledge of complex variable theory.

Groups and Symmetry

Jonathan L. Alperin

The idea of a group is one of the great unifying ideas of mathematics. It arises in the study of symmetries, both of mathematical and of scientific objects. Very surprisingly, the examination of these symmetries leads to deep insights which are not available by direct inspection: while the notion of a group is very easy to explain, the applications of this concept do not at all lie on the surface. In mathematics the concept of a group is fundamental to the fields of differential geometry, topology, number theory and harmonic analysis, while in science this idea is essential in spectroscopy, crystalography, and atomic and particle physics. The importance of abstraction is nowhere more evident than in the concept of a group.

Let us begin with a very simple geometrical object, the equilateral triangle. A symmetry of this triangle is a way of moving the triangle so as not to disturb its appearance; a symmetry thus preserves all the relations of distance and angle between the vertices and edges of the triangle. For example (see Figure 1), if we leave vertex 1 alone and flip the triangle over so that the vertex 2 is interchanged with vertex 3, we have performed a symmetry operation which we shall call operation A. Similarly, if we leave vertex 2 alone and

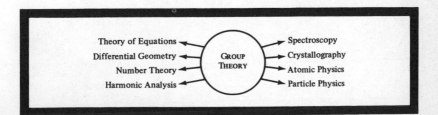

flip the triangle over we get a different symmetry operation which we shall call *B*. Leaving vertex 3 alone and flipping produces operation *C*. Moreover, we can rotate the triangle counterclockwise by 120 degrees, taking vertex 1 to vertex 2, 2 to 3, and 3 to 1; we call this operation *R*. Similarly, we can rotate the triangle clockwise and get a symmetry operation which we call *S*. Finally, we can just leave the triangle alone and we certainly have not disturbed its appearance; we call this operation *E*. These six operations, illustrated in Figure 1, are all the possible symmetries of the equilateral triangle.

Let's look at these six operations more closely. If we perform *R* and then follow this by carrying out *A*, what is the result? Well, *R* takes vertex 1 to

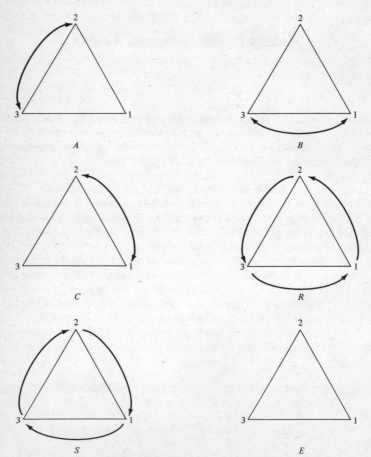

Figure 1. The six possible symmetries of an equilateral triangle. For example, *R* displays a counterclockwise rotation through 120°. These six symmetries are an example of a group.

vertex 2 and then A takes 2 to 3 so the result of performing R and then A carries 1 to 3. Similarly, R takes 2 to 3 and A takes 3 to 2 so this result leaves 2 alone. Also, R takes 3 to 1, which is left fixed by A, so the result carries 3 to 1. Hence, what we have done as a result of these two operations is to perform the operation B. To symbolize this, we write $RA = B$, where the juxtaposition of R and A denotes the result of performing first R and then A.

Let's consider next the product RS. This means the result of performing R and following this by S. But R is counterclockwise rotation and S is clockwise rotation, so $RS = E$, the operation which leaves the triangle alone. In fact, performing any two symmetry operations consecutively produces the same result as some single operation. This information can all be displayed in a table (Figure 2) where the rows and columns are labeled by the operations and the result of performing X and then Y is listed in the X-row under the Y-column.

This table has several important special properties. First, if X is any operation then

$$XE = X = EX$$

since the result of X and then the operation E which doesn't move the triangle, is just X again. Moreover, any one of the operations can be reversed. This follows from the fact that for any operation X there is another one Y such that

$$XY = E = YX.$$

For example, $RS = E = SR$, $AA = E = AA$. Finally, suppose that X, Y, and Z are three of the operations, perhaps not even distinct. Consider the operation $(XY)Z$, which means the product of the operation XY with Z. But XY is the result of performing X and then Y so that $(XY)Z$ is nothing more than the net result of doing X, Y, and Z in turn. Similarly, $X(YZ)$ means doing

	E	R	S	A	B	C
E	E	R	S	A	B	C
R	R	S	E	B	C	A
S	S	E	R	C	A	B
A	A	C	B	E	S	R
B	B	A	C	R	E	S
C	C	B	A	S	R	E

Figure 2. A multiplication table. A display of the results of combining the symmetries illustrated in Figure 1. For example, in the row labelled A under the column headed by B there is an S because the net result of first performing the symmetry A and then following this with the symmetry B is just the symmetry S.

first X and then YZ, so this all means again just doing X, Y, and Z in that order. In other words,

$$(XY)Z = X(YZ)$$

for any symmetry operations X, Y, and Z.

Any system that behaves as does the collection of symmetry operations of the equilateral triangle is a group. More formally, a group is a collection of objects and a rule for multiplying any two of them which has three properties:

1. There is an object e in the group such that $xe = x = ex$ for any object x in the group.
2. For any object x there is an object y in the group with $xy = e = yx$.
3. Whenever x, y, and z are in the group then $(xy)z = x(yz)$.

For another example of a group, let us examine an equilateral pentagon and its symmetries. There are now ten symmetry operations (Figure 3). There is the one operation which leaves the pentagon alone. There are four counterclockwise rotations of the pentagon through angles of 72, 144, 216, and 288 degrees, respectively. There are also five flips corresponding to the five vertices; these are the operations which leave a single vertex fixed and turn the pentagon over along the line from the fixed vertex to the midpoint of the edge opposite the vertex. In general, if we look at an equilateral n-gon, that is, a polygon with n sides, then there will be a symmetry group consisting of $2n$ operations.

Return, for a moment, to the symmetry group of the equilateral triangle. Exactly three of the six operations, namely R, S, and E, do not turn over the triangle. These three operations by themselves also form a group. Notice, by examining the nine entries in the upper left-hand corner of Figure 2, that

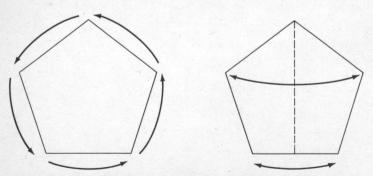

Figure 3. Symmetries of a regular pentagon. These are examples of two of the ten symmetries. The figure on the left represents a rotation, while the symmetry on the right is a flip. All ten symmetries can be constructed by repeated application of these two basic symmetries.

the product of any one of E, R, or S with one of these is again one of these. In general, n of the operations associated with an n-gon do not turn the n-gon over and these n operations form a group by themselves. This group is called the cyclic group of order n and is customarily denoted by Z_n. The objects in this group consist of the operation which leaves the n-gon fixed and the n-1 counterclockwise rotations of the n-gon.

Another important example of a group is the rearrangements of a set, which in mathematical language are called permutations. Let $N(n)$ be the collection of numbers $1, 2, \ldots, n$. A permutation of the set $N(n)$ is just a reordering of these numbers, that is, an operation which just shifts these numbers among themselves. For example, if n happens to be four, and we imagine the numbers 1, 2, 3, and 4 in order, then we can perform the operation which moves 1 to where 2 is, 2 to 4, 3 to 1, and 4 to 3. We denote this operation by the symbol

$$\begin{pmatrix} 1 & 2 & 3 & 4 \\ 2 & 4 & 1 & 3 \end{pmatrix}.$$

As one might expect, if you perform two permutations of $N(n)$ in succession, the result is another permutation. The totality of all the permutations of $N(n)$ is another example of a group, called the symmetry group of degree n and denoted by S_n. It has exactly $n! = n \cdot (n\text{-}1) \cdot (n\text{-}2) \cdot \ldots \cdot (2) \cdot (1)$ permutations in it.

We have seen above that the $2n$ symmetries of the n-gon are of two kinds: there are n symmetries which do not turn over the n-gon; there are n others which do turn it over. The $n!$ permutations of $N(n)$ are also of two kinds. To see this, examine the extent to which the permutation of $N(4)$ given above keeps the numbers 1, 2, 3, and 4 in order. It sends 1 to 2 and sends 2 to 4; the numbers that 1 and 2 are sent to are in the same order. But 2 is sent to 4 and 4 to 3 so that the numbers that 2 and 4 are sent to are in the reverse order. In fact, if we compare any two of the four numbers in this way, and thus make six comparisons, the order is preserved three times and reversed three times.

In general, for a permutation in S_n, we proceed in a similar way. For each pair of numbers i and j between 1 and n, a permutation either keeps them in order or does not. A permutation is called even if the number of such pairs which are not kept in order is even; if the number of pairs not kept in order is odd we say the permutation is odd. It turns out that exactly half the permutations in S_n are even; there are $n!/2$ even permutations and $n!/2$ odd permutations.

In the case of the n-gon we saw that the symmetries that did not turn over the polygon formed a group by themselves. The same sort of thing happens for permutations. The $n!/2$ even permutations in S_n form a group by themselves, called the alternating group and denoted by A_n. As we shall see, the group A_n is very important.

Galois Theory

The solution to quadratic equations is a standard part of elementary mathematical education. For example, if one is given

$$2x^2 + 2x + 5 = 0$$

then by the formula we all have (once) learned, we deduce

$$x = \frac{-2 \pm \sqrt{-36}}{4}.$$

We express the solutions just by using addition, subtraction, multiplication, and division together with the extraction of square roots. Similarly, as was discovered already in the sixteenth century, if one is faced with a cubic equation such as

$$2x^3 + 5x^2 + x - 7 = 0,$$

one can find formulas for its solutions using the four arithmetic operations and the extraction of cube roots, while for quartics, such as

$$x^4 - 3x^3 + 7x^2 - x + 18 = 0,$$

the extraction of fourth roots is also needed. But for centuries, despite great efforts, no such formulas were found for quintic equations, where x^5 also appears. It was only in 1832 with the use of group theory that the young French mathematics student Évariste Galois (1811–1832) was able to explain all of this.

Galois' idea was to associate with any polynomial equation a group in such a way that the properties of the group and the nature of the solutions of the equations are closely related. In particular, he devised groups that reflect the symmetry properties of the roots of polynomial equations. This is indeed one of the most remarkable ideas in the history of mathematics.

Let us illustrate it by looking at a couple of examples, starting with the equation $x^5 - 1 = 0$. This has five solutions, these being the number 1, plus the four numbers

$$\left(-1 + \sqrt{5} + \sqrt{-10 - 2\sqrt{5}} \right) / 4$$

$$\left(-1 - \sqrt{5} + \sqrt{-10 + 2\sqrt{5}} \right) / 4$$

$$\left(-1 - \sqrt{5} - \sqrt{-10 + 2\sqrt{5}} \right) / 4$$

$$\left(-1 + \sqrt{5} - \sqrt{-10 - 2\sqrt{5}} \right) / 4.,$$

If we denote the first of these four numbers by a then $a^5 = 1$ since a is a root of the polynomial. Hence, $(a^2)^5 = a^{10} = (a^5)^2 = 1$ and a^2 is also a root of the polynomial; it is in fact the second of the four numbers just listed. In fact, if

Evariste Galois: Extraordinary Prodigy

Galois wrote down most of his great discovery the night before the duel that resulted in his early death. It took years before his work was published and decades before it was understood. He was born October 25, 1811 in Bourg-la-Reine in France and died tragically on May 31, 1832 in Paris. Besides devoting himself to his great research he was also a student radical and was twice imprisoned because of it.

we label these four numbers a, b, c, and d then $b = a^2$, $c = a^3$, and $d = a^4$. We use the letter e for the other solution, the number 1.

The Galois group associated with this equation consists of certain permutations of the solutions a, b, c, d and e. Previously, when we considered geometrical objects, the associated symmetry group consisted of the operations which preserved *geometric* properties like length and angle. Now we shall consider those permutations of these five solutions which preserve their *algebraic* properties. For example, since $a^2 = b$, if a permutation is to be in the

Galois group and it sends a to the solution x and b to y, then x^2 must equal y. This preserves the relation. In particular, suppose the permutation sends a to c; it must then send b to c^2. But $c = a^3$ so $c^2 = a^6 = a$, as $a^5 = 1$. Thus, if a is sent to c then b must be sent to a. Similarly, since c is the cube of a and a is sent to c we must have c sent to the cube of c, which is $(a^3)^3 = a^9 = a^4 = d$. Also, since d is the fourth power of a, another calculation shows that d must be sent to b. Thus, if the permutation sends a to c then b goes to a, c goes to d and d goes to b; the number e is sent to itself.

More generally, if we know where a permutation in the Galois group sends the root a, then the permutation is determined in a similar fashion. The number a can, in fact, be sent to any one of the numbers a, b, c, or d. There are exactly four permutations in the Galois group of this equation and they are listed in Figure 4. Each row describes one of these permutations by listing in turn where this permutation takes a, b, c, d, and e.

But let's look at another quintic equation, namely

$$x^5 - x - 1 = 0,$$

which doesn't look all that different. Again it has five solutions. But here the Galois group consists of all one hundred and twenty possible permutations of these five solutions, which is a very different situation.

The main result of Galois theory is that the solutions to a polynomial equation can be expressed by the elementary arithmetic operations and the extraction of roots exactly when the Galois group has a certain special property which basically says that the group is made up in a very nice way (see box on p. 73). For our second quintic this implies that it has no solution at all which can be expressed by the arithmetic operations and the extractions of roots: the symmetry group of this equation is just too complex.

Galois theory also explains very well why the quadratic, cubic, and quartic equations behave so well. The idea is, roughly, that a group of permutations of two, three, or four objects just can't be very complicated, so that all such groups are made up in the necessary way. In fact, the methods for dealing with these equations can be derived from the theory. Galois theory is indeed a remarkable and successful theory, bringing in ideas which look to-

a	b	c	d	e
a	b	c	d	e
b	d	a	c	e
c	a	d	b	e
d	c	b	a	e

Figure 4. Symmetries of a quintic equation. The four rows under the heading describe the four elements of the group corresponding to the polynomial $x^5 - 1$. The properties of the polynomial are closely related to this group.

Main Results of Galois Theory

The Set-up

1. Let $f(x) = x^n + a_{n-1}x^{n-1} + \ldots + a_1x + a_0$ be a polynomial with rational numbers $a_0, a_1, \ldots, a_n$ for coefficients.
2. Let $r_1, \ldots, r_n$ be the roots of the equation $f(x) = 0$.
3. Let G be the Galois group of this equation so G is a group of permutations of the n roots $r_1, \ldots, r_n$.
4. Let F consist of the totality of all numbers which can be expressed in terms of the n roots by the arithmetic operations of addition, subtraction, multiplication, and division. F is a field, that is, a number system in which the usual arithmetic operations can be done.
5. We consider number systems contained in F and groups contained in G. To each number system there corresponds a group and to each group there corresponds a number system.

Main Theorem

There is a natural pairing off of the number systems contained in F and the groups contained in G.

How to Use the Result

From a problem about the nature of the roots pass to a problem about the number system F and then pass to a problem about the group G.

tally irrelevant but which lie at the very heart of the matter. This type of unity in mathematics is very common.

Lie Groups

What works once can, of course, be tried again! Why not try to use group theory for all sorts of other equations, like the differential equations that arise in the calculus and its original applications to the physical sciences? Could there be a similar explanation for the successes and failures of elementary methods for finding exact solutions to differential equations? Such questions motivated the Norwegian mathematician Sophus Lie (1842–1899) to develop another sort of group, just over a century ago.

We will restrict ourselves to just giving a couple of examples of such groups; it's much harder to give a precise definition. Let's examine the usual x-y coordinate plane of analytic geometry and calculus. A rotation of this

Sophus Lie: Father of Continuous Groups

Lie was born in 1842 near Bergen, Norway, and died in Oslo in 1899 just after returning to take up a special professorship at a great salary set up for him. He was primarily a geometer, but his work on continuous groups is basic in many areas of mathematics and science.

plane about the origin—the intersection of the coordinate axes—through an angle θ in the counterclockwise direction can be considered as a symmetry of the plane, since it preserves the relations of distance and angle. Using elementary trigonometry it is easy to calculate where a point (x,y) will be moved by this rotation operation: it goes to the point

$$(x \cos \theta - y \sin \theta, \, x \sin \theta + y \cos \theta),$$

where $\cos \theta$ and $\sin \theta$ are the cosine and sine of the angle θ (Figure 5). The totality of these rotations form a group, called a Lie group in honor of its discoverer. Indeed, if we perform the rotation through an angle θ_1 and follow

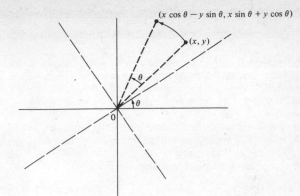

Figure 5. A plane rotation. If the plane is rotated counterclockwise through an angle θ, then the point (x,y) is moved to the point $(x \cos \theta - y \sin \theta, x \sin \theta = y \cos \theta)$.

this with a rotation through the angle θ_2 then the result is just the rotation through the angle $\theta_1 + \theta_2$. This gives a rule for multiplying any two rotations. It is not hard to check that all the properties required for a group are satisfied. For example, the rotation through the angle zero can be used for the element e. If α is the rotation through the angle θ and if β is the complementary rotation through the angle $2\pi - \theta$ (which is the angle which then added to θ gives an angle of 2π, or 360 degrees) then $\alpha\beta = e = \beta\alpha$.

Instead of dealing with rotations in two dimensions, one can do the same thing in three dimensions or in higher-dimensional spaces: in each case the rotations about the origin form a Lie group. Also, for the plane, we can consider the totality of all rigid motions of the plane. These also form a Lie group, larger than the rotation group, that contains (in addition to rotations) the translations of the plane. Similarly, Lie groups of rigid motions can be formed for spaces of higher dimension.

The essential feature of Lie groups is continuity: they are described by continuously varying parameters like the angle θ. This seems quite different from the finite and discrete groups formed from permutations or symmetries of polygons. However, there are very fundamental connections between these continuous Lie groups and the finite groups. They are really just variations on a single theme. This has been known from the very beginning, but only in a superficial way. Only now are we reaching a deeper understanding of these relations.

To indicate this connection let's go back to the group of rotations in the plane and describe it slightly differently. We set $a = \sin\theta$ and $b = \cos\theta$, so a and b are numbers with the property that $a^2 + b^2 = 1$. (This latter formula is just the famous Pythagorean formula applied to a right triangle whose hypotenuse has length 1 and whose legs have lengths a and b.) Conversely, if a

An Algebraic Formula for a Plane Rotation

1. The rotation through the angle θ sends the point (x, y) to the point $(bx - ay, ax + by)$ where $a = \sin\theta$ and $b = \cos\theta$.
2. The rotation through the angle θ' sends the point (x', y') to the point $(b'x' - a'y', a'x' + b'y')$ where $a' = \sin\theta'$ and $b' = \cos\theta'$.
3. If we set $x' = bx - ay$ and $y' - ax + by$ and then substitute these expressions in the formula for the rotation through the angle θ' we will get the formula for the point that (x,y) is sent to if we first perform the rotation through the angle θ and then perform the rotation through the angle θ'. We get

$$(b'(bx - ay) - a'(ax + by), a'(bx - ay) + b'(ax + by)) =$$

$$((-aa' + bb')x - (ab' + a'b)y, (ab' + a'b)x + (-aa' + bb')y).$$

4. If we now set

$$a'' = ab' + a'b, \qquad b'' = -aa' + bb'$$

then we have sent (x, y) to $(b''x - a''y, a''x + b''y)$. However, the result of performing the two rotations is just a rotation through an angle of $\theta + \theta'$ so that we must have

$$\sin(\theta + \theta') = ab' + a'b$$

$$\cos(\theta + \theta') = -aa' + bb'.$$

These formulas are, of course, just the addition formulas for sine and cosine.

5. Thus, if the algebraic notation for the first two rotations is $r(a, b)$ and $r(a', b')$, then the algebraic notation for the result of performing them in turn is $r(ab' + a'b, -aa' + bb')$.

and b are numbers with the property that $a^2 + b^2 = 1$, there is a unique rotation to which they belong: there is one and only one angle θ such that $a = \sin\theta$ and $b = \cos\theta$. Therefore, we can denote a rotation by the symbol $r(a,b)$.

But now if $r(a,b)$ and $r(a', b')$ are two rotations, then the result of performing them in turn is again a rotation so there should be a formula expressing this fact. Indeed,

$$r(a,b)\, r(a',b') = r(ab' + a'b, -aa' + bb').$$

But this is merely an algebraic formula (derived in the box on this page) that does not involve transcendental functions like sine and cosine. This makes possible a simple but great idea. Up until now a and b have been just ordi-

Addition and Multiplication Tables
for a Number System With Five Elements

+	0	1	2	3	4
0	0	1	2	3	4
1	1	2	3	4	0
2	2	3	4	0	1
3	3	4	0	1	2
4	4	0	1	2	3

×	0	1	2	3	4
0	0	0	0	0	0
1	0	1	2	3	4
2	0	2	4	1	3
3	0	3	1	4	2
4	0	4	3	2	1

A number system with five objects, denoted by 0, 1, 2, 3, 4, works like "clock arithmetic" for a clock which shows only five hours on its face. The two tables show how to add and multiply these objects so that usual properties of addition, subtraction, multiplication and division hold.

nary numbers, just restricted by the equality $a^2 + b^2 = 1$. But there are lots of different numbers systems in mathematics, systems in which one can add, subtract, multiply, and divide. For example, there is the system of the complex numbers, comprising all numbers of the form $u + iv$ where u and v are ordinary real numbers and i is a square root of -1. And there is the system of rational numbers, that is, the totality of all quotients m/n of an integer m by a nonzero integer n. If we take any such number system and consider the symbols $r(a,b)$ for those numbers a and b in the system that satisfy $a^2 + b^2 = 1$, and then *define* multiplication by the above formula, we get a group! After all, such a system is a collection of objects with a rule for multiplication and it is not hard to prove that all the required properties are satisfied.

Among the many number systems in mathematics there are some that are entirely finite. For example, we can form a system with five objects, denoted by 0, 1, 2, 3, and 4, but where these are not the usual numbers. We can define addition and multiplication on these five objects so that subtraction and division is possible and so that all the usual rules of calculation hold. (The box on this page shows how this can be done.) Now if the number system is finite then so is the group formed by all the corresponding symbols $r(a,b)$. This yields a finite group which is a version of the continuous Lie group of rotations of the plane.

By varying the number system used in this construction, and by selecting different Lie groups for the background structure, one can obtain many different groups which are of interest in various branches of mathematics. This approach makes it possible to study a large number of different groups at the same time. Even better, an intuition gained from one particular case can be used in a general range. The finite groups that arise in this and related ways are called groups of Lie type. As we shall see, they are the most important of the finite groups.

Classification

It is very common in science to try to classify the objects under examination. The usual approach is to determine certain fundamental objects that can be put together to produce all the objects of interest. Well-known examples include the periodic table of the elements, which is at the very foundation of chemistry, and the fundamental particles of physics. This type of study is also very common in mathematics: prime numbers, for instance, are the fundamental building blocks of all numbers. Other examples abound, especially in algebra and topology. In group theory the fundamental objects, the building blocks, are the so-called simple groups. But simple to understand and find they are not.

Instead of explaining what a simple group is, it is easier to show what it means for a group to be not simple. To do this, we re-examine our first example of a group, the symmetry group of an equilateral triangle (Figures 1 and 2). We have already seen that the three symmetries E, R, and S which do not turn over the triangle form a group by themselves: the multiplication table for this smaller group is given, as we noted above, by the nine entries in the upper left-hand corner of the table in Figure 2. The elements which do turn over the triangle do not form a group (since the essential symmetry E is not one of them). Nevertheless, they have some remarkable multiplication properties.

Let X_1 be the set of the three symmetries E, R, and S and let X_2 be the other three symmetries A, B, and C. If we multiply any two elements of X_2 we always get an element of X_1 and if we multiply an element of X_1 by an element of X_2 the result is always an element of X_2. (These observations can be easily verified by routine, albeit tedious, examination of the multiplication table given in Figure 2.) In general, if we multiply an element of X_i by an element X_j, where i and j can each be 1 or 2, the result always lies in the same one of the two sets X_1 or X_2. This allows us to define a multiplication on the set whose two elements are the object X_1 and the object X_2 (Figure 6) in such a way as to make this set into a group! Thus the group of the equilateral triangle, a group with six elements, is made up of two smaller groups, the group with three elements which are E, R, and S and the group with two elements X_1 and X_2. For this reason the group of the equilateral triangle is not simple, and neither is any finite group that can be similarly decomposed. Groups that are not made up in this way of smaller groups are called simple.

At the end of the last century and the beginning of this one, the simple Lie

	X_1	X_2
X_1	X_1	X_2
X_2	X_2	X_1

Figure 6. The rule of multiplication for a group with only two elements.

groups were completely classified. On the whole they turn out to be connected with various kinds of geometry. The complete list (see box on p. 80) is very satisfying from an esthetic point of view and also extremely useful in further research. In recent times, analogous work has been done on another kind of group that we haven't explicitly mentioned, the algebraic groups. Amazingly, the groups that turn up in this classification are groups of Lie type—those obtained from Lie groups by changing the number system used! Furthermore, in other branches of algebra where classifications have been achieved, the answers are closely related to these. There seems to be a short list of all possible building blocks for groups. Yet even today we really don't have a real understanding of this phenomenon.

The simple finite groups have been the subject of an enormous amount of research over the last quarter century. It's interesting that although the Lie groups were first discovered by motivations from the theory of finite groups, they were completely classified long before the same problem was seriously studied for finite groups. We shall see in a moment reasons for this occurrence. But first let us give some examples.

The most basic simple finite groups are the cyclic groups Z_p with a prime number p of elements. (These are the symmetry groups of polygons with p sides which do not turn over the polygon.) They have an intimate connection with Galois theory, since the Galois groups that are built up from these simple groups are exactly those that correspond to equations whose solutions can be expressed in terms of the arithmetic operations and the extraction of roots. Another type of simple groups are the alternating groups A_n of even permutations for integers $n \geq 5$. There are also a number of kinds of simple finite groups of Lie type, associated with the Lie groups. In fact, there are sixteen families of such simple groups, so that with the cyclic groups and the alternating groups there are eighteen infinite families of finite simple groups.

To what extent is this the complete list? The evidence is mounting that this list is very nearly complete and that we are close to demonstrating that this is true. As we mentioned already, this is the result of a vast project. With present trends, if the whole classification is completed, the work will require a total of about ten thousand pages in journals to do all the necessary proofs. This will be very formidable achievement, one of a scale never before seen in mathematics.

The reason we believe that the list is very nearly complete rather than ac-

Simple Complex Lie Groups

Notation	Name	Dimension
$A_n, n \geq 1$	Linear	$n^2 + 2n$
$B_n, n \geq 2$	Orthogonal	$2n^2 + n$
$C_n, n \geq 3$	Symplectic	$2n^2 + n$
$D_n, n \geq 4$	Orthogonal	$2n^2 - n$
G_2	Exceptional	14
F_4	Exceptional	52
E_6	Exceptional	78
E_7	Exceptional	133
E_8	Exceptional	248

This is the complete list of all the simple complex Lie groups which consists of the four families of geometrical ones plus five other individual groups.

tually complete is that there are a few dozen other simple finite groups, called sporadic groups, that do not fit into any infinite family of simple groups. There are twenty-four of these sporadic groups already known and there is overwhelming evidence for the existence of two more. The first five of the sporadic groups were discovered over a century ago by Émile Léonard Mathieu. The smallest of the ones he uncovered has exactly 7920 elements in it; it consists of certain permutations on an eleven-element set. Until the last decade the five Mathieu groups were the only sporadic groups known, but there have been a whole string of recent discoveries of such groups in the context of the work on the classification of all the simple finite groups. We list all twenty-six sporadic groups in the box on p. 81.

It is not really clear exactly how many more sporadic groups exist. At least two more than we presently know should exist. One of these has already been given a name, the monster, reflecting its great size; considering its size, its properties are remarkably nice. The number of elements in the monster will be

$$2^{46} \cdot 3^{20} \cdot 5^9 \cdot 7^6 \cdot 11^2 \cdot 13^3 \cdot 17 \cdot 19 \cdot 23 \cdot 29 \cdot 31 \cdot 41 \cdot 47 \cdot 59 \cdot 71$$

which is approximately 8×10^{53}, that is, an eight followed by fifty-three zeroes. This group seems to be a group of certain rotations in a space of dimension 196,883.

Perhaps now it is clear why the classification of simple finite groups has appeared to lag behind work on other classifications. The answer is much more involved! Even though the Lie groups (in their finite versions) dominate the scene, new phenomena do occur. Nearly 150 years after group

Sporadic Finite Simple Groups

Group	Number of elements
M_{11}	$2^4 \cdot 3^2 \cdot 5 \cdot 11$
M_{12}	$2^6 \cdot 3^3 \cdot 5 \cdot 11$
M_{22}	$2^7 \cdot 3^2 \cdot 5 \cdot 7 \cdot 11$
M_{23}	$2^7 \cdot 3^2 \cdot 5 \cdot 7 \cdot 11 \cdot 23$
M_{24}	$2^{10} \cdot 3^3 \cdot 5 \cdot 7 \cdot 11 \cdot 23$
J_1	$2^3 \cdot 3 \cdot 5 \cdot 7 \cdot 11 \cdot 19$
J_2	$2^7 \cdot 3^3 \cdot 5^2 \cdot 7$
J_3	$2^7 \cdot 3^5 \cdot 5 \cdot 17 \cdot 19$
J_4 (?)	$2^{21} \cdot 3^3 \cdot 5 \cdot 7 \cdot 11^3 \cdot 23 \cdot 29 \cdot 31 \cdot 37 \cdot 43$
HS	$2^9 \cdot 3^2 \cdot 5^3 \cdot 7 \cdot 11$
MC	$2^7 \cdot 3^6 \cdot 5^3 \cdot 7 \cdot 11$
Sz	$2^{13} \cdot 3^7 \cdot 5^2 \cdot 7 \cdot 11 \cdot 13$
C_1	$2^{21} \cdot 3^9 \cdot 5^4 \cdot 7^2 \cdot 11 \cdot 13 \cdot 23$
C_2	$2^{18} \cdot 3^6 \cdot 5^3 \cdot 7 \cdot 11 \cdot 23$
C_3	$2^{10} \cdot 3^7 \cdot 5^3 \cdot 7 \cdot 11 \cdot 23$
He	$2^{10} \cdot 3^3 \cdot 5^2 \cdot 7^3 \cdot 17$
F_{22}	$2^{17} \cdot 3^9 \cdot 5^2 \cdot 7 \cdot 11 \cdot 13$
F_{23}	$2^{18} \cdot 3^{13} \cdot 5^2 \cdot 7 \cdot 11 \cdot 13 \cdot 17 \cdot 23$
F_{24}	$2^{21} \cdot 3^{16} \cdot 5^2 \cdot 7^3 \cdot 11 \cdot 13 \cdot 17 \cdot 23 \cdot 29$
Ly	$2^8 \cdot 3^7 \cdot 5^6 \cdot 7 \cdot 11 \cdot 31 \cdot 37 \cdot 67$
O	$2^9 \cdot 3^4 \cdot 5 \cdot 7^3 \cdot 11 \cdot 19 \cdot 31$
R	$2^{14} \cdot 3^3 \cdot 5^3 \cdot 7 \cdot 13 \cdot 29$
F_5	$2^{14} \cdot 3^6 \cdot 5^6 \cdot 7 \cdot 11 \cdot 19$
F_3	$2^{15} \cdot 3^{10} \cdot 5^3 \cdot 7^2 \cdot 13 \cdot 19 \cdot 31$
F_2	$2^{41} \cdot 3^{13} \cdot 5^6 \cdot 7^2 \cdot 11 \cdot 13 \cdot 17 \cdot 19 \cdot 23 \cdot 31 \cdot 47$
F (?)	$2^{46} \cdot 3^{20} \cdot 5^9 \cdot 7^6 \cdot 11^2 \cdot 13^3 \cdot 17 \cdot 19 \cdot 23 \cdot 29 \cdot 31 \cdot 41 \cdot 47 \cdot 59 \cdot 71$

The two groups not yet proved to exist are indicated by (?).

theory began and after all sorts of other types of groups have been understood, mathematics has finally reached the threshold of a great success with the finite groups.

Bibliography

F. J. Budden, *The Fascination of Groups*. Cambridge University Pr, Cambridge, 1972.

Lee Dembart, Theory of groups: a key to the mysteries of math. *New York Times* (May 17, 1977) 35, 55.

Arthur L. Loeb. *Color and Symmetry*. Interscience, New York, 1971.

Arthur L. Loeb, *Space and Structures*. Addison-Wesley, Reading, 1976.

Carolina H. MacGillavry, *Fantasy & Symmetry, The Periodic Drawings of M. C. Escher*. Abrams, New York, 1976.

George W. Mackey, Group theory and its significance for mathematics and physics. *Proceedings of the American Philosophical Society* **117** (1973) 374–380.

Marjorie Seneschal and George Fleck (Eds.). *Patterns of Symmetry*. University of Massachusetts Pr, Amherst, 1977.

The Geometry of the Universe

Roger Penrose

One of the most fruitful sources of mathematical intuition is physical space. For not only does physical space provide us with the basic concepts of Euclidean geometry, but it also gives us a pictorial framework for visualizing the very much more general types of space that occur continually throughout mathematics. Moreover, it was the picture of physical space that led to those key ideas of mathematical analysis: continuity and smoothness. Indeed, even the very basic mathematical notion of real number originated from measurement of spatial separation—and of time intervals too, these being, as Albert Einstein's relativity has told us, geometrical quantities again, whose measurement is essentially bound up with that of space. So it comes as a shock when we also learn from relativity that our now cherished notion of Euclidean geometry does not, after all, describe physical space in the most accurate way. Yet, from these Euclidean beginnings, a more subtle and flexible geometry, known as differential geometry, has grown to maturity. It is in terms of this geometry that Einstein's theory finds expression. And now, more than sixty years after general relativity was first put forward as a daring original view of the world, the theory stands in excellent agreement with observation. So if we wish to understand how the world is shaped, we must come to terms with this theory.

In this article I shall first introduce the simplest models of non-Euclidean geometry and then indicate how the powerful general machinery (manifolds, tensors, tangent spaces, geodesics) of differential geometry may be constructed. In so doing, I shall use some of the more modern ideas for this that were not available to Einstein in his day. Finally, I shall outline the basic physical notions of both special and general relativity, describing the geometric content of Einstein's equations and taking the discussion as far as black holes and cosmology.

The physical theory of general relativity could not have evolved were it not for the work of many generations of mathematicians (in particular, Carl Friedrich Gauss and Bernhard Riemann) who were able to free geometry from its earlier imprisonment in Euclidean rigidity. But this debt to pure geometry that relativity had owed has now been amply repaid. For many of the ideas of the modern subject of differential geometry received their initial stimulus from concepts arising from Einstein's general relativity.

Thus we have, in the physics of relativity and in the mathematics of differential geometry, a supreme example which illustrates how physics and mathematics can enrich one another. Here we can begin to see something of that profound interplay between the workings of the natural world and the laws and sensitivity of thought—an interplay which, as knowledge and understanding increase, must surely ultimately reveal a yet deeper interdependence of the one upon the other.

Non-Euclidean Geometries

But what is differential geometry? I shall begin by describing what was historically the first of the non-Euclidean geometries—Lobachevskian (or hyperbolic) geometry—which deviates the least from that of Euclid. Just one of the Euclidean postulates is abandoned, namely the fifth: "If a straight line falling on two straight lines make the interior angles on the same side less than two right angles, then the two straight lines if produced indefinitely meet on that side on which the angles are less than two right angles." Lobachevskian geometry was, in effect, explored in considerable detail by the Jesuit Geralamo Saccheri in a work published in 1733, though his aim had really been to show that Euclid's fifth postulate was consequent upon the others. But it was the great German mathematician Carl Friedrich Gauss (1777–1855) who first clearly realized that in fact a consistent alternative geometry to that of Euclid could be obtained in which the fifth postulate of Euclid was simply denied. Gauss did not publish his findings, and his main results in this area were independently rediscovered first (apparently) by his contemporary Ferdinand Schweikart, then by the Russian Nikolai Ivanovich Lobachevski (1793–1856) and then again by the Hungarian János Bólyai (1802–1860). As is not infrequently the case in mathematics, the name which becomes most commonly attached to an idea need not belong to the person who first thought of it. (But I wish to emphasize that, contrary to the sentiments apparently expressed in Tom Lehrer's witty satirical song, there is absolutely no suggestion that Lobachevski had heard of these ideas beforehand.)

Consider the two-dimensional case first. On way to proceed is to model the Lobachevskian plane in terms of ordinary Euclidean concepts but where the basic notions of "distance" and "straight" are interpreted in a new way. We shall name this model L^2, where the superscript "2" (and "1" in "S^1",

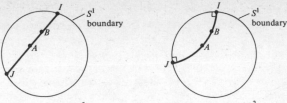

Projective model of L^2 Conformal model of L^2

Figure 1. The Lobachevski Space L^2. "Lobachevskian distance AB" $= \epsilon \log_e (IA \cdot JB / IB \cdot JA)$ where IA, JB etc. denote *Euclidean* distances, and where $\epsilon = \frac{1}{2}$ for the projective model and $\epsilon = 1$ for the conformal model.

etc.) refers to the dimension of the space. (Read "L^2" as "L-two", not "L squared.") There are two standard representations of L^2 in the Euclidean plane E^2, the projective model and the conformal model. In each case, L^2 is represented as a disc in E^2—the interior of a circle S^1 of unit radius (see Figure 1). According to the projective model a "straight line" of Lobachevskian geometry is depicted as a chord, i.e., as a line segment drawn across the disc which is *straight* also with respect to the Euclidean geometry of E^2. According to the conformal model, a Lobachevskian "straight line" is depicted as an arc drawn across the disc which is *circular* with respect to the geometry of E^2 and which meets S^1 at right-angles. The Lobachevskian "distance" between two points A, B of the disc is given by a simple formula (see Figure 1) which, remarkably, has almost precisely the same form in each model.

While the projective model has the obvious advantage that the Lobachevskian "straightness" is correctly depicted, the conformal model has the slightly more subtle and perhaps more significant advantage that Lobachevskian "angles" are correctly depicted. This has the effect that small shapes are represented without distortion, only an appropriate magnification having to be applied at each point. Furthermore, Lobachevskian "circles" are always correctly depicted as circles in this model. Some very elegant woodcuts by the Dutch artist Mauritus C. Escher illustrate these facts in a striking way (see Figure 2), being accurate representations of the conformal model of L^2. (The idea of making such illustrations was suggested to Escher by the Canadian geometer H.S.M. Coxeter.) Notice that though the figures become very small and crowded towards the edge, they remain roughly true in their shape. The smaller the portion of each figure that is examined, the truer its shape remains. For that is what is meant by a conformal representation: arbitrarily small figures are accurately represented, with only a change in scale.

The Lobachevski world is an infinite world. For example, each devil in Escher's second woodcut is of equal Lobachevskian area, and there are infinitely many devils. The boundary circle S^1 represents infinity for L^2.

The conformal and projective models can be directly related to one

Figure 2. Escher's view of L^2.

another. If the two diagrams are superimposed, then the point P_p of the projective model is obtained from the corresponding point P_c of the conformal model by moving it radially outwards from the centre O by the amount shown in Figure 3. The fact that we have two different ways of representing L^2 in the Euclidean plane (there are in principle many more) illustrates an important point. It is the internal (or *intrinsic*) geometry of L^2 that really concerns us. Many different ways of representing this same geometry are possible, but the differences have to do with the representation only.

For another type of two-dimensional geometry, distinct from that of both L^2 and E^2, consider an ordinary spherical surface S^2 in three-dimensional Euclidean space E^3. Again Escher has depicted his vision of life in such a world in terms of interlocking angels and devils, this time in the form of a wooden model (see Figure 4). For a specific mathematical description we can take the sphere S^2 to have unit radius. The "straight lines" with respect to the intrinsic geometry are the great circles (i.e., circles on S^2 whose centers coincide with that of S^2). The measure of "distance" on S^2 is simply arc length along a great circle. The geometry of S^2 differs from Euclidean geometry a little more seriously than does that of L^2. For the "straight" lines of S^2 are closed up on themselves (they have, in mathematical terminology, the same *topology* as circles) and intersect one another in *pairs* of points rather than in single points. This last difference can be removed if we pass to the so-called elliptic space P^2, which is derived from S^2 by the identification of antipodal points. Each point of P^2, therefore, corresponds to *two* points of S^2, which lie

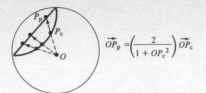

$$\overrightarrow{OP_{\mathrm{p}}} = \left(\frac{2}{1 + OP_{\mathrm{c}}{}^2}\right) \overrightarrow{OP_{\mathrm{c}}}$$

Figure 3. The relation between the two L^2 models.

on a straight line through its center. We may think of each point of P^2 to be defined abstractly as such a pair of antipodal points of S^2. (Such abstract representations of spaces are common in mathematics. What is defined as a "point" in one space may just be some more elaborate structure in another. We shall see more examples of this later.)

If we try to represent the points of S^2 or P^2 globally on the Euclidean plane we have a difficulty. This is because the topology of S^2 and P^2 is different from that of E^2 or of any subset of E^2. But we shall see in a moment how to obtain a valid representation by adjoining some "points at infinity" to E^2. The topology of P^2 can be described as a "sphere with a cross-cap," as depicted in Figure 5. In Figure 6 we see how to get from the first picture of Figure 5 to the second. Since each point of P^2 appears twice in S^2, we consider, say, just the southern hemisphere of S^2, but where antipodal points on the equator must be identified.

Figure 4. Escher's model of an S^2 world.

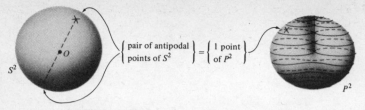

Figure 5. Spherical and elliptic geometry.

A standard conformal representation of S^2 on the plane is given by *stereographic projection*. Regard E^2 as the equatorial plane of S^2 and let N be the north pole. A point X on the sphere is projected from N to the point Z on E^2 (so N, X, and Z are all in line) as shown in Figure 7. Notice that as X approaches N, the line becomes a tangent of S^2 at N and is therefore parallel to E^2. The point Z has gone to "infinity" on E^2. Thus we have realized S^2 not as a subset of E^2 but as the whole of E^2 together with one extra point, adjoined at infinity, to represent N.

This stereographic correspondence is an important one. It is conformal in the same sense as was the conformal model of L^2. An angle between curves on S^2 projects to an equal angle between the corresponding projected curves; a small shape on S^2 projects to a similar small shape on E^2 (perhaps expanded or contracted). Furthermore (over and above conformality) any circle of any size on S^2 will project to an exact circle on E^2—except that if the circle on S^2 happens to pass through N, the projection is a straight line in E^2 (see Figure 8). A great circle on S^2 (i.e., a "straight line" for its intrinsic geometry) is represented on E^2 as a circle (or straight line) whose two intersections with the equator are diametrically opposite.

Let us now imagine N and the plane E^2 to be held fixed, but the sphere S^2 to be moved upwards until its center is the point N, so that it now touches E^2 at the point O. The projection, from N, of S^2 to E^2 now carries a *pair* of antipodal points X, X' on S^2 to a single point Z of E^2, except that the points of the entire equator are now carried to infinity (see Figure 9). Great circles on S^2 are now carried to straight lines on E^2. The model is projective, rather than conformal and represents P^2, rather than S^2 (since X and X' give the same point). Now E^2 is to have the points of an additional *line* adjoined at infinity (representing the equator of S^2).

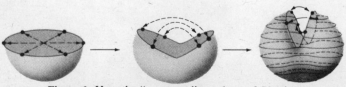

Figure 6. How the "cross-cap" topology of P^2 arises.

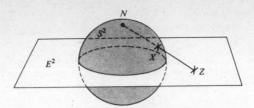

Figure 7. Stereographic projection of S^2.

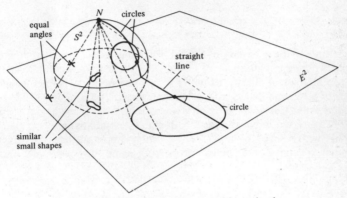

Figure 8. Effects of the stereographic projection.

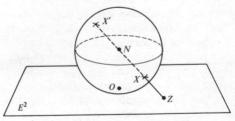

Figure 9. Projection of P^2 to E^2.

As in the case of our two models for L^2, these representations of S^2 and of P^2 can be related to one another by simple geometry (see Figure 10). Note that the formula relating the two models differs by only a sign change from the Lobachevskian one of Figure 3.

We have been considering particular ways of representing the geometries L^2, S^2, and P^2 in terms of a Euclidean description. But each of these geometries can be developed in its own way in an entirely "synthetic" manner, as

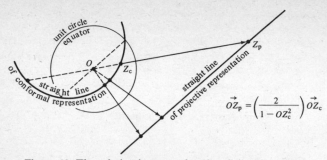

Figure 10. The relation between the S^2 and the P^2 models.

an alternative to the geometry of Euclid. Each has its own body of theorems, sometimes agreeing with those of Euclid, sometimes more complicated, occasionally simpler. A striking example in this last category is the formula for the area of a triangle discovered by Johann Heinrich Lambert (1728–1777) even before non-Euclidean geometry was established! If the angles are α, β, and γ (measured in radians, where 180° equals π radians) and the sides are "straight" with respect to the relevant geometry, then the area Δ is given by how much the sum of the angles differs from π, i.e., by

$$\Delta = \begin{cases} \pi - (\alpha + \beta + \gamma) & \text{for } L^2, \\ (\alpha + \beta + \gamma) - \pi & \text{for } S^2 \text{ or } P^2. \end{cases}$$

No such simple formula exists in the Euclidean case, for there we always have $\alpha + \beta + \gamma = \pi \; (= 180°)$ and a side length must be specified also in order that the area can be determined. This brings out another feature of the geometries L^2, S^2, and P^2, which is distinct from the Euclidean case, namely that there is now an absolute scale of length defined. The Euclidean property of similarity does not now hold. A large triangle is now not really like a small one, since the difference between π and the sum of its angles is a measure of its size. In fact, in the preceding formulas, units of length have been tacitly assumed to concur with this absolute scale. Thus S^2 was taken as a *unit* sphere, its total area being 4π; similarly, the area of P^2 is 2π. The scale of L^2 shows up in the fact that π is the upper bound of all possible areas of triangles. By recalibrating the unit of length we obtain Lobachevskian, spherical, or elliptic geometries whose quantitative departure from that of Euclid is as large or as small as we please.

These geometries are of especial interest not only because they represent the simplest deviations from the Euclidean case but also because they describe, according to mainstream modern cosmology, the best candidates for the large-scale spatial structure of the actual universe. But since physical space is three-dimensional rather than two-dimensional, we must consider the extensions to three dimensions of the models just introduced. In fact, this extension is quite straightforward. We can model the three-dimensional

Lobachevskian space L^3, again either by a projective or conformal representation, as the interior of a sphere S^2 in ordinary Euclidean three-space E^3, the Lobachevskian definition of "straightness" and of "distance" being the same as in the two-dimensional case. Thus, for example, in the conformal representation, a Lobachevskian "straight line" is depicted as a circular arc, in the 3-space inside S^2, which is perpendicular to S^2 at the two points I, J where it meets it; and the formula for the Lobachevskian distance between points A, B on the arc is as given in Figure 1. In fact it is the Lobachevski geometry L^3 that is most favoured by present (albeit somewhat indirect) observations. So why is it that we are not directly aware that the geometry of the spatial world probably accords more with that of Lobachevski than with that of Euclid? The reason lies in the fact that the magnitude of the absolute unit of length (the "radius of curvature of the spatial universe") is of the order of 10,000,000,000 light years which is approximately

60,000,000,000,000,000,000,000 miles!

Compared with this, the measurement of any terrestrial distance must be regarded as essentially infinitesimal. And the *infinitesimal* geometry of L^3 is Euclidean.

The S^3 and P^3 models must also be considered seriously in this connection. Cosmological observations do not yet have a good record of reliability; and the balance of informed opinion could easily shift from its present preference, namely L^3, to S^3 (or P^3). Again we can represent these geometries in terms of E^3, using a conformal or a projective description. The discussion is just as before—though it is harder actually to visualize the projections which achieve the representation, since this would require *four* Euclidean dimensions!

Two new features which do arise are, perhaps, worthy of note here. The first concerns the *orientability* of P^3, as opposed to the nonorientability of P^2. The question is whether a consistent sense of local "handedness" can be defined over the whole space. Consider S^2 first. At each point we can define a handedness (or orientation) in terms of a rotation sense in the surface about that point—say clockwise as we view the sphere from the outside (see Figure 11). But the antipodal correspondence is incompatible with preserv-

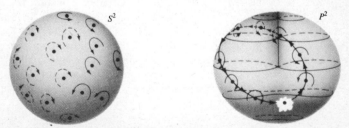

Figure 11. Orientability of S^2 and nonorientability of P^2.

ing this sense of orientation, so that when we identify antipodal points to produce P^2 we get a nonorientable surface (Figure 11) which cannot have a consistent continuous orientation assigned to it: by travelling around certain closed paths on P^2 we can return to our starting point but with the sense of handedness reversed.

The new feature that arises with P^3, however, is that this is *not* the case in three (or any odd number of) dimensions; that is, the antipodal correspondence on S^3 is here compatible with the orientation, so that P^3 is orientable. Thus, a space-traveller who travels from his home planet, in a P^3 world, and finally returns, will never find (as happened to the school-teacher in H.G. Wells's "The Plattner Story") that his sense of handedness is reversed in relation to the standard defined on his home planet. There are, however, other (nonhomogenous) three-dimensional models which are not orientable, so that if the space traveller in such a world left his right glove behind on the planet and took his left glove with him, he could find, on his return, that he had two identically rather than oppositely shaped gloves!

The second new feature arising is a more detailed one, which may seem (initially) like a curiosity. I mention it because it is actually important in a number of different contexts and is a good example of a recurring theme in differential geometry, as we shall see. Recall that the geometries L^2 and P^2 (or S^2) differ from that of Euclid with regard to the concept of parallel lines. Thus in the case of L^2, two straight lines may appear to start out parallel to one another, but as we move out along the lines, they begin to diverge away from one another more and more, farther and farther along the lines. On the other hand, with S^2 or P^2 the situation is opposite: two lines which start out seeming to be parallel will begin to converge towards one another and finally meet. If we pass to the three-dimensional geometry L^3, or to S^3 and P^3, this situation is exactly corresponding; but for S^3 (or P^3) a slightly different concept of "parallel" can be introduced for which the lines do *not* get closer together!

First consider a straight line l in ordinary Euclidean space E^3. Imagine l to be displaced parallel to itself by moving it out in a direction $\vec{d}$, perpendicular to itself. Now rotate the line slightly about the direction $\vec{d}$ as we move it out. The result is a line m which is *skew* to l. The distance between l and m will increase if we move outwards along them in either direction. Now apply the same procedure in the space S^3 (or P^3). By carefully adjusting the rate at which we rotate as we move m away from l, we can balance this *diverging* effect arising from the skewness against the *converging* effect of the non-Euclidean geometry in such a way that the lines l and m now remain the *same* distance apart all the way along. Such pairs of lines are called *Clifford parallels* on S^3 (or P^3) (see Figure 12) and can be of a right-handed or left-handed sort. (William Kingdon Clifford (1845–1879), an English geometer, had forcefully suggested that matter curves space!)

More remarkably, it is possible to fill up the whole of S^3 (or P^3) with a system of lines (great circles on S^3), all of which are "parallel" to one another in

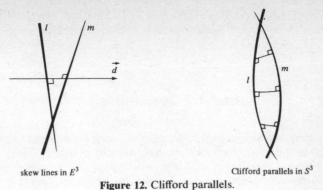

skew lines in E^3 Clifford parallels in S^3

Figure 12. Clifford parallels.

this sense, where there is precisely one through each point of the space and where every pair is linked. Figure 13 illustrates such a complete system of Clifford parallels. The S^3 is represented in ordinary Euclidean space E^3 (together with one point at infinity) according to the conformal representations discussed above. The Clifford parallels now appear as a system of mutually linked circles (and one straight line). The circles lie in a twisted fashion on a system of "nested" tori. (A torus is a surface shaped like an inner tube or the surface of a doughnut; "nested" means that these tori lie

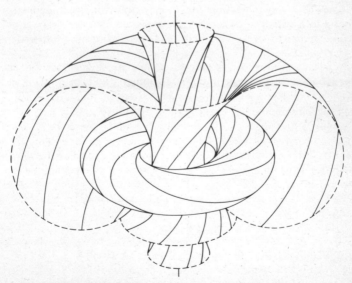

Figure 13. Clifford parallels filling the whole of S^3.

threaded within one another in continuous succession.) As can be seen, the configuration is somewhat complicated, indicating something of the intriguing possible situations that arise in higher-dimensional non-Euclidean geometry.

Irregular Geometries

So far, we have been concerned with only the simplest and most symmetrical possible non-Euclidean geometries. According to Einstein's theory, however, such models can, at best, represent only the very large scale averaged-out spatial geometry of the universe. Local irregularities in geometric structure (i.e., local "curvature") will arise on a much smaller scale, corresponding to the presence of local gravitational fields. In the immediate neighborhood of the surface of the earth this local curvature would have, as it so happens, a radius of the general order of the distance from earth to sun, or about 100,000,000 miles. (A large radius of curvature means a small curvature and *vice versa*.) As compared with the very rough value of approximately 10^{22} miles for the universe as a whole, we see that the *local* curvatures in our neighborhood dominate, locally, by a factor of something like 10^{14}. But these curvatures are still very small by any normal standards of terrestrial geometry and do not disturb the accurate employment of Euclid's geometry in the ordinary way. Even in the most extreme situation arising at the surface of a neutron star we can expect curvature radii of 100 miles or so (see Figure 14). But these small deviations from Euclidean geometry are, according to Einstein's view, *totally* responsible for the effects of gravity, so we must come to terms with them in some detail if we are to understand how gravity acts.

How, in fact, are we to give precise mathematical description to such irregular geometries, where the notions of symmetry and congruence, so basic

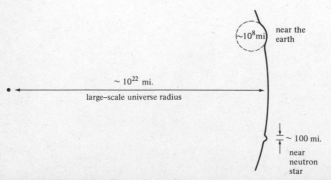

Figure 14. The rough orders of magnitude of the curvatures of our universe (not drawn to scale!).

to Euclidean arguments, are now totally absent? In the solution to this problem lies the essence of differential geometry.

One way that we might envisage proceeding would be to imagine the curved space that we wish to describe embedded as some kind of surface in a Euclidean space of higher dimension. This is actually a legitimate procedure and represents an older approach to differential geometry. But if, as is the case with relativity theory, we are concerned with *intrinsic* (that is, "internal" geometric) properties only, this procedure is rather indirect and is less satisfactory than certain other approaches. Any particular embedding or representation that may be employed is something extraneous to the geometry under study. We have seen this kind of thing already in our examination of L^2, S^2, and P^2, where various *different* representations in terms of E^2, or E^3, were given to describe the *same* space. It is often hard for people to grasp, at first, that such an embedding space is actually not needed. To picture a curved universe one may *feel* a need for it to be sitting inside some kind of flat background, so ingrained are the Euclidean concepts in our minds. But such an embedding space, when it is used, is for purposes of convenience only and need have no "reality" in terms of the curved geometry under consideration. (In this respect, pictures such as that given in Figure 14 can be rather misleading.)

To proceed with an intrinsic approach to differential geometry we need first the mathematical concept of a *manifold*. This generalizes the concept of a curve or surface and consists of a set of "points" together with some "local structure." The objects which are called the points of the manifold are usually undefined entities. There is no such Euclidean requirement that they be objects of infinitesimal size, devoid of physical extent. In fact we can construct examples of manifolds were the "points" themselves represent finite, extended objects. For example, each point of our manifold might represent a single location (i.e., position and orientation) of a rigid body in ordinary Euclidean 3-space, the manifold so constructed then being what is called the *configuration space* of the body (see Figure 15). Since six parameters are required to describe the location of the body (three for the position of its

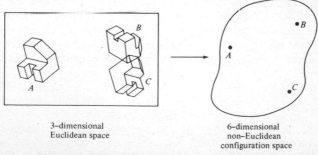

3–dimensional
Euclidean space

6–dimensional
non–Euclidean
configuration space

Figure 15. The configuration space of a rigid body.

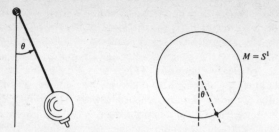

Figure 16. Configuration space of a simple pendulum.

center of mass and three for its orientation in space) the manifold we obtain will be six-dimensional (and, in fact, non-Euclidean—closely related to P^3). The word "point," when used in such a context to denote a rigid body location, for example, serves mainly to channel ones thinking in a useful direction: in the picture conjured up in ones mind's eye, such a "point" is, indeed, devoid of physical extent, unlike the original picture of the rigid body itself.

In this example it is hard to visualize the configuration space in detail, since it is six-dimensional. This does not detract from the usefulness of the configuration-space concept, however, and configuration spaces of much larger dimension are often considered. (For example, a tiny pea-sized box containing a gas with 10^{18} molecules—considered as point particles—would have a 3,000,000,000,000,000,000-dimensional configuration space!) But to fix ideas at this stage, let us consider a simpler situation, namely that of a rigid body (a "simple pendulum") which is constrained to move only in one plane and which is pivoted at one point, so that it has only one degree of freedom. The configuration space, in this case, is one-dimensional and is a circle S^1, the points of which may be labelled in terms of an angle θ determining the orientation of the body (Figure 16). Note that even in this simple case, the configuration space is non-Euclidean, since a one-dimensional Euclidean space is a straight line, which has a different topology from a circle.

Configuration spaces are not the only manifolds that arise naturally in mechanics. For example, velocity or momentum variables may be incorporated in addition to position variables. In the case of the constrained rigid body just considered, the angular momentum about the pivot would provide such an additional variable, and the resulting space (called a "phase space") would then be two-dimensional, having the topology of a cylinder (see Figure 17), the height along the cylinder representing the magnitude Ω of the angular momentum. Again this space is non-Euclidean (for obvious topological reasons).

Consider now the question of "local structure" for a manifold. The most primitive local structure is *topology*. Roughly speaking, the topology of a space provides a way of telling which points are very "close" to one another and which are not, or, a little more accurately, which sequences of points approach other points indefinitely closely. It does not provide any concept of an actual "distance" between points. Topology does provide us with a

Figure 17. Phase space of a simple pendulum.

way of telling whether a curve drawn on the space is continuous (i.e., unbroken) or discontinuous, but it does not provide a distinction between a smooth curve and a kinky or angled one and it does not provide any concept of length along a curve. Topology tells us what the "continuity" structure of the space is, but not "smoothness" and certainly not "size." Thus the surface of a cube has the same topology as the surface of a sphere (S^2), though one is angled and the other smooth. But a cube or 2-sphere has a different topology from a torus (T^2) or a projective plane (P^2), or a 3-sphere (S^3), for example, or from a 2-sphere with a single point removed ($S^2 -$ {point}), this last having the same topology as a Euclidean plane (E^2) and as a Lobachevskian plane (L^2). The cylinder of Figure 17 has the same topology as a 2-sphere with *two* points removed, i.e., as $E^2 -$ {point}.

A topological space is called an *n*-dimensional (topological) *manifold* if the topology in the neighborhood of any point is the same as in an *n*-dimensional Euclidean space E^n. For example, L^2, S^2, P^2 are two-dimensional manifolds, while S^1 and S^3 are one- and three-dimensional manifolds, respectively. Thus, we can envisage piecing together a number of local portions of a Euclidean *n*-space (with regions of overlap specified) to form a manifold M; in Figure 18, this procedure is applied to produce a sphere and a torus. The rules for how this piecing together is to be achieved in any particular case

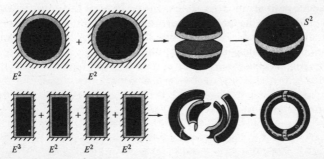

Figure 18. Piecing together a sphere and a torus out of Euclidean portions.

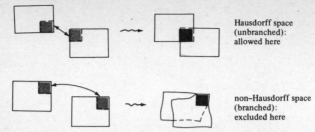

Figure 19. Hausdorff and non-Hausdorff manifolds.

can be given in an entirely intrinsic manner without reference to any embedding space: portion A of piece number 1 is to be identified with—or "glued" against—portion B of piece number 2, etc. (Hermann Weyl, 1885–1955).

(One has to be slightly careful, when specifying how these gluings are to be arranged, that the resulting space does not possess leaves or branches—that is, that the resulting space is what is called a Hausdorff manifold (see Figure 19). In fact, non-Hausdorff manifolds, in which such branching occurs, form legitimate mathematical objects worthy of study. But in normal mathematical parlance, the word "manifold," taken by itself, means a Hausdorff manifold. I shall not consider the other more general possibility in this article.)

This patching, so far, enables us to construct a manifold as a topological space only. And while we certainly do not wish to retain so rigid a structure as, say, the concept of Euclidean distance inherited from that in the individual patches, topology, by itself, is too weak a structure to support differential geometry. What we need is a *differentiable structure* for a manifold M. Such a differentiable structure may be thought of as a means of specifying whether or not a (real-valued) function on M is to be regarded as "smooth."

Let us consider an example of what "smoothness" is to mean. Take the unit sphere S^2 and choose, for example, the function h which measures the vertical height of a point on S^2 above the equator (see Figure 20a). (The value of h below the equator is then negative.) This would be an example of a smooth function of S^2. Now consider another function $|h|$, defined on S^2, which measures the *absolute* vertical distance from the equator (so the value below the equator is now positive). This function is *not* smooth along the equator, as it suddenly changes its direction of increase from that given just below to that given just above the equator (see Figure 20b). On the other hand the square $h^2 (= |h|^2)$ of the vertical distance *is* smooth on S^2. This is because the change in the increase of the function below and above the equator occurs suitably gradually (see Figure 20c). To make these notions adequately precise would require entering into the intricacies of differential calculus and analysis. I do not propose to do this in any detail here, assuming that an intuitive picture of the notion of smoothness may have been supplied to the reader by the above examples. But, in fact, the whole point of assign-

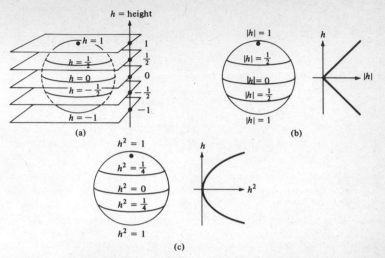

Figure 20. The smoothness of a function on S^2.

ing a differentiable structure to a manifold M is that it enables the ideas of calculus to apply on M. For this, a kind of algebra, known as tensor algebra, must be developed.

Vectors and Tensors

One way of proceeding is to try to pin things down more precisely by introducing *coordinates* into the manifold. This entails labelling, in a smooth manner, the various points of an n-dimensional manifold M with different n-tuples $(x^1, x^2, \ldots, x^n)$ of real numbers. (The indices $1, 2, \ldots$ in "x^1," "x^2," $\ldots$ do not here indicate that a power of x is to be taken, nor are they dimensions of spaces; they are just markers distinguishing the different coordinates.) But it may not be possible to do this all at once for the whole manifold. For example, in the cases S^2, P^2, S^3, P^3, the topology rules out such a global coordinate system. The procedure in such cases is to cover the manifold with a patchwork of overlapping coordinate systems (see Figure 21), so that each point lies in at least one patch. But some points will then lie in more than one patch and are correspondingly assigned more than one set of coordinates, so coordinate transformations must then be presented, which tell how to translate from one coordinate system to another. In this way we can provide a labelling for all points of the manifold.

This introduction of a coordinate system for M does not, however, pin things down as definitely as one might think. For the coordinate system is itself something quite arbitrary in most cases. Sometimes (as is the situation

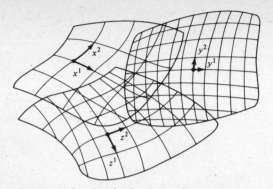

Figure 21. Overlapping local coordinate systems.

with Cartesian coordinates for the Euclidean plane) particular types of coordinates suggest themselves as more "natural" than the general curvilinear type and then it is helpful to employ such "natural" coordinates. But when M is curved in an arbitrary way, it tends to be inconvenient to try to single out special types of coordinate systems in this way. Instead, one asks for properties of the manifold M which can be ascertained equally well *whichever* system of coordinates is selected.

It may seem surprising at first that such coordinate-independent properties exist at all. But in fact they do—and the whole classical approach to differential geometry rests on this: the coordinates are introduced into the formalism only to be eliminated again, as an irrelevance, from those quantities which one is finally concerned with. There is much that is ingenious and much that is powerful in this classical approach, but it is often complicated. More recently, largely owing to the influence of the remarkable French mathematician Élie Cartan (1869–1951), a more economical coordinate-free approach has evolved, which leads directly to the required calculus of "coordinate-independent" properties. Though a detailed discussion of either the classical or coordinate-free approach is well beyond the scope of this article, it will be necessary to discuss briefly some of the key concepts that are involved.

The idea is to introduce a whole hierarchy of objects and operations, forming a kind of algebra or calculus—sometimes referred to as the *absolute differential calculus*. The most primitive objects (after the points of M themselves) are the *scalar fields*, these being simply the "smooth functions" on M that we have just been considering. A coordinate system may, if desired, be invoked to provide the initial notion of what "smoothness" means on the manifold: a function on M is then deemed to be smooth if, when expressed in terms of the coordinates, it is differentiable (in the ordinary calculus sense) as a function of these coordinates. Next, objects called *vector fields, covector fields, tensor fields,* and *connections* may be introduced in turn (and, if

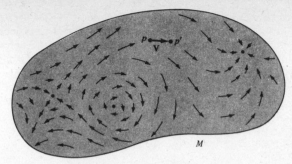

Figure 22. A vector field on M.

required, further more complicated objects too). In the classical approach, the coordinate system is called forth again and again in order to define each new class of entity in terms of the transformation properties of their explicit descriptions when the coordinate system is changed. But the economy of the coordinate-free approach is that no further mention of coordinate systems is needed once the concept of a scalar has been established.

Let us examine how this works for the next most complicated object after a scalar field, namely a *vector field*. To picture a vector field **V** (or contravariant vector field, in the older terminology) we think of a "field of arrows" on M (see Figure 22) where, crudely, each arrow is envisaged as connecting some point p of M to another one p' which lies infinitesimally away from p. In the modern viewpoint, **V** is interpreted as a *differential operator*, which is a function that acts on a scalar field ϕ to produce a new scalar field $\mathbf{V}(\phi)$, where the value of $\mathbf{V}(\phi)$ at a point p is the rate of increase of ϕ at p in the direction indicated by the arrow at p.

The set of different possible vectors (or "arrows") at a single point p of M constitutes what is called the *tangent space T_p* to M at p (see Figure 23). To picture what this tangent space means, we envisage expanding the immediate neighborhood of p in M uniformly outwards by a large amount so that

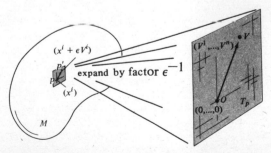

Figure 23. The tangent space T_p at a point p of M.

what used to be an infinitesimal displacement $\overrightarrow{pp'}$ on M now becomes a finite displacement $\mathbf{V} = \overrightarrow{OV}$ in the resulting space T_p. (Here p has become the origin point O of T_p and p' has become V.) In terms of a coordinate system for M, the vector $\mathbf{V}$, at p, may be specified in terms of n real numbers V^1, $V^2, \ldots, V^n$ called its components at p. Their precise relation to $\mathbf{V}$ is given in the box on p. 103. These components may be used as (Cartesian) coordinates for the point V in the n-dimensional space T_p. The point O, which represents the zero vector $\mathbf{V} = 0$, is the origin of coordinates in T_p (but, of course, p need not be the origin of x^i-coordinates in M).

Though the manifold M may be "curved," it always turns out that each T_p is, in a well-defined sense, *flat*. In fact it is a *vector space*—which means that elements of T_p can be added together (as represented geometrically by the parallelogram law) and mulitplied by numbers. However T_p should *not* be regarded (at least in the most general cases) as a Euclidean space E^n. The reason for this is that the concept of Euclidean distance between points of T_p—e.g., the distance OV which, by Pythagoras' theorem (in n-dimensions), would be given by

$$OV^2 = (V^1)^2 + (V^2)^2 + \cdots + (V^n)^2$$

—is not "invariant" under change of coordinates on M. However, it is possible to impose further structure on M which does supply such an invariant concept of Euclidean distance in each T_p. But to describe this, we must look a little further into the absolute differential calculus.

Having the concept of a vector field on M it is not hard, using a purely algebraic procedure, to pass to the next step and define a *covector field* on M. A covector field $\mathbf{A}$ (or covariant vector field, in the older terminology) may be thought of, according to the coordinate-free approach, as a certain type of function (called a *linear* function) which takes a vector field $\mathbf{V}$ to a scalar field $\mathbf{A(V)}$. A rough geometrical picture of a covector field $\mathbf{A}$ is a field of plane-elements (actually $(n-1)$-dimensional plane-elements) on M (see Figure 24). The plane element at p contains all the directions of all the arrows representing vectors $\mathbf{V}$ at p for which $\mathbf{A(V)} = 0$. This enables us to proceed

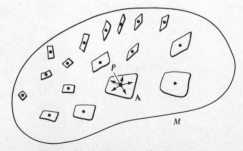

Figure 24. A covector field on M.

Vectors

A *vector field* **V** is a map which takes a scalar field ϕ to another scalar field $\mathbf{V}(\phi)$ such that:

$$\mathbf{V}(\phi + \psi) = \mathbf{V}(\phi) + \mathbf{V}(\psi)$$

$$\mathbf{V}(\phi\psi) = \phi\mathbf{V}(\psi) + \psi\mathbf{V}(\phi)$$

$$\mathbf{V}(\kappa) = 0 \quad \text{whenever } \kappa \text{ is constant.}$$

In terms of coordinates x^1, $x^2, \ldots, x^n$, **V** has components $V^1, V^2, \ldots, V^n$, each dependent upon $x^1, \ldots, x^n$, where

$$\mathbf{V}(\phi) = V^1 \frac{\partial\phi}{\partial x^1} + V^2 \frac{\partial\phi}{\partial x^2} + \cdots + V^n \frac{\partial\phi}{\partial x^n}$$

$$= V^i \frac{\partial\phi}{\partial x^i} \quad \begin{array}{l}\text{using the repeated-index}\\ \text{summation convention.}\end{array}$$

to the general concept of a *tensor field* on M, which can be viewed as a similar type of map (called a *multi-linear* map) from collections of vectors and covectors to scalars (or else to vectors, covectors, or other tensors, if desired).

Perhaps the most important type of tensor field is that which defines the structure of a *Riemannian manifold*—named after the great German mathematician Bernhard Riemann (1826–1866), who, following the important work of Gauss in the two-dimensional case, virtually initiated the subject of differential geometry. This tensor field is called the *metric tensor* and is generally denoted **g**. It may be viewed as a map which takes a pair of vector fields **U**, **V** to a scalar field $\mathbf{g}(\mathbf{U},\mathbf{V})$; in terms of a coordinate system, **g** is represented as a square $n \times n$ array of components g_{ij}. (The relation between $\mathbf{g}(\mathbf{U},\mathbf{V})$ and g_{ij} is displayed in the box on p. 104, as are the two conditions of symmetry and positive-definiteness that the tensor **g** must satisfy in order to qualify as a Riemannian metric.)

The geometrical significance of **g** is that it actually does give a Euclidean metric structure (E^n) to each tangent space T_p. Recall that the Pythagorean sum-of-squares expression for OV^2 that was mentioned above will not do this because it gives a result which is dependent on the choice of coordinate system. But the expression $\mathbf{g}(\mathbf{V},\mathbf{V})$ may now be used instead:

$$OV^2 = \mathbf{g}(\mathbf{V},\mathbf{V}) = g_{ij}V^iV^j.$$

Its value is independent of coordinates simply because the quantity $\mathbf{g}(\mathbf{V},\mathbf{W})$ (in the coordinate-free approach) is something which makes no reference to a coordinate system. Furthermore, this new definition of OV can be shown

Metric Tensor

$$\mathbf{g(U,V)} = g_{11}U^1V^1 + g_{12}U^1V^2 + \cdots + g_{1n}U^1V^n$$

$$+ g_{21}U^2V^1 + g_{22}U^2V^2 + \cdots + g_{2n}U^2V^n$$

$$+ g_{n1}U^nV^1 + g_{n2}U^nV^2 + \cdots + g_{nn}U^nV^n$$

$$= g_{ij}U^iV^j \text{ using the summation convention}$$

$$\mathbf{g(U, V)} = \mathbf{g(V, U)}, \quad \text{i.e., } g_{ij} = g_{ji}.$$

$\mathbf{g(V, V)} > 0$ unless $\mathbf{V} = 0$, i.e., (g_{ij}) is a positive-definite matrix.

to satisfy all the properties that an ordinary Euclidean "distance from the origin" has to satisfy in a vector space. (This follows from a classical theorem which states that a coordinate system can always be chosen such that at one *particular* point p the components g_{ij} are zero for $i \neq j$ and unity for $i = j$. The expression $g_{ij}V^iV^j$ is then equal to the Pythagorean sum of squares in T_p and the required properties are satisfied. In other coordinates the expression for OV^2 would look more complicated, as is to be expected since most coordinate systems would give rise to "oblique" coordinate axes for T_p in which the Pythagorean form would not hold.)

Let me recapitulate what has been established. First there is the general concept of a manifold M, with its topology and "smoothness" structure. This is the place where the objects of absolute differential calculus can live. But at that stage no specific concept of distance is singled out, even in the tangent spaces. If we wish to introduce such a distance concept, we must do so by specifying a particular metric tensor field **g**. This enriches the structure of M and turns it into what is called a *Riemannian manifold*. The fact that the smooth manifold structure *alone* does not give us a distance concept is graphically illustrated in Escher's woodcuts (see Figure 2); the Euclidean metric on the unit disc is very different from the Riemannian ("Escher") metric of L^2, since according to the latter, the devils are all the same size!

In geometrical terms, the assignment of such a Riemannian structure means the assignment of a particular concept of *distance* between any pair of infinitesimally separated points p, p'. Hence, stringing such points together into a (smooth) curve on M we obtain a concept of distance measured along the curve, called the (Riemannian) *length* of the curve (see Figure 25). Figure 25 also illustrates another Euclidean-type concept which a Riemannian structure defines, namely the angle between two vectors, or the angle of intersection between two curves.

The "curve-length" concept on M completely defines the Riemannian

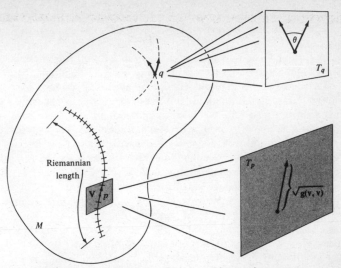

Figure 25. Lengths and angles in a Riemannian space.

structure on M: the two are equivalent. On the other hand, the "angle" concept does not quite define the Riemannian structure, but defines something weaker, known as a *conformal* structure. Recall the conformal representations of L^2 and S^2 in E^2. These are examples in which one Riemannian space (L^2 or $S^2 - \{\text{point}\}$) is mapped into another (E^2) with a different Riemannian structure (so distances are not faithfully represented) but where the conformal structure is preserved in each case (so angles are faithfully represented).

Curvature

A Riemannian manifold is something that looks *locally* like a Euclidean space—in the sense that the tangent spaces are each given a Euclidean structure. Euclidean space E^n is itself a particular example of a Riemannian manifold. So also are the non-Euclidean spaces L^n, S^n, P^n, as well as the more general irregular spaces that we are concerned with now. The *deviations* from Euclidean geometry can be described, on a local scale, in terms of a certain tensor field **R** known as the (Riemann–Christoffel) *curvature* tensor. The geometrical meaning of the tensor **R** is best understood in terms of a certain operation called *covariant differentiation*, denoted ∇, a notion which was first introduced by Elwin Christoffel (1829–1900) and studied as an operation in its own right by Tullio Levi-Civita (1873–1941). The operation ∇ carries a tensor field Φ into another slightly more complicated one, denoted $\nabla \Phi$, which measures the rate at which Φ is varying. (Scalars, vectors, and covectors may be regarded as special cases of tensors, so ∇ applies to them too.)

This operation ∇, as first introduced by Christoffel, is something which arises in a unique and rather remarkable way once a Riemannian structure g has been assigned to the manifold. But Levi-Civita showed how a ∇ could be defined and interpreted (though not uniquely) even when *no* specific metric g was given on the manifold. Such an assignment of a ∇ to M makes M a *manifold with connection*. By Christoffel's original result, any Riemannian manifold is, in a natural way, also a manifold with connection. But not every manifold with connection is Riemannian. That is only a special case, though an important one. Another important special case, which we discuss later, is the pseudo-Riemannian case needed for relativity.

To understand the geometric meaning of ∇ and of curvature we may think in terms of an essentially equivalent concept: *parallel propagation*. Suppose γ is some smooth curve on M, and let the vector $\mathbf{U}$ be tangential to γ all the way along it (see Figure 26). An operation $\nabla_{\mathbf{U}}$ is defined by ∇, called the propagation derivative, which measures the "rate of change" of any tensor field Φ along the curve γ. (In the special case when Φ is a scalar ϕ, this is our old friend the vector operator $\mathbf{U}$, i.e., $\nabla_{\mathbf{U}}\phi = \mathbf{U}(\phi)$.) We now define *parallel propagation* of Φ along γ by the condition $\nabla_{\mathbf{U}}\Phi = 0$, which means that the "rate of change" of Φ along γ is zero. Now consider the case when Φ is a vector $\mathbf{V}$. The parallel propagation of $\mathbf{V}$ along γ defines a system of "parallel directions" in M along γ. This is illustrated in Figure 26.

But there is a crucial respect in which this general notion of "parallel" differs from the familiar particular case of it that occurs in Euclidean space: for if γ is a *closed* loop, we may now find that a net *rotation* occurs when γ is traversed. Consider, for example, the case when M is a sphere S^2. Now parallel propagation of $\mathbf{V}$ along γ does *not* correspond to keeping $\mathbf{V}$ parallel to itself in an embedding in Euclidean space E^3 in which S^2 may be visualized as sitting, since $\mathbf{V}$ must always be kept pointing tangentially to S^2. Suppose that the vector $\mathbf{V}$ starts at a point p situated on the equator of S^2 and is moved by

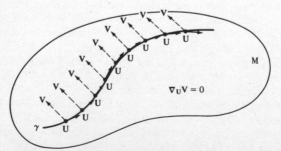

Figure 26. Parallel propagation of a vector along a curve.

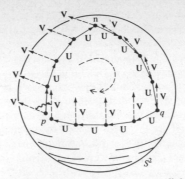

Figure 27. The effect of S^2's curvature on parallel propagation.

parallel propagation along a meridian (great circle) until it reaches the north pole (see Figure 27). If **V** begins by pointing due west, then by symmetry we can expect that it remains pointing due west until it is about to reach the north pole. Suppose it is next moved back by parallel propagation to the equator along a different meridian, in the direction opposite to the one in which **V** now points. Clearly **V** points due north all the way down until the equator is reached at q. Finally, it is returned to p by parallel propagation along the equator—pointing north all the time. But on return, we find there is a considerable discrepancy! Though the vector **V** has been moved by parallel propagation throughout, we find that it has undergone a resulting *rotation* (here through a right-angle) when it finally returns to the starting point.

This phenomenon is a direct manifestation of the presence of *curvature*, which is described quantitatively by the *curvature tensor* **R**. We may regard **R** as a map which takes three vectors **X**, **Y**, and **V** into a fourth vector **W** = **R**(**X**,**Y**;**V**), where **W** (scaled up suitably) represents the difference between the final and initial states of a vector **V**, when **V** is carried by parallel propagation around a small parallelogram whose two sides at a point p are defined by vectors **X** and **Y** (see Figure 28). Thus, **R** measures an infinitesimal discrepancy which results from parallel propagation around an *infinites-*

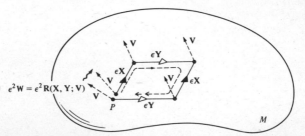

Figure 28. The geometrical meaning of the curvature tensor.

imal closed loop. The finite discrepancy for a finite loop would then be obtained by appropriately combining together these infinitesimal effects. The box on p. 109 contains details of the analytic definition of **R**.

Next, we can define *geodesics*, which are the natural analogs of straight lines for M. A geodesic is a curve with tangent vectors **U**, at different points of the curve, related to one another by parallel propagation. Hence a geodesic moves along M "parallel to itself." There is a unique (maximally extended) geodesic through a point p in each direction at p (see Figure 29).

The geodesics on S^2 are the great circles; those on each of L^2, S^2, P^2, S^3, etc. are what were referred to as the "straight lines" of those geometries. Recall also the concept of a triangle made up of such lines, and the fact that the angles summed to a quantity whose difference from π $(= 180°)$ was a measure of the area of the triangle. This last fact is actually an illustration of the above-mentioned property of parallel propagation, in the special case when the curvature is constant. The integrated discrepancy, for parallel propagation of a vector around the triangle, is then just proportional to the area enclosed. Thus, as we pass around the triangle we encounter a net rotation effect which is proportional to the area of the triangle. The total angle that the sides of the triangle rotate through, as the triangle is traversed, differs from the Euclidean value by an amount proportional to this area (see Figure 29).

I have been discussing geodesics and curvature as concepts arising solely from the properties of ∇. But a concept of "geodesic" arises also directly from a metric **g**. For we could define a geodesic which connects points p and q (not too far from one another) as the curve from p to q of *minimum Riemannian length*. In fact, this definition concurs precisely with the one given in terms of ∇ provided that we choose our ∇ correctly in the first place, as the one naturally defined by **g** according to Christoffel's prescription. It is a remarkable and very basic fact of Riemannian geometry that this ∇ is in fact uniquely defined from **g** by two simple requirements. The first is that the length $(= [\mathbf{g}(\mathbf{V},\mathbf{V})]^{1/2})$ of any vector **V** remain constant under parallel propagation along any curve (or, equivalently, that $\nabla \mathbf{g} = 0$). The second is that ∇ be what is called *torsion-free* (see the box on p. 110). In geometrical terms,

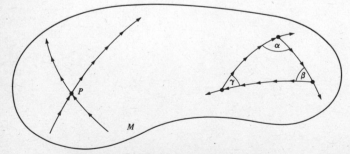

Figure 29. Geodesics and geodesic triangles.

Curvature Tensor

$R_{ijk}{}^l$ denotes the components of **r**, so $W^l = R_{ijk}{}^l X^i Y^j V^k$ stands for $\mathbf{W} = \mathbf{R(X,Y;V)}$. Then

$$\mathbf{R(X,Y;V)} = \{\nabla_X \nabla_Y - \nabla_Y \nabla_X - \nabla_Z\} \mathbf{V}$$

defines **R**, where $\mathbf{Z}(\phi) = \mathbf{X}(\mathbf{Y}(\phi)) - \mathbf{Y}(\mathbf{X}(\phi))$ for scalar field ϕ.

---

the condition that ∇ be torsion-free is, essentially, that infinitesimal parallelograms effectively "close" (see Figure 30). This point was glossed over in the geometrical interpretation of curvature just given, but that discussion is not seriously affected by it. Though the torsion-free condition is a natural one to impose on a choice of ∇, it may be remarked that there are interesting circumstances in which it is desirable *not* to do so. For example, the concept of "parallel" that gives rise to the Clifford parallels on S^3 that were described earlier (see Figures 12 and 13) is something which arises from a ∇ *with* torsion.

Space-time

Riemannian geometry is one of the most important parts of differential geometry, but it is only a part. There are many other types of local structure that can be defined and which give rise to rich and interesting alternative geometries. One such alternative—though differing only minimally from the Riemannian case—is the *pseudo-Riemannian* geometry of Einstein's general relativity.

To appreciate the nature of this particular geometry, recall, first, some of the basic tenets of special relativity. Space and time are combined together to give a 4-dimensional picture of the world—the *space-time* of Minkowski. (Hermann Minkowski (1864–1909), a Russian-born mathematician, had lec-

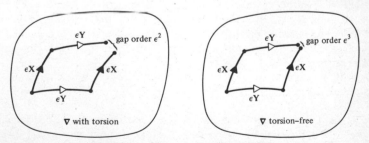

Figure 30. The torsion-free condition.

Torsion-Free Connection

∇_i denotes the components of ∇. ∇ is torsion-free if $\nabla_i \nabla_j \phi = \nabla_j \nabla_i \phi$ (ϕ a scalar field); equivalently, $\nabla_U V(\phi) - \nabla_V U(\phi) = U(V(\phi)) - V(U(\phi))$. In this case, $R_{ijk}{}^l V^k = (\nabla_i \nabla_j - \nabla_j \nabla_i) V^l$.

---

tured to the student Einstein while at the Zürich Polytechnic in Switzerland. Later, it was he who put forward the geometric space-time view of Einstein's special relativity.) This picture may be conveniently described in terms of four global coordinates t, x, y, z, where t is an ordinary time coordinate and x, y, z are ordinary Cartesian space coordinates. It simplifies matters to choose units in such a way that the speed of light (conventionally denoted c) is chosen to be equal to unity ($c = 1$). This means that units of length and time become convertible into one another according to the scheme:

$$1 \text{ second} = 299{,}792{,}458 \text{ metres} = 186{,}284 \text{ miles}$$

$$1 \text{ year} = 1 \text{ light year}.$$

Then the *light cone* of the origin is the locus

$$t^2 = x^2 + y^2 + z^2 .$$

It has two portions, namely the *future* light cone (given when $t > 0$), representing the history of a spherical light pulse travelling outwards from an initial flash at the center at $t = 0$, and the *past* light cone ($t < 0$) which is the history of a spherical light pulse converging inwards to the center at $t = 0$ (see Figure 31).

We may think of light to be propagated as particles, called photons; then the *generators* of the light cone (i.e., straight lines on the cone through the origin) represent the histories of individual photons of the light flash. The same picture would also hold for any other type of "massless" particle since all such particles travel with the speed of light. A massive particle, however, must always travel with a speed less than that of light, so the history of a free massive particle emitted in the same explosive event that produced the photons would be a straight line from the origin O lying in the interior of the future light cone (see Figure 32).

Any other particle in free (unaccelerated) motion would be likewise depicted as a straight line in Minkowski space. For a massless particle the line would be tilted at 45° to the t-axis; for a massive particle, the tilt would be less than 45°. A massive particle not in free motion would be described as a curve—the *world-line* of the particle—whose tangent vectors are every-

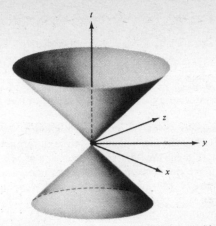

Figure 31. The light cone of the origin in Minkowski space-time.

where tilted at less than 45° to the *t*-axis, i.e., which everywhere lie inside the local light cone (see Figure 32). Such a world-line is called *timelike*.

Minkowski space M^4 has a natural geometry which is *pseudo-Euclidean* rather than Euclidean. The *Minkowskian distance OQ* of a point Q, with co-ordinates t, x, y, z, from the origin is given by

$$OQ^2 = t^2 - x^2 - y^2 - z^2 ,$$

which differs from the Pythagorean expression (for Euclidean 4-space E^4) by the presence of minus signs. (Sometimes the form $x^2 + y^2 + z^2 - t^2$ is used, but this would not conform with the particular interpretation given here.) Notice that $OQ^2 = 0$ if Q lies on the light cone of O and $OQ^2 > 0$ if Q lies inside it. Now the significance of this "distance" OQ is that (for Q within the light cone of O) it measures the time-interval experienced, between the

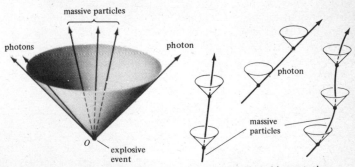

Figure 32. The world-lines of particles in Minkowski space-time.

events O and Q, by a clock whose world-line is the straight path OQ. There is a corresponding result for world-lines not passing through O. If R has co-ordinates t', x', y', z', then the Minkowski distance RQ is given by

$$RQ^2 = (t - t')^2 - (x - x')^2 - (y - y')^2 - (z - z')^2.$$

We have $RQ^2 > 0$ if Q lies within the light cone of R, and RQ then measures the time-interval experienced by a clock in free motion from R to Q.

The fact that measurement of time accords with such an expression, rather than with the Newtonian "absolute" time difference $t - t'$, is the key to special relativity. Time is the "distance" measure of Minkowskian geometry. And like ordinary distance in Euclidian geometry, the time measure along a timelike curve (world-line) becomes a *path-dependent* concept. If R and Q are two space-time points connected by several timelike curves (see Figure 33) then the time interval between R and Q will differ, in general, along the different curves.

Though this seems strange at first, it should be emphasized that there is overwhelming experimental support for such nonabsoluteness of time measurements (e.g., lifetimes of cosmic ray particles created in the upper atmosphere, accurate time measurements made in airplanes, the behavior of particles in high energy accelerators). We are familiar with the fact that ordinary Euclidean distance is a path-dependent concept. But our intuitions about time are built up from experiences in which the path-dependence effect is very minute—because ordinary velocities are so much less than that of light.

In Euclidean geometry, the "straight" path RQ would be the one along which the distance measure is a minimum. There is a curious reversal of this

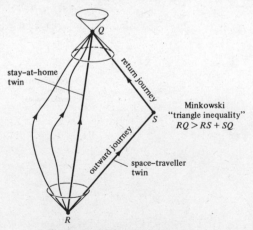

Figure 33. The "clock paradox."

for Minkowskian geometry: among all time-like curves from R to Q the straight (unaccelerated) world-line has *maximum* length (i.e., time interval). This relates to the famous so-called clock paradox of special relativity (see Figure 33). Two twins start from R, one of whom remains unaccelerated on his home planet, while the other travels outwards at a sizable fraction of the speed of light to a distant planet at S, whereupon he turns back for home. When they are reunited at Q, it is found that the traveller has aged less than the stay-at-home! The stay-at-home's time which elapsed during the journey of his brother is just RQ—and this must be *greater* than that measured along the traveller's nonstraight world-line.

It is worth remarking at this juncture that the Minkowskian "unit sphere," given by the set of points Q at unit Minkowskian distance from the origin O ($OQ = 1$), is actually a Lobachevskian space L^3 (or, rather two copies of such a space). The situation is depicted in Figure 34 in the case of two space-dimensions. (The conformal and projective representations of L^3 that were described earlier can be obtained by stereographic projection from $O' = (-1, 0, 0, 0)$ to the E^3 hyperplane $t = 0$, and from $O = (0, 0, 0, 0)$ to $t = 1$, respectively. This is in close analogy with the projections of S^3 and P^3 to E^3 illustrated in Figures 7 and 9.)

Velocities in special relativity are often represented as Minkowski (future-pointing) unit vectors, that is, by points on the Minkowskian unit sphere. Thus the space of velocities is really *non*-Euclidean, being a Lobachevskian space L^3. This relates to the fact that, in relativity, velocities do *not* add together in the familiar Galilean–Newtonian way.

We now pass to *general* relativity. The flat Minkowskian space-time M^4 is now to be replaced by a *curved* manifold M. In the small, the structure of M is to resemble Minkowski space, and we need a local structure which expresses this fact. So we choose a "metric" tensor $\mathbf{g}$ for M which bears the same relation to Minkowskian geometry as does a Riemannian metric tensor to Euclidean geometry. The only difference from the Riemannian case is that the positive-definiteness requirement ($\mathbf{g}(\mathbf{V}, \mathbf{V}) > 0$ unless $\mathbf{V} = 0$) is now

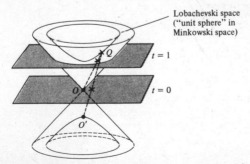

Lobachevski space ("unit sphere" in Minkowski space)

$t = 1$

$t = 0$

Figure 34. The Minkowsian unit sphere is a Lobachevski space.

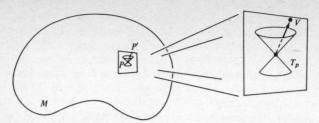

Figure 35. A curved space-time has a locally Minkowskian metric.

dropped and replaced, suitably, so that each tangent space T_p now acquires a Minkowskian rather than a Euclidean structure. This means that, at any point p, a coordinate system can be found so that, in T_p, the component form of $OV^2 = g(V, V)$ reduces to the standard Minkowskian expression $(V^0)^2 - (V^1)^2 - (V^2)^2 - (V^3)^2$. The vanishing of this expression (i.e., of the coordinate-independent expression $g(V, V)$) defines a "light cone" in each tangent space T_p—often called the *null cone* at p (see Figure 35). Such a g defines a pseudo-Riemannian structure for M of the kind referred to as *Lorentzian* (after the Dutch physicist Hendrik Antoon Lorentz (1853–1928)).

Vectors for which $g(V, V) > 0$ are called *timelike* and are possible tangent vectors for world-lines of massive particles; vectors for which $g(V, V) = 0$ are called *null* and can be tangents to world-lines of photons; those for which $g(V, V) < 0$ are *spacelike* and are not permitted as tangents to physical particles. As in a Riemannian space a measure of length is defined, but now along timelike or null curves only. This "length" measures the time experienced by a particle having the curve as its world-line.

The null cones at the various points of the spacetime M provide the most important part of its structure. (They are defined by the metric g up to proportionality and constitute, in effect, the conformal structure of M.) The main physical importance of the null cones is that they determine the *causal relations* between space-time points. If a point p can be connected to a point q by a future-directed null or timelike curve then it is possible for a signal to get from p to q, but not otherwise. And it is precisely the null cone structure that determines which these curves are.

In a general Lorentzian manifold M, the causal relations may differ greatly from the situation in Minkowski space (see Figure 36). It is even possible to construct models with timelike curves which form closed loops—but such models are normally excluded on the clear physical ground that in such a universe it would in principle be possible for an astronaut to travel (or send signals) into his own past, thereby leading to the possibility of a paradoxical situation.

An example of a space-time model with unusual and interesting (though *not* paradoxical) causal properties is that representing gravitational collapse

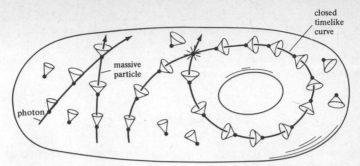

Figure 36. Null cones in a Lorentzian manifold (with a closed timelike curve).

to a *black hole* (see Figure 37). Physical theory predicts that the end-point of the evolution of a massive star—of, say, twenty times the mass of our own sun—would be such a black hole. The most noteworthy feature of this model is the presence of a space-time 3-surface, called the *absolute event horizon*, with the property that signals or particles emitted inside it cannot escape to the outside world. Signals or particles can, however, get from out-

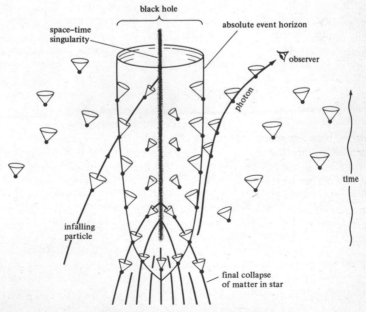

Figure 37. Collapse to a black hole.

side the horizon to inside it. The matter which collapsed to form the black hole can itself not communicate signals to the outside, once it has fallen within the horizon. Instead, this matter falls inwards to the vicinity of the center, at which it encounters a *space-time singularity* where the magnitude of the curvature of space-time approaches infinity!

Einstein's Equations

The reader may reasonably ask, however, why he should accept that space-time is curved at all, let alone why it should be curved to the extreme degree encountered in the black-hole model. In fact small deviations from the flat Minkowskian form of the metric have been experimentally detected, these deviations occurring by virtue of the presence of a gravitational field. In the 1960 experiment of Pound and Rebka, the ratio between clock rates at the top and the bottom of a 22.6 meter tower was measured and found to be the value 1.0000000000000025 that had been predicted by general relativity. Since clock rates measure space-time geometry, this may be taken as a direct measurement of the deviation of the actual geometry of space-time from the Minkowskian form. There is now a host of rather more conclusive tests of the theory available, and the theory must now be regarded as very well confirmed. For example, a fairly direct test of the non-Minkowskian nature of the null cone structure is afforded by the phenomenon of bending of light by the gravitational field of the sun—since light rays (in vacuum) follow the null cones (Figure 38). The bending effect was observed by Sir Arthur Eddington's expedition in 1919; but much more accurate measurements with radio signals have been made quite recently and conform closely to the predictions of general relativity.

These tests all measure rather minute deviations from Minkowskian space-time geometry. Large-scale deviations could occur either in a cosmological context or else with extreme collapse situations such as in the case of a black hole. It is very probable that a black hole is responsible for the X-ray source Cygnus X-1, the X-rays being emitted when matter is dragged into the hole. So it seems that such extreme situations do actually occur on a comparatively local scale. (The black-hole horizon in Cygnus X-1 would be something of the order of 40 miles across and its distance away, about 8,000 light years.) But the expected detailed properties of such a black hole depend heavily on the exact form of the equations of general relativity. These will be considered next.

So far, the only structure of the space-time that we have looked at is its metric **g**. But **g** determines a unique torsion-free ∇ (with $\nabla g = 0$) in the Lorentzian case just as well as in the Riemannian one. Thus we have a notion of space-time geodesic defined, and also a curvature tensor **R**. Analogously to the Riemannian situation, timelike curves that are (locally) of maximal length turn out to be geodesics with respect to this ∇. And it is a

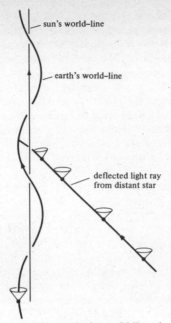

Figure 38. The earth's geodesic world-line about the sun.

hypothesis of the theory that massive test particles falling freely under gravity have world-lines which are timelike geodesics; and correspondingly, light rays are null geodesics. (Later on, one finds that it is not necessary to postulate this hypothesis; it turns out to be a consequence of the Einstein field equations.) A physical basis of this geodesic hypothesis is Einstein's *principle of equivalence*. This is essentially the observation of Galileo Galilei (1564–1642) that all bodies, whatever their constitution, fall at the same speed (neglecting air-resistance). Thus the motion of a body under free fall is determined once its initial position and velocity have been fixed, no parameters defining its mass or internal constitution being required. This is necessary for the geodesic hypothesis since a geodesic is correspondingly determined once its initial point and space-time direction at that point (i.e., spatial velocity) are known. This has the effect that freely falling objects have no local way to detect "gravitational force"—now a familiar phenomenon of the space age. Astronauts in orbit (examples of free fall under gravity) feel no gravitational force even though the intensity of the earth's gravitational field in their vicinity may differ but little from that at the earth's surface. The astronauts, together with their space-craft, are following a geodesic path in space-time. It is *we* who feel a "force," corresponding to the fact that our space-time paths are *not* geodesics. (The earth's surface is continually "pushing" us away from a geodesic, i.e., from free-fall orbit). On the other hand, the earth itself can be regarded as a "particle" with a geodesic world-line about the sun (see Figure 38). The same applies to any other small body in orbit. None of them "feels" any gravitational force.

How is it, then, that gravitation can be detected at all, if its effects can always be locally eliminated by free fall? The answer lies in the phenomenon of *geodesic deviation*. For it is not quite true that (extended) bodies in free fall cannot locally detect the effects of gravity. There are slight differences between the motions of neighboring particles, and these relative motions *can* be detected locally. These small effects can accumulate over large volumes and become very substantial.

Imagine, first, an observer falling freely in the earth's field, where he starts off surrounded by a small sphere of particles initially at rest with respect to him. The particles nearer to the earth than himself will accelerate towards the earth slightly faster than he will; those farther from the earth, slightly more slowly; and those at the same distance, slightly inwards towards the earth's center (see Figure 39). Thus, relative to the observer, there is a slight distortion effect and the sphere accelerates unevenly about him, becoming deformed essentially into an ellipsoid. This is the *tidal effect* of gravity. (Replace the earth by the moon and our sphere by the earth, and we derive the standard explanation of the tides.) Now, if we just work within Newtonian theory, a remarkable feature of the inverse-square law shows up here: the volume of the ellipsoid is initially equal to that of the original sphere. In fact, this property characterizes the inverse-square law completely! The effect assumes that no gravitating matter is enclosed by the sphere. If matter *is* enclosed, then the volume is initially slightly reduced—by an amount proportional to the total mass of this matter. If we change our picture and imagine the "observer" at the center of the earth and the sphere of particles to be

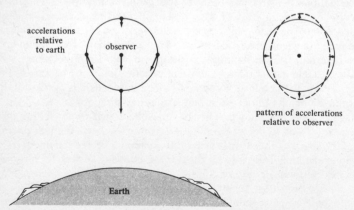

Figure 39. The distortion of a spherical volume in free fall towards earth.

the earth's surface, then we obtain the accumulative effect that there is acceleration inwards at every point of the sphere, leading to the familiar effect of gravity that we all feel! The volume effect is, in fact, a way of stating Newton's law of gravity.

Let us now pass to Einstein's theory. It turns out that if we state things in the above way, there is almost no change whatever from the Newtonian result. The only essentially new feature is that the volume change now is not just proportional to the mass surrounded (in the sense of mass-energy) but to the mass plus the sum of the pressures (in three perpendicular directions). But this extra effect due to the pressures is absolutely minute in all "normal" circumstances. It can only become significant when the gravitating matter becomes "relativistic," i.e., when the internal motions approach the speed of light. Such circumstances can actually begin to arise in a star which is in an extreme situation, near collapse. Then instead of tending to halt collapse, extra pressure simply serves to make matters worse owing to the increased gravitational pull. Effects such as this can contribute to the final demise of a star towards a black-hole state.

But what has this to do with space-time geometry? Let us examine our observer and his surrounding particles from the space-time point of view. We have, now, a bundle of timelike geodesics which start out parallel to one another, but where distortions are introduced as we move along the geodesics. The surrounding geodesics start by being equidistant from the central ray, but after a short while the relative accelerations between the geodesics begin to show their effects (see Figure 40). There is a formula—called the *Jacobi equation*, or *geodesic deviation equation*—which relates this relative acceleration directly to the curvature tensor. If **T** is a vector tangential to the geodesics and **V** connects a point on the central ray to a

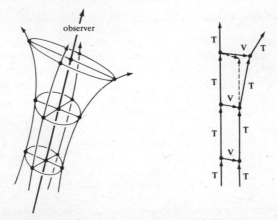

Figure 40. Space-time picture of tidal distortion.

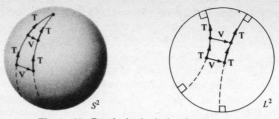

Figure 41. Geodesic deviation in S^2 and L^2.

neighboring one, then the acceleration of $\mathbf{V}$ is $(\nabla_{\mathbf{T}})^2\mathbf{V} = \mathbf{R}(\mathbf{T}, \mathbf{V}; \mathbf{T})$. In fact this result is closely related to the one depicted in Figure 28, from which it may be derived. One can also see directly this effect taking place in the geometries L^2 and S^2 that we considered at the beginning. Two geodesics which start out parallel to one another cease to be so as they are followed along for a short while (see Figure 41).

The Jacobi equation tells us that tidal distortion is a direct measurement of space-time curvature. And when we incorporate into this the statement about volumes described above, we can quite quickly obtain *Einstein's field equations*. These state the equality between two tensors, the first being a geometrical curvature quantity called the Ricci tensor, which measures the "volume change" part of the geodesic deviation, and the second describing the matter density ("mass-energy" plus "sum of pressures") being effectively the so-called *energy-momentum tensor*. (Einstein's equations are illustrated in the box on p. 121.)

The discussion of black holes is based mainly on the Einstein vacuum equations (given when the matter density is zero). When one considers cosmological models, however, it is necessary to come to terms with the structure of the space-time where matter *is* present. For this we represent a smoothed out average of the matter density as a suitable approximation to the actual contents of the universe. Local velocities are generally small (compared with that of light) and pressures are likewise small; consequently an adequate approximation is obtained by inserting the so-called energy-momentum tensor of "dust" into Einstein's equations. Furthermore, there is strong evidence that, on a large enough scale, the universe is essentially *isotropic* (that is, rotationally symmetric) about the earth's location. If, in addition, it is assumed that the earth does not occupy any special privileged position with respect to the universe as a whole, then it follows that space-time must accord with one of the famous *Friedmann models* of the universe, named after their discoverer, the Russian Alexander Alexandrovitch Friedmann (1888–1925).

There are three essentially different Friedmann models, distinguished by the value of a discrete parameter k $(= \pm1$ or $0)$ which describes the spatial curvature. In the case $k = -1$ (as of now, the observationally most favored case), there are spatial sections which are all L^3 geometries; when $k = 0$ the

Einstein's Equations

The argument in the text gives Einstein's equations in the form:

$$R_{ij} = -8\pi G\left(T_{ij} - \frac{1}{2}T_k{}^k g_{ij}\right),$$

where

$R_{ij} = R_{ikj}{}^k$ = components of Ricci tensor

G = gravitational constant

T_{ij} = components of energy-momentum tensor.

An equivalent but more familiar form is:

$$R_{ij} - \frac{1}{2}R_k{}^k g_{ij} = -8\pi G T_{ij}.$$

Sometimes an extra (very small) "cosmological term" λg_{ij} is added on the left.

sections are all E^3, and when $k = 1$ they are all S^3 (or all P^3) geometries. In each case there is an initial *singularity* at which space-time curvatures approach infinity—called the *big-bang* origin of the universe. The universe expands (according to the model) very rapidly at first, but then the expansion slows down somewhat. In the case $k = 1$, the expansion actually reverses to become a collapse, and the whole model terminates in a final singularity. When $k = 0$ or -1 the expansion continues indefinitely, though rather more rapidly in the case $k = -1$ (see Figure 42).

When $k = 1$, the graph of the universe size, as plotted against time, is one branch of a cycloid (a curve traced out by a point on the circumference of a circle which here rolls on a vertical straight line). A complete cycloid is a periodic curve which repeats the same pattern again and again. This has led people to refer to the $k = 1$ model as an oscillating universe. However this is misleading because there is no way, within accepted physical laws, of getting the universe through the singularities at zero radius.

At one time it had been thought that these singularities might just be the result of the assumptions of exact spherical symmetry that have been made, since all the matter is concentrated together at one point when the singularity occurs. Perhaps if this symmetry is not required to hold exactly, people argued, the different bits of matter might "miss" one another, so that although large curvatures might be present, infinite curvatures might hopefully be avoided. This same suggestion was made as a possible means of avoiding the singularity at the center of a black hole. However, later work has shown

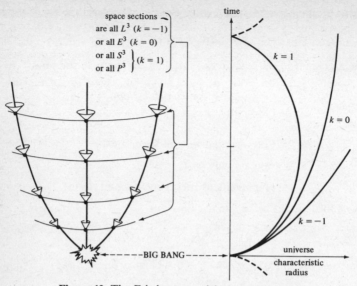

space sections
are all L^3 ($k = -1$)
or all E^3 ($k = 0$)
or all S^3 } ($k = 1$)
or all P^3

time

$k = 1$

$k = 0$

$k = -1$

universe
characteristic
radius

——————BIG BANG——————

Figure 42. The Friedmann models ($\lambda = 0$ assumed).

these suggestions to be untenable. Using methods of global differential geometry it has been possible to show that the singularities are not just a feature of high symmetry but must be present in *all* general-relativistic space-times which qualitatively resemble the Friedmann or black-hole models. Thus, it appears that the resolution of the singularity problem is not to be found within classical general relativity. But I do not regard this as a sign of weakness in the theory since the physical conditions near singularities are not remotely like those familiar to physicists here on earth. Rather, it is the theory's strength that its essential simplicity and elegance allows arguments of such general character to be applied to it, so that rigorous deductions concerning its own limitations may be made.

New Directions

In the preceding pages, we have seen something of the power and scope of the ideas of differential geometry. But we have really witnessed only a small fraction of this power, even when restricted to applications which relate to the structure of the physical world. For example, there is another type of manifold, called a *symplectic* manifold, whose structure is somewhat like that of a (pseudo-) Riemannian manifold, but where a *skew-symmetric* tensor s (i.e., such that s(**U**, **V**) = −s(**V**, **U**)) takes the place of the symmetric tensor **g**. The phase spaces of classical systems (like that depicted in Figure 17) are always symplectic manifolds. Yet again, there are important types of manifolds called *fibre bundles*, introduced by pure mathematicians to handle problems within differential geometry itself, but now finding significant application within theoretical physics, such as to the so-called gauge theories of elementary particle interactions.

However, to me, the most exciting new applications of geometry to the structure of the world lie within the realm of *complex* geometry. Recall the definition of a complex number: $a + ib$, where a and b are ordinary real numbers and $i = \sqrt{-1}$. Such numbers may be used in place of the real numbers of ordinary real differential geometry. In this way we may define a *complex manifold*, where complex coordinates and complex "smooth" functions (called *holomorphic* functions) take the place of the previous real ones. Recall also the representation of complex numbers as points of an ordinary plane—called the Argand (or Gauss) plane (Figure 43). This shows how, in "real" terms, every complex coordinate describes *two* dimensions. Each complex n-dimensional coordinate patch is actually $2n$-dimensional in real terms, and a complex n-dimensional manifold is $2n$-dimensional as a real manifold (with a special type of local structure).

Now, the very simplest of all closed complex manifolds—a complex analogue of a real 1-dimensional closed loop—is the (conformal) sphere S^2. This is covered by two complex 1-dimensional coordinate patches (Argand planes) which may be obtained by stereographic projection from the north and south poles of S^2, respectively (recall Figure 7). Thus S^2 is a closed complex manifold of *one* dimension or a *complex closed* curve!

Now think of S^2 in a different way—as the *celestial sphere*, or *sky* of an observer in space. Suppose a second observer, momentarily coincident with the first, moves relative to the first and is rotated with respect to him. Each

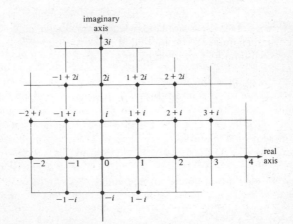

Figure 43. The Argand plane.

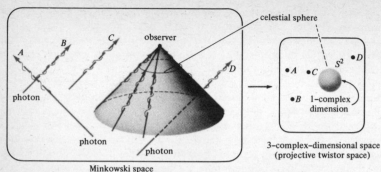

Figure 44. The complex geometry of spinning photons.

observer may use a quite different description of his "sky" in terms of complex coordinates but, remarkably, the two descriptions will give S^2 a completely equivalent structure as a "complex curve." Even more remarkably, the *Lorentz group*, whose elements represent the freedom of passing from one observer to another, corresponds precisely to the group of complex self-transformations of this complex curve. The most fundamental of symmetry groups in physics is, therefore, just the symmetry of a simple complex closed curve!

Many properties of the Lorentz group and Lorentz transformations may be very rapidly obtained using this viewpoint. But this is just the beginning. The celestial sphere may be thought of as a section through the past light cone of the observer—since the light rays entering his eye generate this cone. But the system of *all* light rays, not just those that enter his eye, can also be naturally represented as points of a complex manifold—this time 3-complex-dimensional—provided that we restrict attention to Minkowski space and that we correctly treat the energy and spin of the photons (see Figure 44). This leads us into *twistor* theory, in which complex manifolds play a key role in the description of physics.

Complex geometry is a remarkably rich, elegant and powerful area of mathematics. A great deal of intricate structure can be subsumed into logically simple concepts. For example, even the flat tangent space of a 2-complex-dimensional manifold reveals, upon examination, our now familiar friend, the system of Clifford parallels on S^3. And this, in its turn, has direct relevance to twistor theory, to the structure of spinning massless particles and to the architecture of quantum mechanics. Clearly there is much yet to be revealed in such interplay as this between mathematics and physics, or between the constructions of the mind and the works of nature.

Suggestions for Further Reading

General

Kaufmann, William J. *Cosmic Frontiers of General Relativity*. Little, Brown, Boston, 1977.

Rindler, Wolfgang. *Essential Relativity*. D. van Nostrand, New York, 1969; Springer-Verlag, New York, 1977.

Taylor, Edwin F. and Wheeler, John A. *Spacetime Physics*. Freeman, San Francisco, 1963, 1966.

Technical

Guillemin, Victor and Pollack, Alan. *Differential Topology*. Prentice-Hall, Englewood Cliffs, 1974.

Hawking, S.W. and Ellis, G.F.R. *The Large Scale Structure of Space-Time*. Cambridge University Pr, New York, 1973.

Misner, Charles W., Thorne, Kip S., and Wheeler, John A. *Gravitation*. Freeman, San Francisco, 1973.

O'Neill, Barrett. *Elementary Differential Geometry*. Academic Pr, New York, 1966.

The Mathematics of Meteorology

Philip Duncan Thompson

In its modern sense, meteorology is the science that deals with the structure and behavior of the atmosphere or, more precisely, that part of the gaseous envelope that extends upward from the earth's surface to an altitude of about 100 kilometers. The latter limit is rather arbitrary, but corresponds roughly to the altitude below which electromagnetic forces and photochemical reactions are presumed to be relatively unimportant, and whose effects are therefore assumed to have little influence in the course of events in the underlying atmosphere. The name was evidently taken from the first "scientific" treatise on weather, the *Meteorologica*, written by Aristotle in the fourth century B.C. Although Aristotle's early work was concerned with a wider variety of subjects (including the qualitative description of various astronomical, oceanographic, and geologic phenomena), it appears likely that "meteorology" is derived from the Greek word "meteoros," meaning "something that falls from the sky"—rain, snow, hail, or hard-rock meteors.

Although Aristotle dubbed meteorology a science, it would be difficult to describe his studies as "hard science" today. To quote one example, Aristotle says (through his protegé, Theophrastus):

> We must now show that each wind is accompanied by forces and other conditions in due and fixed relation to itself; and that such conditions in fact differentiate the winds from one another.

There are, of course, a few grains of sense in this statement: some of the great persistent seasonal and regional wind patterns are clearly governed by a few simple and dominant physical processes. But to suppose that the peculiarities of Aristotle's "eight winds" are determined by totally different and distinct causes is virtually a denial of the universality of physical law. Theophrastus says further: " ... the Etesian Wind (monsoon) ... is caused by the melting of the snow." This is a declaration that most modern scientists would consider implausible for its very appearance of oversimplification, but might be considered valid as a statement about two different effects with a common cause.

Exactly when the northwestern Europeans became preoccupied with the vicissitudes and vagaries of weather is buried in prehistory. Presumably their dependence on weather mounted as tribes took up fixed abodes, planted and harvested crops, and sailed to catch fish or explored overseas for richer prizes. Unlike the relatively mild Greek squalls, however, their weather was worse and more changeable, and their attitudes were accordingly less didactic and more pragmatic. They (notably the British and Norse) were not so much concerned with explaining why weather happened as they were in predicting it, and gradually built up a huge folklore and literature of portents—the unseasonable migration of birds, hibernation of wild beasts, unusual sexual behavior of farm animals, color of the sunset, the type and sequence of cloud forms, direction of the wind, aches in the joints, migraines, and so on. Discounting simple numerology, such as the number of crows in a single flight, these homely rules of thumb were completely nonquantitative. In fact, even the most sophisticated description of the state of the atmosphere and its behavior was entirely qualitative, clear up through the Middle Ages to the time of Galileo Galilei (1564–1642).

The New Mechanistic View

Meteorology did not become a mathematical science in a single moment. It began to assume a quantitative flavor when the state of the atmosphere was described by measurements of such physical variables as temperature, pressure, wind speed and direction, quantities that are now measured at a worldwide network of observing stations on a routine basis.

With Galileo's invention of the thermometer in about 1631 and his student Evangelista Torricelli's discovery of the principle of the barometer (a device for measuring atmospheric pressure) in 1643, man's conception of the atmosphere as a coherent physical entity began to take shape. There was increasing international exchange of observations and ideas among scholars and gentlemen amateurs during the seventeenth century. These, together with reports and accounts of winds and ocean currents from seafarers plying the newly opened trade routes across the Atlantic and Pacific, revealed a regular and recurring pattern of the world's weather. Some of its main features are reflected in old and rather vivid nautical terms for various climatic zones and their characteristic seasonal or regional types of weather, e.g., "tradewinds," "monsoon," "horse-latitudes," or "roaring forties."

Measurement and quantitative description was and is indispensable, but theory is guided by and tells what to do with it. Observation and experiment

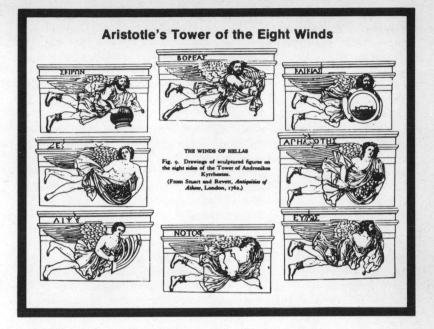

Aristotle's Tower of the Eight Winds

THE WINDS OF HELLAS

Fig. 9. Drawings of sculptured figures on the eight sides of the Tower of Andronikos Kyrrhestes.

(From Stuart and Revett, *Antiquities of Athens*, London, 1762.)

were essential ingredients of the new Galilean view of the natural world. An equally important development, however, was the evolution of a quantitative theory to account for what was observed. In fact, the germs of a mathematical theory of the motion of *solid* bodies are contained in Galileo's heretical works of the sixteenth century. Perhaps his best-known work concerned the motion of falling bodies, whose distance of fall he found to be proportional to the square of time elapsed from the moment of release. This was the first experimentally confirmed law of motion of bodies accelerated in the earth's gravitational field.

Great as were the achievements of Galileo in Italy, Nicolaus Copernicus (1473–1543) in Poland, and Johann Kepler (1571–1630) in Germany, the culmination of the new mechanistic view of the universe was Isaac Newton's *Principia Mathematica*, in which he enunciated certain universal physical principles governing all matter and stated them in *mathematical* form. The best known is perhaps the law of gravitation, from which Newton (1642–1727), Karl Friedrich Gauss (1777–1855), and present-day space scientists have predicted the orbits of planets and satellites with incredible accuracy. A more general and important statement is Newton's Second Law of Motion, relating the acceleration of *any* material body to the forces acting on it. The concepts contained in *Principia Mathematica*, especially Newton's "method of fluxions" (now known as differential and integral calculus), rapidly gained acceptance in the western scientific world, and are still the keystones in the study of the behavior of physical systems in natural environments encountered on earth.

The consequences of Newton's laws of motions were not lost on his contemporaries on the continent. Daniel Bernoulli (1700–1782) and Leonhard Euler (1701–1783), both Swiss mathematicians, were quick to apply Newton's physical concepts and mathematical methods to the study of *nonsolid* bodies (i.e., gases and liquids), a special case of which is the earth's atmosphere. With the advantage of long hindsight, it is now easy to see that many present-day problems in meteorology could have been solved in the early eighteenth century: they were not solved at that time, however, principally through lack of observations and an understandable timidity in formulating "working hypotheses" in the absence of clearcut tests. Moreover, the theoretical framework at that time was still not complete, for reasons that are explained in detail in the box on p. 131.

Lacking a complete mathematical formulation of the laws of atmospheric dynamics, Newton and his contemporaries did not essay a systematic attack on the problems of meteorology and atmospheric motion. Crucial connections between the variables were still missing. Nevertheless, Newton's colleague Edmund Halley for whom the famous comet is named, suggested in 1686 a qualitative explanation for the northeast trade-winds, based on some empirical ideas about convection (vertical motion generated by differential heating of air), coupled with a misguided notion of the effects of the earth's rotation. It took another 50 years for his fellow Englishman John Hadley to give the correct description of the earth's rotational effect on the direction of the "trades." The moral of this story is that qualitative arguments are of little use in trying to understand something as complicated and interactive as the inner workings of the atmosphere.

In view of the rather slow march of classical physics, it should not be surprising that the history of mathematical applications in meteorology has followed the gradual filling-in of the fundamental physical framework of the theory of atmospheric motions. We shall next briefly review some of the major developments in atmospheric physics and thermodynamics that have led to a fairly complete formulation of the total mathematical problem.

The Emergence of Physical Hydrodynamics

The mathematical formulation of the theory of fluid motions was incomplete in the mid-eighteenth century because there was no known physical relation between mass density and ambient pressure. Robert Boyle's discovery (in 1662) of a linear proportionality between pressure and density provided part of the answer, but this simple relationship was shown to hold only if temperature were held fixed. Presumably, the factor of proportionality could be variable, since temperature is not necessarily constant in naturally occurring flows.

Hydrodynamical Equations

For those who have been initiated into some of the mysteries of fluid mechanics, we summarize here the mathematical theory of fluids as it stood in the age of Newton, Euler, and Bernoulli. It can be compressed into four equations, expressing the three components of acceleration in terms of the forces per unit mass, and the condition for conservation of mass:

$$\frac{\partial u}{\partial t} + u\frac{\partial u}{\partial x} + v\frac{\partial u}{\partial y} + w\frac{\partial u}{\partial z} - fv + \frac{1}{\rho}\frac{\partial p}{\partial x} = 0,$$

$$\frac{\partial v}{\partial t} + u\frac{\partial v}{\partial x} + v\frac{\partial v}{\partial y} + w\frac{\partial v}{\partial z} + fu + \frac{1}{\rho}\frac{\partial p}{\partial y} = 0,$$

$$\frac{\partial w}{\partial t} + u\frac{\partial w}{\partial x} + v\frac{\partial w}{\partial y} + w\frac{\partial w}{\partial z} + g + \frac{1}{\rho}\frac{\partial p}{\partial z} = 0,$$

$$\frac{\partial \rho}{\partial t} + \frac{\partial}{\partial x}(\rho u) + \frac{\partial}{\partial y}(\rho v) + \frac{\partial}{\partial z}(\rho w) = 0.$$

Here u and v are the components of fluid velocity in two horizontal and mutually perpendicular coordinate directions (x and y) and w is the component of velocity in the vertical or z-direction. The local mass density of fluid is ρ, and p is the ambient pressure. The so-called Coriolis parameter f takes into account the earth's rotation, and g is the gravitational acceleration.

The variables that characterize the state of the fluid are u, v, w, ρ, and p, all of which may vary from one time or place to another. The fundamental problem in the theory of fluids—including the atmosphere—is to deduce the structure and evolution of the flow by solving the equations listed above, given the initial state (or other constraints) on the flow. (Such constraints reflect, for instance, the fact that air does not flow freely into or out of the ground or escape outward to space.)

Counting the physical variables in these equations, one immediately notices that there are five variables but only four equations, so the solution of the equations is therefore not completely determined. But it is also true that, in dealing with many problems in meteorology and fluid dynamics, one may assume that the density is constant or depends almost entirely on pressure, and conversely; in this case the system of equations enumerated above is formally complete. Each equation then expresses the instantaneous time rate of change of one of the four basic variables u, v, w, and p in terms of the current values of the same variables. If anyone in Euler's time had been bold enough to make this assumption, the theory of atmospheric motions might have been advanced by a century: there were not, however, enough quantitative observations to make anyone comfortable with such an enormous conceptual leap.

The next salient experimental fact did not emerge until 1802, when Jacques Charles (1746–1823) in France found that pressure is linearly proportional to temperature, with volume (or density) held fixed. This result, when combined with Boyle's Law, leads to a remarkably simple relationship between the thermodynamical variables of pressure, density, and temperature (see the box on p. 133). But this did not completely solve the problem. Although the combination of Boyle's and Charles's Laws provided one new relationship between the variables that characterize the state of a fluid, it also introduced one new variable, namely temperature. The number of variables was still one greater than the number of independent equations.

The missing piece of the puzzle was another and independent connection between the thermodynamic variables. Beginning early in the seventeenth century, a number of mathematicians and experimental physicists began to look more deeply into the nature of "heat" which, up to that time, was considered a special form of matter, called "phlogiston." The American-British scientist Sir Benjamin Thompson, Count Rumford, by boring out the barrel of a brass cannon and keeping careful account of the weight of the filings and the heat engendered by the boring, showed in 1798 that "heat" did not have perceptible mass and therefore did not have the usual properties of matter. It appeared rather to be associated with the motion of the boring process. Not long afterward, James Joule in England carried out more systematic experiments which established an equivalence between "heat energy" and the "mechanical energy" of stirring, by which the heat was produced.

All of these isolated results may be summarized by the First Law of Thermodynamics. Put in plain language, this law says that the heat energy added to a fluid is equal to the change in its internal energy (a measure of the kinetic energy of motions on molecular scale) and the work which the fluid does in expanding against pressure or "normal" forces. (The mathematical formulation of this principle is given in the box on p. 133.)

The discovery of the First Law of Thermodynamics dropped the last major piece of the puzzle into place. There is, however, one important *caveat*: the formulation of the problem is still not complete unless the rate at which the fluid is heated is in some way specified, or related to other variables in the system. Speaking in rather general terms, the rate at which heat is gained or lost by a unit mass of the atmosphere depends on several very complicated processes, which we cannot hope to elaborate on here. Specifically, the local gain or loss of heat is effected by diffusive exchange between the vapor, liquid, and ice phases of water, by absorption of incoming radiation from the sun (which is almost exactly balanced by emission of infrared radiation from the earth and its atmosphere), and by turbulent exchange of air between the earth's surface and the atmosphere at higher levels. At this point we merely remark that the general radiative properties of all materials were fairly well-established by the studies of Gustav Kirchhoff (1824–1887) in Germany and Ludwig Boltzmann (1844–1906) in Austria. Moreover, the laws for the change of phase of water, with concomi-

The Boyle—Charles Law

Under the assumption that the thermodynamic variables p, ρ, and T are connected by a single relationship, Boyle's Law may be written as $p = \rho f(T)$ and Charles's Law as $p = Tg(\rho)$ in which T is the absolute temperature, the quantity $f(T)$ depends only on T, and $g(\rho)$ depends only on ρ. Equating these two independent expressions for p, we get

$$\frac{f(T)}{T} = \frac{g(\rho)}{\rho}.$$

But the left-hand side of this equation depends only on T, whereas the right-hand side depends only on ρ. This is possible if and only if both sides are equal to a constant that is independent of p, ρ, and T. In other words, $f(T)/T = R = $ constant. This implies that

$$p = R\rho T.$$

This is the Boyle—Charles Law, the equation of state for an ideal gas. The gas constant R is related to Avogadro's number.

First Law of Thermodynamics

In mathematical symbolism, the First Law of Thermodynamics is expressed as:

$$\frac{dq}{dt} = C_v \frac{dT}{dt} + p \frac{d}{dt}\left(\frac{1}{\rho}\right)$$

where dq/dt is the rate of heat addition to a unit mass of fluid (such as air), C_v is the specific heat at constant volume, and the symbol d/dt stands for the time rate of change experienced by a material element of fluid, following its motion. The remaining quantities p and ρ are, as defined in the box on p. 131, the fluid density and ambient pressure.

The equation expressing the First Law explicitly involves only variables appearing in the five equations listed in previous boxes, i.e., the three Newtonian equations of motion, the equation for conservation of mass, and the Boyle—Charles equation of state. Thus, with an independent equation that introduces no new unknown variables, the system of hydrodynamical equations becomes formally complete.

tant release or uptake of the "heat of condensation" or "heat of fusion," have been known for two centuries (John Dalton, 1766–1844). The remaining enigma was the transfer of heat by highly irregular and quasi-random motions. Considerable heat is exchanged between the earth and atmosphere by turbulent vertical motions engendered by heating of the ground during the day. The scale of such motions is small enough that they cannot be observed in detail and cannot be predicted in any greater detail. This is still a piece of the puzzle. Despite these uncertainties, however, for purposes of predicting atmospheric behavior over relatively short times (i.e., a few days), the atmosphere may be viewed as a system that receives little or no heat.

With these qualifications in mind, let us now take stock of the situation at mid-nineteenth century in the time of Hermann von Helmholtz (1821–1894). By 1844, all of the basic dynamical and thermodynamical laws that govern the behavior of a fluid like the atmosphere were widely known and assumed to apply under fairly general conditions. To recapitulate, these are Newton's three laws of motion, the principle of mass conservation, the Boyle–Charles Law, and the First Law of Thermodynamics. The corresponding equations (the so-called hydrodynamical equations) are six in number and involve six physical variables. Their detailed form is displayed in the fragmentary pieces of mathematical digression set apart in the boxes on pp. 131 and 133.

In 1858 Helmholtz discovered some rather general properties of solutions of the hydrodynamical equations, notably, that the "vorticity," that is, the spin or whirling motion, of an identifiable "blob" of fluid tends to persist as time goes on. Thirty years later, he applied the hydrodynamical equations explicitly to the study of atmospheric motions: in particular, he investigated the stability and propagation of small wave-like departures from a state of equilibrium, in which the wind and density vary with altitude. Two other papers of this period by the British scientists Lord Kelvin and Lord Rayleigh, both dealing with wave motions of the atmosphere, opened the way to a systematic treatment of a rather general class of solutions of the hydrodynamical equations.

The mathematical techniques used in the works of Helmholtz, Kelvin, and Rayleigh had not previously been exploited in meteorology, but were an outgrowth of a method developed by the astronomers in the calculation of orbits. This is the "perturbation" method, in which one considers small departures from an unperturbed orbit, assuming that those departures are due solely to the influence of some body whose presence is not accounted for in the calculation of the unperturbed orbit. The virtue of this scheme is that, if the gravitational attraction of the perturbing body is small, the corrections for its influence are governed by *linear* equations, which are mathematically much more tractable than the *nonlinear* equations governing the exact motions of three or more interacting bodies. (The distinction between linear and nonlinear equations is elaborated in the box on p. 140.)

The applicability of the perturbation method is clearly suspect if the influence of the perturbing body is fully as large as that of the bodies accounted for in the calculation of the unperturbed orbit. So it is also with the application of this method in meteorological theory: if the amplitudes of departures from equilibrium become large, the approximations of the perturbation method fail. Wave motions of small amplitude are particularly amenable to this treatment, whereas tornadoes, hurricanes and the like are not. I hasten to point out, however, that many features of atmospheric motions have been studied successfully by these methods. It is no exaggeration to say that Helmholtz, Kelvin, and Rayleigh laid much of the groundwork for modern theoretical meteorology.

The Central Problem

The next turning point was the realization that the hydrodynamical equations could be solved in their general form, in principle and if necessary, by sheer brute force, using *approximate* methods of solution. In 1904, in a remarkable manifesto and testament of faith in determinism, the Norwegian meteorologist Vilhelm Bjerknes (1862–1951) stated the central problem of weather prediction: to solve the general form of the hydrodynamical equations, given observations of the state of the atmosphere at a *single* time. In principle, he said, the state of the atmosphere at any time in the future is determined by its state at a single instant in time. This was, in fact, the first explicit assertion that the future state of the atmosphere is determined by its past. (The mathematical and physical basis for this assertion is outlined in the box on p. 137.)

The mathematical methods by which Bjerknes proposed to calculate rates of change would be considered crude and cumbersome by present-day standards. Briefly described, they depended on the *graphical* multiplication of vectors (such as wind velocity), an operation that must be carried out by hand and eye at a finite number of discrete points. This approach would be too time-consuming, if not too tedious, for human patience. The importance of Bjerknes's work, however, does not lie so much in his analytic or numerical methods, but in his original statement of the prediction problem, the fact that he did persuade the Norwegian government to establish a denser network of observing stations, and for his creation of two institutes where three succeeding generations of theoretical meteorologists have cut their teeth. His contributions to the "polar-front" theory and the study of atmospheric stability will be more appropriately discussed in the next section.

One of the most singular and original figures in the field of mathematical meteorology and numerical weather prediction was the British scientist Lewis Fry Richardson (1881–1953). Trained in physics, mathematics, and meteorology, he took an early interest in discrete variable representations, notably the finite-difference method. Briefly described, this method consists

Vilhelm Bjerknes

in characterizing a continuous variable by its values at a large number of fixed and discrete points, and in approximating the continuous changes of a variable by its differences between discrete, but closely-spaced points. This approach to the approximation of derivatives has an intrinsic advantage over the graphical methods of Bjerknes: it is simply that the finite-difference approximation to a given partial differential equation leads to a sequence of algebraic equations, whose solution could presumably be carried out by an automaton. In most other respects, however, Richardson's view of the physical aspects of the prediction problem was that of Bjerknes, as outlined earlier.

Richardson, a Quaker and humanitarian, did his service during World War I as an ambulance driver. But between trips to the front, he carried out a long hand-calculation of a solution of the complete nonlinear hydrodynamical equations, beginning with estimated initial values of the basic variables at the centers of squares on a layered checkerboard. Anticlimactically, Richardson's papers disappeared in the general confusion of war, but were later found under a coal heap in Belgium and sent back to England. This was the first genuine attempt to apply mathematical methods in weather prediction.

The results of Richardson's calculation and the description of his method

Bjerknes's Meteorological Manifesto

The roots of mathematical meteorology can be traced directly to Vilhelm Bjerknes's 1904 proposal that the state of the atmosphere at any time in the future is determined by its state at present. The basis for Bjerknes's reasoning is summarized in the following equations, which are just the Newtonian equations of motion, taken together with three independent equations that are easily derivable from the Boyle-Charles equation of state, the principle of mass conservation and the thermodynamic energy equation for flows to which no heat is added:

$$\frac{\partial u}{\partial t} = -u\frac{\partial u}{\partial x} - v\frac{\partial u}{\partial y} - w\frac{\partial u}{\partial z} + fv - \frac{1}{\rho}\frac{\partial p}{\partial x},$$

$$\frac{\partial v}{\partial t} = -u\frac{\partial v}{\partial x} - v\frac{\partial v}{\partial y} - w\frac{\partial v}{\partial z} - fu - \frac{1}{\rho}\frac{\partial p}{\partial y},$$

$$\frac{\partial w}{\partial t} = -u\frac{\partial w}{\partial x} - v\frac{\partial w}{\partial y} - w\frac{\partial w}{\partial z} - g - \frac{1}{\rho}\frac{\partial p}{\partial z},$$

$$\frac{\partial \rho}{\partial t} = -u\frac{\partial \rho}{\partial x} - v\frac{\partial \rho}{\partial y} - w\frac{\partial \rho}{\partial z} - \rho\left(\frac{\partial u}{\partial x} + \frac{\partial v}{\partial y} + \frac{\partial w}{\partial z}\right),$$

$$\frac{\partial p}{\partial t} = -u\frac{\partial p}{\partial x} - v\frac{\partial p}{\partial y} - w\frac{\partial p}{\partial z} - \frac{C_p p}{C_v}\left(\frac{\partial u}{\partial x} + \frac{\partial v}{\partial y} + \frac{\partial w}{\partial z}\right),$$

$$\frac{\partial T}{\partial t} = -u\frac{\partial T}{\partial x} - v\frac{\partial T}{\partial y} - w\frac{\partial T}{\partial z} - \frac{RT}{C_v}\left(\frac{\partial u}{\partial x} + \frac{\partial v}{\partial y} + \frac{\partial w}{\partial z}\right).$$

The symbols in these equations are precisely those introduced in previous boxes, except for C_p which is the coefficient of heat at constant pressure. These six equations involve only the six dependent variables u, v, w, ρ p, and T, and express the instantaneous *time* rate of change of each of these variables at any fixed point in terms of *spatial* derivatives of variables in the same set.

Now suppose that the values of u, v, w, ρ, p, and T were known at every point in the atmosphere at one arbitrary initial time. Thus, one could compute the right-hand sides of the equations above and could thus calculate the instantaneous time derivatives of each variable at the initial time. But, knowing the initial value and initial time rate of change of each variable, we could then extrapolate over a very short time interval to obtain the predicted values of each variable at a short time later than the initial time. We then simply regard the *predicted* values of u, v, w, ρ, p, and T as a new set of *initial* values and repeat the process. In this way, one can (in principle) build up a prediction over a long period of time from a sequence of successive predictions over short intervals of time.

were published in a single monograph, *Weather Prediction by Numerical Process* issued by the Cambridge University Press in 1922. It is an oddly quixotic work, being a candid report of an admitted but glorious failure. The computation predicted enormously rapid changes of pressure, more nearly

Richardson's "Checker-board" Grid and Notation

consonant with those expected from the propagation of sound waves, rather than those associated with the rather slow movement of large-scale cyclones and anticyclones.

Near the end of this monograph he describes a phantasmagorical vision of the "weather factory"—a huge organization of specialized human computers, housed in something like Albert Hall, directed by a mathematical conductor perched on a raised pulpit, and communicating by telegraph, flashing colored lights, and pneumatic conveyor tubes (the kind you used to see in large department stores). In this fantasy, he estimated that, even using the newfangled mechanical desk calculators, it would take about 64,000 human automata just to predict weather as fast as it actually happens in nature. Richardson's preface concludes with a rather wistful but prophetic statement: "Perhaps some day in the dim future it will be possible to advance the computations faster than the weather advances and at a cost less than the saving to mankind due to the information gained. But that is a dream." That dim future came only 25 years later.

In the meantime, meteorologists were understandably skeptical about the applicability of Richardson's method or, for that matter, any prediction scheme based on mathematical theory. The deficiencies were compounded by the fact that any such method obviously could not be applied in the routine practice of daily weather forecasting, simply owing to the staggering requirements on raw computing speed.

The Retreat to Linearity

The equations expressing the dynamical laws that govern the behavior of the atmosphere are nonlinear in a most horrible, but essential way. The hydrodynamical equations (in the box on p. 137) involve terms that are both linear and quadratic in their dependent variables. Thus it is not usually possible to find a solution general enough to satisfy the equations with initial conditions and boundary constraints. There are, in fact, no known methods by which the nonlinear hydrodynamical equations can be solved under general conditions, and few cases in which they can be solved exactly, even under very special conditions.

Faced with the apparent failure of numerical methods (such as Richardson's), theoretical meteorologists found themselves caught between two opposing difficulties. On the one hand, their analytical methods were not powerful enough to deal with the problem of nonlinearity; on the other hand, purely numerical techniques were not far advanced nor were they feasible from the standpoints of economics and engineering. Taking the only possible way out, they changed the problem, tailoring it to fit their modest mathcmatical means. Specifically, they "linearized" the equations using the perturbation methods mentioned above (see the box on p. 140).

Now, one may cavil about the legitimacy of this approach, since the amplitude of velocity fluctuations in the atmosphere are often fully as large as the equilibrium velocity. Moreover, much of the kinetic energy found in very large and very small scale motions is transferred from intermediate scales by *nonlinear* interactions between motions of different scales. So it is clearly impossible to construct an accurate description of all aspects of atmospheric behavior from the solutions of linear equations. Nevertheless, they are still extremely valuable in classifying the various *kinds* of motions (especially "wave" motions) that can exist under a wide range of equilibrium conditions, and in understanding the dominant physical processes that account for their existence. (Simple wave motions are oscillations due to restoring forces that tend to restore the atmosphere to a state of relative rest, in a manner analogous to that in which a weight hanging at the end of a coiled spring oscillates around its rest position.) In this respect, the weakness of linear methods is also their greatest strength, since they enable us to separate out each class of wave motions and to study them in isolation from all of the others. Some of these different types of motion will be discussed briefly later in this section.

Although Bjerknes and his students of the Bergen school did not fulfill their dream of predicting weather by graphical solution of the nonlinear equations, they made monumental contributions to the theory of meteo-

Linearization by the Perturbation Method

To illustrate the fundamental ideas behind the process of "linearization," let us specialize the hydrodynamical equations of the box on p. 137 for the special case when $v = w = 0$. If, in addition, we consider motions neighboring the state of rest, both $u(\partial u/\partial x)$ and $u(\partial p/\partial x)$ are *products* of deviations from zero, whereas $\partial u/\partial t$ and $\partial p/\partial t$ depend *linearly* on the deviations. Hence, for very small deviations, the first and fifth equations reduce to

$$\frac{\partial u}{\partial t} = -\frac{1}{\rho}\frac{\partial p}{\partial x},$$

$$\frac{\partial p}{\partial t} = -\gamma p \frac{\partial u}{\partial x},$$

where $\gamma = C_p/C_v$. By ignoring other products of small deviations from equilibrium, we achieve the further simplification

$$\frac{\partial u}{\partial t} = -\frac{1}{\bar{\rho}}\frac{\partial p}{\partial x},$$

$$\frac{\partial p}{\partial t} = -\gamma \bar{p} \frac{\partial u}{\partial x},$$

in which $\bar{\rho}$ and $\bar{p}$ are the equilibrium values of density and pressure, respectively. It is easily verified by direct differentiation that the solutions of this system are of the form $u = u_0(x-ct)$ and $p = p_0(x-ct)$, where the functions u_0 and p_0 are defined by $u_0(x) = u(x,0)$ and $p_0(x) = p(x,0)$. These expressions above are solutions of the linearized equations provided that

$$c = \pm\sqrt{\frac{\gamma\bar{p}}{\bar{\rho}}} = \pm\sqrt{\gamma R\bar{T}}.$$

This will be recognized as the familiar formula for the speed of sound, one possible type of wave motion in the atmosphere.

rologically important wave motions in the atmosphere, most notably, unstable waves feeding on the potential energy of a sloping interface between two rotating fluids with different densities and equilibrium angular velocities. Their work culminated in the so-called polar front theory for the origin of middle- and high-latitude storms. In the five decades following their work, there have been several amendments, but the basic mathematical formulation has remained essentially unchanged.

Since the heyday of Bjerknes's Bergen school, there have been innumerable applications of linear theory in the study of the stability and propagation of atmospheric waves. One of the first systematic treatments was a small monograph by Bernhard Haurwitz of Colorado State University on internal gravity waves, which are analogous to waves travelling over the surface of the sea. In the atmosphere, internal gravity waves are often manifested as "billow" clouds—long parallel bands of condensed water vapor lying along the crests of the waves, where the upward displacement of air is the greatest. The mechanism of propagation of these waves is the buoyancy or "Archimedean" restoring force due to the weight of the overlying fluid: this force is directed upward when the fluid is displaced below its equilibrium position, and conversely. Thus, the fluid oscillates up and down around its equilibrium altitude.

Around 1939, Carl-Gustav Rossby at M.I.T. applied linear methods to a strange type of wave motion, in which the particles of fluid move in horizontal planes, back and forth from north to south; the associated "Rossby waves," however, are propagated from east to west. These odd wave motions correspond to special solutions that had been known for two centuries, but Rossby was apparently the first to recognize their importance in meteorology and to isolate them in pure form. It turns out, in fact, that (of all the various types of atmospheric wave motions) Rossby waves most resemble the kinds of large-scale motions actually observed. In this case, the mechanism of propagation is due to the earth's rotation. Haurwitz later gave a more rigorous solution for Rossby waves, with spherical geometry, but still using linear methods.

Near the end of this rather narrowly defined era, Jule Charney of M.I.T. attacked what was considered to be one of the burning questions of the day, namely, the growth or stability of waves in a rotating fluid, with continuous vertical variations of density and velocity in the equilibrium state. Although he employed the essentially classical linear approach to the problem, by examining the stability of waves with different horizonal wavelengths, the degree of mathematical sophistication required was considerably greater than that displayed in many earlier works.

Since Charney's work of 1946, in which he determined the equilibrium conditions under which large-scale perturbations (with wavelengths of several thousand kilometers) would amplify, there have been several new developments in the mathematical theory of linear atmospheric instability. These could all be said to be genuine contributions to mathematics, although the original motivation grew out of meteorology or, more generally, geophysical fluid dynamics. These methods are clearly applicable to many other linear systems.

The Renaissance of Numerical Methods

The most discouraging aspect of Richardson's proposed method for solving the main problem of weather prediction was the sheer volume of calculation required. His pilot computations were clearly in error, but there might have been some hope of isolating and analyzing the various sources of inexactitude if it were possible to carry out a series of controlled numerical experiments, in which a number of critical features of the physical formulation, numerical method and initial data were varied systematically. Owing to the exigencies of hand calculation, however, this was plainly impossible in Richardson's day.

An absolutely necessary ingredient in the success of any practical scheme of "numerical weather prediction" was a computing device that was capable of calculating a one-day forecast in less than 24 hours. In retrospect (recognizing that even Richardson probably underestimated the "administrative overhead" in dealing with automata), we now see that a one-day prediction requires on the order of at least a billion numerical and logical operations. (A billion is approximately the number of ping-pong balls required to form a continuous string around the equator of the earth.) The requirements on total data storage (memory) capacity and rate of data transfer to and from storage and processor were equally severe. What was needed was a computing organism capable of performing something like 10,000 operations per second.

During World War II and the years immediately following, substantial progress was made in designing numerical processors in which the switching elements were not cogged wheels or electro-mechanical relay switches, but consisted essentially of radio tubes (or electronic switches), which have virtually no mechanical inertia. By 1945, in fact, J.P. Eckart (Sperry Rand Corp.) and John Mauchly (Carnegie Institute of Technology) had designed and built processors with speeds of the order of one logical or numerical operation per millisecond. These developments brought the basic operating speed of the processor to within one order of magnitude of that required for the routine application of numerical methods to weather prediction.

But there were still two basic deficiencies in the system. First, the data storage device was limited by the density of sound impulses that could be cyclically regenerated and propagated through a mercury-filled tube. Second, and more important, the programming of the processor was completely external and "human-limited" in the sense that the entire sequence of instructions to the processor had to be written out in advance and conveyed to it instruction-by-instruction through a manually wired plug-board or by setting switches.

The big advance, however, was not dependent on sheer hardware development. It arose from John von Neumann's realization that computing machines of this class must be "self-programming," i.e., that it should not be necessary to tell the computer what to do in complete detail. If, as a simple example, the same sequence of operations is to be performed on different sets of data, one may achieve a degree of self-programming capability by storing the operands (numbers to be operated on) at enumerated "addresses" or locations in the machine's memory and, in addition, storing in its memory the execution orders, *which include the address of the operand.* Thus, the basic execution cycle for one set of data (or operands) may be reused merely by concluding the cycle with a sequence of instructions to change the addresses of the operands that appear in the basic execution cycle. One then repeats the whole cycle. This is the essential logical basis of "stored-programming" and the single stroke that broke the human bottleneck of writing out the entire sequence of instructions in advance. Since the number of *kinds* of numerical operations in even large hydrodynamical calculations is fairly small, but are repeated many times on different combinations of data, this feature of the stored-program machine made it ideally suited to the demands of numerical weather prediction.

Early in 1946, von Neumann, one of the most brilliant and versatile mathematicians of this age, singled out the problem of numerical weather prediction for special attention. Although he had a deep appreciation of its practical importance and intrinsic scientific interest, he also regarded it as the most complex, interactive, and highly nonlinear problem that had ever been conceived of—one that would challenge the capabilities of stored-program machines for many years.

At the outset of von Neumann's Meteorology Project at the Institute for Advanced Study in Princeton it was recognized that a machine would only recommit Richardson's errors if it merely duplicated his calculations exactly. Accordingly, considerable effort was put into the analysis of various possible sources of error. One potential source of error (but not one that accounted for Richardson's failure) was found to be computational "instability." It had been shown by Richard Courant, *et al*, in 1928 that the numerical solutions of certain finite-difference equations would not remain stable unless the time increments were taken to be less than the time required for a wave-impulse to be transmitted from one mesh-point to another. Otherwise, numerical solutions would become unstable and "blow up."

A more vexing cause of grief emerged from a simple error analysis of the "divergence" of measured velocities—i.e., the rate of horizontal expansion or vertical contraction of fluid elements. It was alarming to discover that the errors in observational estimates of the "divergence" were fully as large as the divergence itself, or perhaps even larger. This appeared to account for the extremely rapid displacement of disturbances in Richardson's calculation.

But what to do about it? In 1946, computers still fell short of the point when the problem could be explored by even brute machine force. In 1948 Charney, following a lead that he saw in his own thesis work, published a remarkable paper in which he suggested that a derived form of the hydrodynamical equations be modified in such a way that solutions corresponding to high-speed sound and gravity waves (both of which may lead to computational instability) are excluded, but such that solutions corresponding to the large-scale "meteorological" modes are retained almost intact. It turned out that exactly those features of the atmospheric motions that make the *gen-*

eral problem difficult—namely, the almost exact balance of mechanical forces—also characterize the large-scale motions and provide the physical basis for a theory of approximation. Briefly, Charney showed that the fundamental dynamical law governing the large-scale motions of the atmosphere is the principle of absolute vorticity conservation, which states that the product of the vorticity (or spin) of a fluid element around its vertical axis and the area of its horizontal cross-section remains constant with time. (This is, in fact, the same mechanical principle by which ballet dancers and figureskaters go into a rapid spin, starting out with their arms and legs extended and then pulling them in to nearly vertical positions.)

By systematic but selective introduction of the conditions for mechanical balance, Charney derived a single equation with a single unknown. The resulting equation had no solutions corresponding to sound or gravity waves and left the remaining solutions virtually untouched. This approach, generally known as "filtering," does not have the disadvantages of dealing with the original (or "primitive") hydrodynamical equations, in that the equation for the rate of change of "spin" depends only on forces that exert torque on the fluid. Since the pressure force, being a normal force, can exert no torque, there is no compensation between the pressure and inertial torques. Charney's formulation of 1948 skirted two difficulties: first, the solutions of his equation were not sensitive to errors in the initial wind field and, second, the characteristic speed of wave propagation was considerably less than that of sound or gravity waves. These advantages, taken together with the fact that the equation contained only one unknown, brought the problem into the realm of computational tractability.

There are numerous earlier instances, of course, in which filtering or some other modification of the governing equations has been invoked to simplify hydrodynamical problems, sometimes unwittingly or without conscious knowledge of the real basis for such methods. By considering water as an idealized incompressible fluid, the oceanographer modifies his version of the hydrodynamical equations so that they have no solutions corresponding to sound waves, whose existence depends on the compressibility of the medium: he knows perfectly well that water is actually compressible, but also knows that its slight compressibility has very little effect on the propagation of gravity waves on the sea surface. Similarly, by disregarding gravitational forces, the acoustical engineer modifies his version of the hydrodynamical equations so that solutions corresponding to gravity waves are excluded, without significantly changing the solutions corresponding to sound waves.

These were deliberately chosen as exceptionally clear-cut examples, in which the nature of the desired modification and its consequences can be grasped immediately, and where the validity of the procedure is so obvious as to require no further justification. In other instances, however, the question of filtering is a much more subtle one. With regard to the atmosphere's large-scale behavior, the existence of gravitational forces is very important from some points of view but relatively unimportant from others. In this case, the problem of discarding certain features of the system while retaining the meteorological essence is not so simple as merely omitting terms from one of many possible formulations of the general system of equations. It might be expected that similar questions would arise in the analysis of any system which is capable of displaying widely different modes of behavior under different external conditions or which, under normal conditions, displays a dominant mode of behavior. In such cases, the methods of filtering developed by the meteorologist may suggest systematic approaches to problem simplification in other fields. In this respect, meteorology and weather forecasting, through their indigenous methods of multiple scale analysis, have undoubtedly contributed to applied mathematics and the general area of continuum mechanics.

In the fall of 1948, Charney and Arnt Eliassen (of the University of Oslo), who had independently arrived at a scale analysis similar to Charney's, joined von Neumann's Meteorology Project and immediately started a preliminary study of general solutions of a linear system that included the effects of the earth's topography. In the spring of 1950, Charney, Ragnar Fjörtoft of the Norwegian Meteorological Institute, and von Neumann carried out a numerical experiment that had been designed over the previous year. It was about the simplest non-trivial test of numerical methods that could be devised, being based on the equation for conservation of absolute vorticity or spin. The numerical methods were straightforward: centered differencing in both space and time, but satisfying the Courant–Friedrichs–Lewy condition for computational stability. The initial state was taken from observations over North America. The calculations were programmed for the computer ENIAC (Electronic Numerical Integrator and Calculator) at Aberdeen Proving Ground by George Platzman of the University of Chicago and Klary von Neumann. The operation of the machine was rather complicated, requiring the wiring of plugboards and the setting of many switches. Nevertheless, the programmers performed successful calculations early in April of 1950, the first in the history of numerical weather prediction.

The results of these experiments showed that numerically calculated predictions of the atmosphere's large-scale motions were fully as accurate as those based on subjective experience, despite the fact that the numerical model was the simplest imaginable. Later generalizations of the model—including the effects of vertical motion, density stratification, vertical wind shear and thermodynamical processes—were tested on the Princeton machine (the MANIAC, for Mathematical Numerical Integrator and Calculator, also affectionately known as the JONIAC, for Johnny von Neumann) around 1953. The latter experiments showed that relatively simple three-dimensional models were capable of predicting the development of new storms, and that such models were already advanced enough to justify putting them into the routine of daily weather forecasting.

The application of numerical methods to operational weather prediction was brought about in a relatively brief span of time, between the first tests of

Participants in the Grand Victory Celebration Upon Completion of the First Successful Numerical Weather Prediction

a prototype model in 1950 and the establishment of the Joint Numerical Weather Prediction Unit in the summer of 1954. In large part, the surprising success of this venture was due to the intimate and continued involvement of a small but tightly-knit group of the original scientific instigators—all with a smattering of mathematics, physics, meteorology, numerical analysis, and computer logic—who had the momentum and enthusiasm to see the whole development through to the stage of application. At the present time, one of the most important factors in the judgments of regional weather forecasters is the daily numerical prediction computed by the National Meteorological Center in Washington and disseminated to outlying stations by teletype and radio-facsimile.

Although Charney's prescription for "filtering out" sound- and gravity-waves as solutions of the hydrodynamical equations was demonstrably effective, there was still some lingering doubt that some other alternative might not be more exact. Since Richardson's failure was in the main almost certainly due to errors in the observations of the initial state, a popular candidate was the solution of the unmodified hydrodynamical equations; we start, however, with an initial state which has been modified in such a way that the amplitude of sound- or gravity-waves is very small. If the initial state

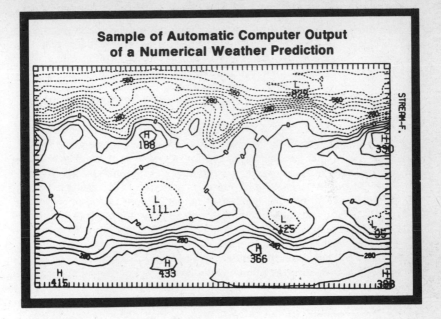

were not so adjusted, errors in observations would imply the existence of sound and gravity waves of spuriously large amplitude.

The problem of adjusting the initial state to remove the unwanted solutions was attacked independently by Philip Thompson of the National Center for Atmospheric Research and Bert Bolin of the University of Stockholm. It was solved by establishing the conditions under which a natural state of balance between the pressure and velocity fields reduced the amplitude of spurious sound- and gravity-waves, but left the Rossby-wave solutions essentially intact. This scheme, based on a detailed analysis of the linear perturbation equations and generalized to establish necessary and sufficient conditions for the exclusion of gravity- and sound-waves as solutions of the nonlinear equations, is the foundation of the prediction methods currently used by the National Meteorological Center.

The Baroque Period

The architecture and basic physical ingredients of mathematical prediction models have not changed drastically in the past two decades, but the facade of numerical procedure has undergone considerable embellishment, some of which one might describe as baroque, high baroque, or even rococo.

The phenomenon of computational instability—exponential growth of error due to finite differencing—has been understood since 1928, and the cure for this particular ailment is known. Yet in the first systematic tests of numerical prediction models it became apparent that finite differencing also introduced a systematic underestimate in the predicted speed at which weather disturbances are propagated. Only recently it has been recognized that the concomitant "truncation" error in the speed of energy propagation is even greater.

The brute force remedy for the truncation error of finite-difference methods is simply to make the distance between adjacent points in the finite-difference grid smaller and smaller, at the considerable expense of increased demands on both storage capacity and volume of computation: decreasing the linear dimensions between grid points by a factor of two results in a sixteen-fold increase in the total number of numerical and logical operations. An ostensibly more palatable alternative is to increase the order of accuracy of the finite-differencing scheme, without changing the resolution of the grid. The usual approach is simply to include a few more terms in the approximate relation between exact continuous derivatives and differences between values at a number of discrete points. Although this procedure presumably reduces the requirements on both memory capacity and total processing time, its basis is somewhat suspect: the functions to be differentiated are not demonstrably smooth, and it is not obvious that higher-order approximations lead to greater accuracy.

Fairly early in the game it was observed that certain quantities should be exactly conserved, or "invariant," but were not conserved in approximate numerical calculations. This defect was traced to the finite-difference scheme. The U.C.L.A. meteorologist Akio Arakawa subsequently discovered that the latter could be modified in such a way that the known invariants are, in fact, conserved. His results are valuable, if only as a program check, but do not appear to prevent anything but widespread computational errors, as opposed to concentrated *local* ones.

In 1956, in the course of analyzing the errors of a meteorological calculation, Norman Phillips of the National Oceanic and Atmospheric Administration turned up a new and unsuspected kind of computational instability arising from the nonlinear interactions between wave motions of different scales. This is a particularly pernicious difficulty, since there is apparently no clearcut cure for the disease. The introduction of a degree of artificial "smoothing" has been found to suppress this kind of instability, but for reasons that seem insufficient and probably specious. It is perhaps evident, however, that problems of meteorology and weather forecasting have provided the motivation for many developments in the analysis of finite-difference methods in fluid dynamics or, more generally, in continuum mechanics.

The finite-difference method is not the only approach to a discrete representation of the state of a fluid. Another is that of representing a continuous variable as a truncated series of superimposed waves, i.e., as a linear combination of elementary waves that are individually independent of time, but whose amplitudes depend only on time. According to this scheme, the time evolution of the atmosphere is described by the changes in the amplitude factors. The latter are determined by a simultaneous system of coupled ordinary differential equations, which are derivable from the original partial differential equations that govern the motion in ordinary geometrical space. These methods (generally known as "spectral" methods) suffer from truncation of the wave representation, but are not subject to precisely the same kinds of errors peculiar to finite-difference methods. They have been made particularly attractive since the development of the so-called Fast Fourier Transform (FFT), a very efficient numerical algorithm for calculating the amplitudes in a wave representation of a function. This feature is especially useful in handling certain quantities that figure prominently in the equations for large-scale atmospheric motions.

In their first applications, spectral methods appeared to require exorbitant amounts of storage space or processing time, even if the number of terms in the wave representation was only moderately large. This is due to the fact that the number of auxiliary constants is roughly proportional to the *square* of the number of terms in the wave representation. With the advent of the FFT, however, it became feasible to carry out some operations in geometrical space, transform to wave-space, carry out other operations in wave-space, and transform back to geometrical space. Over the past decade, these techniques have been widely applied in problems of dynamical meteorology and numerical weather prediction, and in the numerical simulation of fully developed or decaying turbulent flows.

Yet another method of calculation, combining exact analytical methods with a bare minimum of approximate numerical methods, is essentially a Lagrangian scheme, wherein "labelled" material elements of fluid move with the flow, conserving the spin of each element. From the unvarying spin associated with each element and the position of each element at a particular time, it is possible to calculate the velocity of each element at the same time. In principle, therefore, one can compute the evolution of the velocity field from a given initial configuration.

Although Lagrangian methods of this type and generality are new to meteorology (and, for that matter, to any other field) and are still in the experimental stage, there appear to be several advantages that are inherent in such methods. First, physical quantities that should be *locally* conserved are exactly conserved in the calculation and, second, almost all computations are carried out exactly. Finally, and perhaps most important, the size and shape of the elements may be chosen arbitrarily, to provide high resolution in those regions of the flow in which there is considerable fine structure, but lower resolution where less is needed. The latter features are especially suited to the treatment of atmospheric flows, which appear to consist of rather typical and highly localized structures distributed at irregular intervals.

By now it should be apparent that much of the effort in the development of numerical weather prediction over the past twenty years has gone into the improvement of numerical methods. To a considerable degree, this activity has involved only the application or extension of standard or known methods. It is also true, however, that the motivation for several refinements of numerical and mathematical technique came from meteorology, and that some innovations came entirely from within the science, rather than from the mainstream of mathematics.

A New View

Until very recently, virtually all of the people involved in the development of numerical methods took a strictly deterministic view of the prediction problem, that is, that the future state of the atmosphere is completely determined by its present state. Even if this proposition were true, however, the detailed current state of the atmosphere is known only in some probabilistic sense, and the ensuing prediction of the future is also correct only in that same sense. The question then arises: granting that the measured initial value of a variable at each discrete point is its most probable true value, but with a known error distribution, how does the most probable value and its associated error distribution evolve with time through the course of the prediction? If, for example, the error distribution becomes more and more "smeared out," the prediction may be no better than a sheer guess, and possibly even worse than a forecast based only on the climatological mean value.

In 1956, Thompson tried to devise a mathematical technique for dealing analytically with this problem, and slightly more than half succeeded. This was essentially a simple analysis of nonlinear instability around a non-equilibrium state. This work was extended in 1960 by E.A. Novikov of the Institute of Atmospheric Physics in Moscow, who added the mathematical rigor and refinements needed to make this a complete work.

The question was revived again by G.D. Robinson of the Center for Environment and Man in his Presidential Address to the Royal Meteorological Society in 1969. He argued that, assuming a "transfer of uncertainty" from unresolvable small scales of motion to larger scales, all predictive value would be lost after about two days. This argument was not convincing, however, because it led to an estimate of predictability that is lower than the level that is actually achieved in practice. Moreover, Robinson's estimate was based on assumed properties of three-dimensional turbulence, whereas it has been observed that large-scale atmospheric "turbulence" closely approaches two-dimensional or quasi-two-dimensional turbulence. A consequence of the latter is that the "transfer of uncertainty" to larger scales is slower, increasing the range of predictability by a factor of 2 or 3.

Recognizing that predictions should ideally be stated in terms of probability distributions, Edward Epstein of the National Oceanic and Atmospheric Administration proposed to compute at least their low order statistics, i.e., the mean value and the standard deviation around the mean. Incomplete as it is, this development is highly significant and very promising. It has been pushed further by Cecil Leith of the National Center for Atmospheric Research, who has outlined an efficient method for computing the evolution of the probability distribution from a sample of independent predictions. It turns out that the problem of atmospheric predictability is generically similar to certain puzzles in the statistical theory of turbulence, which has recently drawn heavily on some of the techniques and concepts of quantum field theory and statistical mechanics. It is no great exaggeration to say that some of the most advanced methods in the analysis of the statistical behavior of an ensemble of highly nonlinear systems have originated in and were developed in the context of the atmospheric sciences.

To many meteorologists, some of these newfangled statistical-mechanical notions seem rather esoteric. Nevertheless, probabilistic information is precisely what we need in filling out the "payoff table" and in determining the optimum economic strategy in the face of any kind of uncertainty. These ideas are neither new nor strange, but are peculiarly American. The concepts of classical statistical mechanics were laid down in 1880 by Josiah Willard Gibbs at Yale, regarded by his European colleagues as the greatest American scientist of his time. After almost a hundred years, we are still rediscovering the beauty and power of his physical insight and mathematical methods. Anyone who ignores the few small fragments of determinism or discernible statistical order in the universe is losing his grip on an increasingly complicated and chaotic world.

Epilogue

We see, then, that meteorology has become a mathematical science—not by a single stroke, but gradually, following the erratic development of classical physics over a period of some three centuries. It is a difficult science, beset with complications of both interacting physical processes and the purely mathematical complexities of an inherently nonlinear dynamical system. But as new science, new technology, new mathematics have arisen, there have always been a few intrepid and foolhardy meteorologists who have tried to capitalize on the new gains. They were quick to apply anything that came to hand: new theory, perturbation methods, numerical techniques, computers, data-processing systems, or anything else to get on with the problem.

One can also make a case, however, that meteorology has given back about as much as it got from the mainstream of mathematics. Meteorological research has not only provided the motivation for mathematical innovations, but has contributed directly to certain branches of applied mathematics —most notably in the fields of multiple scale analysis and filtering, numerical analysis, linear stability analysis, and the statistical treatment of an ensemble of highly nonlinear dynamical systems.

Having come this far, there is no turning back.

Suggestions for Further Reading

Descriptive

Shaw, Sir Napier. *Manual of Meteorology*, Vol. I. Cambridge University Pr, Cambridge, 1942.

> An introduction to descriptive meteorology: theory is not emphasized, but there are many fascinating fragments of the history of meteorology. It is easily read by anyone with a general scientific background.

Thompson, Philip D. and O'Brien, R. *Weather*. Time-Life Science Libr., 1965.

> This is a general description of weather phenomena and an elementary introduction to the physical processes that govern weather, written for the well-read layman.

Theoretical

Holton, J. *An Introduction to Dynamic Meteorology*. Academic Pr, New York, 1972.

> An excellent beginning text in theoretical meteorology, including many examples of applications of mathematical techniques to the solution of meteorological problems.

Thompson, Philip D. *Numerical Weather Analysis and Prediction*. Macmillan, New York, 1961.

> This deals specifically with the application of the principles of dynamic meteorology to the mathematical prediction of weather.

Applications

Richardson, Lewis F. *Weather Prediction by Numerical Process*. Cambridge University Pr, Cambridge, 1922.

> The pioneer work on the use of numerical methods in meteorology.

Thompson, Philip D. A history of numerical weather prediction in the United States. In *History of Meteorology in the United States: 1776–1976*. American Meterology Society, 1978.

> A chronological account of developments in numerical methods as applied to meteorological problems from about 1945 to 1976.

The Four-Color Problem

Kenneth Appel
Wolfgang Haken

In 1976, the Four-Color Problem was solved: every map drawn on a sheet of paper can be colored with only four colors in such a way that countries sharing a common border receive different colors. This result was of interest to the mathematical community since many mathematicians had tried in vain for over a hundred years to prove this simple-sounding statement. Yet among mathematicians who were not aware of the developments leading to the proof, the outcome had rather dismaying aspects, for the proof made unprecedented use of computer computation; the correctness of the proof cannot be checked without the aid of a computer. Moreover, adding to the strangeness of the proof, some of the crucial ideas were perfected by computer experiments. One can never rule out the chance that a short proof of the Four-Color Theorem might some day be found, perhaps by the proverbial bright high-school student. But it is also conceivable that no such proof is possible. In this case a new and interesting type of theorem has appeared, one which has no proof of the traditional type.

Early History

Despite the novel aspects of the proof, both the Four-Color Problem and its proof have deep roots in mathematics; to see this, one must examine the history of the problem. In 1852, Francis Guthrie (1831–1899), who had been a student in London with his brother Frederick, wrote to Frederick to point out that it seemed that the countries of every map could always be colored with only four colors in such a way that neighboring countries had different colors. By "neighboring" countries he must have meant countries adjacent along a borderline rather than at a single point (or even a finite number of

Announcement of Success

FOUR COLORS

SUFFICE

points), for otherwise a map whose countries looked like the wedges of a pie (see Figure 1) would require as many colors as there were countries. By a "country" he certainly must have meant a connected region, for if a country is allowed to consist of more than one region it is not hard to construct an example of a map with five countries each of which is adjacent to each of the other four (see Figure 2). Francis asked Frederick if there was a way of proving mathematically whether or not this "Four-Color Conjecture" was true. Frederick Guthrie was still at University College, London, where both

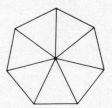

Figure 1. Why countries bordering at a single point are not called neighbors.

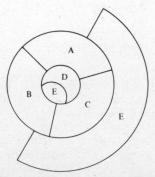

Figure 2. Why a country must consist of a single region.

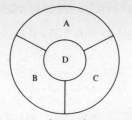

Figure 3. Why four colors are needed.

brothers had attended the lectures of Augustus DeMorgan (1806–1871), one of the major mathematicians of his time. Frederick, unable to answer his brother's question, asked DeMorgan, who could not find any method to determine the truth or falsity of the conjecture.

Guthrie and DeMorgan certainly realized that the map in Figure 3 requires four colors since each of the countries is adjacent to the other three. This means that a "Three-Color Conjecture" is false: three colors will not suffice to color all maps. Moreover, DeMorgan proved (see box on p. 156) that it is not possible for five countries to be in such a position that each of them is adjacent to each of the other four. This result led him to believe that one would never need five different colors and thus that the Four-Color Conjecture was true. But the argument that five mutually adjacent countries cannot exist in a map does not constitute a proof of the Four-Color Conjecture. Many amateur mathematicians, not understanding this, have independently discovered proofs of DeMorgan's result and have then thought that they had proved the Four-Color Conjecture. The difficulty is illustrated in Figure 4: among the six countries of this map there is no collection of four in which each member is adjacent to the other three; yet the map requires four colors—three for the countries in the outer ring and a fourth for the country in the center.

Figure 4 shows that it is not legitimate to conclude that the number of colors required for a map is the same as the maximum of mutually adjacent countries. More powerful mathematical methods are required for a proof of the Four-Color Conjecture.

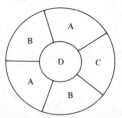

Figure 4. Why it is misleading to try to generalize from Figure 3.

DeMorgan's Argument

Augustus DeMorgan, the first contributor to the theory of the Four-Color Conjecture, showed that no five regions in the plane can be mutually adjacent. The basic idea of his proof is to examine a hypothesized set of five regions which border one another, and derive an internal (geometric) contradiction. Start by assigning numbers to the regions as follows: Choose any region as Region 0, and any other as Region 1. From some point on the common boundary of Regions 0 and 1 proceed clockwise around the boundary of Region 0 until the boundary of a new region is encountered; call this Region 2. Number the next (new) region encountered with 3 and the last region with 4. Various possibilities are illustrated below:

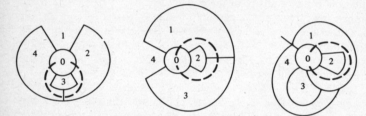

One is then easily able to show that either there is a closed curve (dashed line in the above illustrations) which separates Region 1 from Region 3 or else one which separates Region 2 from Region 4. The two regions separated by this curve cannot border one another since, roughly, one is located inside the curve, the other outside it.

This argument depends on the fact that any closed curve in the plane which looks more or less like a circle bent out of shape (technically, any simple closed curve) has an interior and an exterior. That is, any other curve in the plane that contains a point in the interior and a point in the exterior must cross the given curve. (A precise statement of this idea is given in what is called the Jordan Curve Theorem. Although the idea seems almost obvious, it is not true for every surface; for example on the surface of a doughnut, a closed curve that loops through the hole does not have a distinct interior.) The contradiction that arises in DeMorgan's argument is based entirely on the ability of a simple closed curve to separate its interior from its exterior.

In 1878, the eminent mathematician Arthur Cayley (1821–1895), unable to determine the truth or falsity of the conjecture, proposed the problem to the London Mathematical Society. Within a year after Cayley's proposal, Arthur Bray Kempe (1849–1922), a London barrister and a member of the

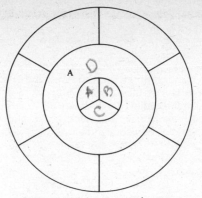

Figure 5. A non-normal map.

London Mathematical Society, published a paper that claimed to prove that the conjecture was true. Kempe's argument was extremely clever, and although his "proof" turned out not to be complete it contained most of the basic ideas that eventually led to the correct proof one century later.

A map is called "normal" if none of its countries encloses other countries (see Figure 5) and no more than three countries meet at any point; for example, the maps in Figures 3 and 4 are normal. Figure 6 gives a larger example of part of a normal map (in cylindrical projection) that was constructed in 1977 by Edward F. Moore of the University of Wisconsin. (Figure 6 has further interesting properties that will be discussed later, but at this point it serves primarily as an example of a map that is not nearly as easy to color with only four colors as are the other examples.)

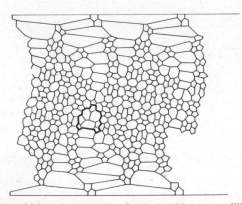

Figure 6. Part of Moore's example of a map with no "small" reducible configuration. (A reducible 12-ring configuration is indicated by heavy edges.)

Normal Maps

A normal map is one in which no more than three regions meet at any point, and in which no region entirely encircles another one. The states in the eastern half of the United States form a normal map, but the entire continental United States does not—because Utah, Nevada, Arizona and New Mexico all meet at a single point:

Since every map can be associated with a normal map which requires at least as many colors, it is sufficient to prove the Four-Color Conjecture for normal maps, for if it is true for these maps then it will be true for all maps. One then shows that any normal map (in a plane) satisfies the formula

$$4p_2 + 3p_3 + 2p_4 + p_5 - p_7 - 2p_8 - 3p_9 - \cdots - (N - 6)p_N = 12,$$

where p_n is the number of countries of the map that have exactly n neighbors and N is the largest number of neighbors that any country has. (Note that $n = 0$ and $n = 1$ cannot occur in a normal map since no enclaves or islands can occur in normal maps; thus the formula begins with p_2.) Now each p_n is either positive or zero and occurs in the formula with a positive sign only if n is less than 6. Thus for the formula to have a positive sum on the left (to match the positive number on the right) at least one of p_2, p_3, p_4, or p_5 must be positive. In other words, some country must have either two, three, four, or five neighbors.

It is easy to modify a non-normal map to produce a normal map which requires at least as many colors. Thus, if there were a map that required five colors, a so-called "five-chromatic map" then there would have to exist a normal five-chromatic map. Hence to prove the Four-Color Conjecture it would be sufficient to prove that a normal five-chromatic map is not possible. Kempe's argument, slightly modified (see box on p. 158), shows that any normal map must have some country with five or fewer neighbors. Kempe noted that if there were a normal, five-chromatic map then there would have to be such a map with a smallest number of countries, a "minimal normal five-chromatic map." Then, proceeding by the classical method of *reductio ad absurdum*, he presented an argument that if a minimal normal five-chromatic map had a country with fewer than six neighbors—which, as he had just shown, every normal map must have—then there would have to be a normal map with fewer countries that was also five-chromatic. (A sketch of Kempe's argument is given in the box on p. 160.) So, if the argument were totally correct up to this point, there could be no number that was the number of countries of a minimal five-chromatic map; hence no minimal five-chromatic map was possible. However, since this meant that there could not be any five-chromatic map at all, the proof would have been complete.

Eleven years later, in 1890, Percy John Heawood (1861–1955) pointed out that Kempe's argument that no minimal five-chromatic map could contain a country with five neighbors was flawed, and that the error did not appear easy to repair. Heawood, in trying to attack the problem, investigated a generalization of the original Four-Color Conjecture. The maps studied by Guthrie and Kempe were maps in a plane or on a sphere. Heawood also considered maps on more complicated surfaces containing "handles" (Figure 7) and "twists" (Figure 8). He was able to obtain an elegant argument, applicable to all surfaces except the sphere and the plane, that provided an upper

Figure 7. A "pretzel" surface with two handles.

Kempe's Argument

The crux of Kempe's purported proof of the Four-Color Conjecture is that a minimal five-chromatic normal map (a smallest normal map that requires five colors) cannot contain any country with exactly two, three, four, or five neighbors. Since Kempe knew that every normal map must contain such a country, he concluded that there is no smallest normal map requiring five colors. Hence there can't be any map that requires five colors. To outline Kempe's argument, we will examine in detail how his proof went for countries with three or four neighbors.

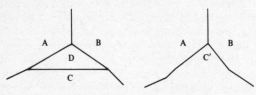

Suppose a minimal five-chromatic map had a country with exactly three neighbors (like country D in the figure above on the left). If that country is amalgamated with one of its neighbors (as in the figure on the right, where countries C and D are united to form country C′), then the resulting map has fewer countries than a minimal five-chromatic map. Hence it is colorable with four colors.

Now if all the countries except the amalgamated country (D) of the original map are assigned the colors of the corresponding countries in the map obtained by the amalgamation, the amalgamated country may be colored with the color not assigned to any of its three neighbors. Thus the original map must have been four-colorable, contradicting the assumption that it was five-chromatic. (Essentially the same argument suffices to show that no country in a minimal five-chromatic map can have exactly two neighbors.)

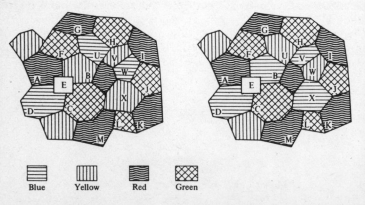

The corresponding argument for four neighbors is an idea of major importance in Kempe's work. Suppose that a minimal five-chromatic map had a country with exactly four neighbors. As before, one can amalgamate such a country and color the rest of the map with four colors, leaving the amalgamated country uncolored (as in the figure above where the country with four neighbors is E). Now if the four neighbors of the uncolored country are colored with fewer than four distinct colors, a color may be chosen for the remaining country. Otherwise, the following argument of Kempe suffices.

Consider the colors of a pair of countries on opposite sides of the uncolored country (for example, the red of country A and the green of country C). Either there is a path of adjacent countries colored with those two colors leading from one of the countries to the other, or there is not. (In the figure above, the path consisting of countries A, F, G, H, I, J, K, L, M, C has only countries colored red and green and leads from A to C. On the other hand there is no path consisting of countries colored yellow and blue leading from B to D.) In honor of Kempe, such two-colored paths are now called Kempe chains.

If both pairs of opposing countries were joined by paths of the corresponding pairs of colors, the two paths would then have a country in common, which is certainly impossible. Thus there is some pair (B and D in our example) not joined by a Kempe chain. Choose one country (say B) of the pair and list all of the countries that are colored by one of the two selected colors (in our example these would be yellow and blue, the colors of B and D) and are joined by a (yellow-blue) path to the chosen country. (In our example, countries B, U, V, W, X form the list.) Now interchange the colors of the countries on the list. (The figure on the right above results from the figure on the left by interchanging blue and yellow on countries B, U, V, W, X.) Now the uncolored country has neighbors of only three colors since the list of countries whose colors were interchanged cannot include more than one (in our case, B) of the four neighboring countries. Thus the uncolored country (E) may be colored with the fourth color (yellow), again leading to a contradiction with the hypothesis that the map required five colors.

bound for the number of colors required to color maps on these surfaces. If the method he used had been applicable to the plane, it would have provided a proof of the Four-Color Conjecture.

Figure 8. A "Moebius strip" with a twist.

Why Are Mathematicians Interested in the Problem?

Heawood continued to work on the problem for the next sixty years. During that time many other eminent mathematicians (as well as countless numbers of amateur mathematicians) devoted a great deal of effort to the Four-Color Conjecture. In fact, much of what is now known as Graph Theory—the geometry of wiring diagrams and airline routes—grew out of the work done in attempting to prove it. It is interesting to ask why so many mathematicians would spend so much time on what appeared to be a question of so little practical significance. To understand the answer to this question is to understand the motivation of pure mathematicians.

Toward the end of the nineteenth century, mathematicians were able to build many powerful theories that enabled them to settle many difficult questions. The feeling grew that any question that could be reasonably posed in the language of mathematics could be answered by the use of sufficiently powerful ideas. Moreover, most mathematicians felt that such questions could be answered in such a way that a competent mathematician could check the correctness of such an answer in a reasonable amount of time. The Four-Color Conjecture was certainly such a problem. Easily stated in the language of mathematics, it could be understood by any intelligent layman. If one could not settle this problem, then one had not developed the appropriate mathematical tools.

In the 1930's a small cloud appeared on the horizon. In logic, the branch of mathematics in which the idea of proof was most precisely stated, the work of Kurt Gödel and Alonzo Church led to some rather disturbing results. First, in what seemed like the most natural logical system, there are statements that are true but not provable in the system. Second, there must be theorems in the system with relatively short statements whose shortest proofs are too long to be written down in any reasonable length of time. In the 1950's it was discovered that the same difficulties affected branches of mathematics other than logic. Some mathematicians thought that since the Four-Color Conjecture had been studied without resolution for such a long time, it might be one of those problems for which neither a proof of correctness nor of incorrectness could be found. Others felt that if a proof existed it might be too long to write down. Still others felt that the disease of unsolvability could not spread to this area and that an elegant mathematical argument must be possible either to prove or disprove the Conjecture.

We now know that a proof can be found. But we do not yet (and may never) know whether there is any proof that is elegant, concise, and completely verifiable by a (human) mathematical mind.

Unavoidable Sets and Reducible Configurations

So many areas of mathematics have been involved in various attempts to prove the Four-Color Conjecture that it would be impossible to discuss them all here; the interested reader could consult references on p. 180 for further historical information. We shall restrict our attention to the work that led directly to the proof.

Kempe had shown that in every normal map there is at least one country with either two, three, four, or five neighbors; there are no normal maps (on a plane) in which every country has six or more neighbors. This may be expressed by the statement that the set of "configurations" (see Figure 9) consisting of a country with two neighbors, a country with three neighbors, a country with four neighbors, and a country with five neighbors is *unavoidable* in the sense that every normal map must contain at least one of these four configurations. Unavoidability is one of the two important ideas that are basic to the theory. Throughout this essay when we say that a set of configurations is unavoidable we shall mean that every normal map must contain some configuration in the set.

The second important idea is *reducibility*. Intuitively, a configuration is reducible if there is a way of showing, solely by examing the configuration and the way in which chains of countries can be aligned, that the configuration cannot possibly appear in a minimal five-chromatic map. The methods of proving configurations reducible grew out of Kempe's proof that a country with four neighbors cannot occur in a minimal five-chromatic map. The use of the word "reducible" stems from the form of Kempe's argument; he proved that if a five-chromatic map contains a country with, say, four neighbors, then there is a five-chromatic map with a reduced number of countries. The reader who understands that argument (in the box on pp. 160–161) has grasped the essential idea of reducibility proofs.

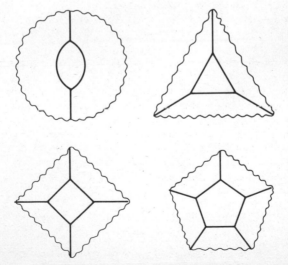

Figure 9. Kempe's small unavoidable set.

In the century since Kempe first introduced the idea of reducibility, certain standard methods for examining configurations to determine whether or not they are reducible have been developed. To use these methods to show that large configurations are reducible requires examination of a large number of details and appears feasible only by computer. We may describe Kempe's attack on the Four-Color Conjecture as an attempt to find an unavoidable set of reducible configurations: finding such a set is sufficient for proving the Four-Color Conjecture.

From 1900 to 1970

In 1913, George David Birkhoff (1884–1944) of Harvard, one of the first eminent American mathematicians, examined Kempe's flawed proof and developed much of the basis for later arguments. Birkhoff, using Kempe's idea and some new techniques of his own, was able to show that certain larger configurations were reducible, for example, the configuration of Figure 10. Using these results and some similar ones of his own, Philip Franklin (1898–1965) of MIT proved that a five-chromatic map (one that, hypothetically, requires five colors) must contain at least 22 countries. The methods Birkhoff developed were used and improved by many mathematicians between 1913 and 1950. Although this fine work established that a large number of configurations were reducible, the set of all configurations that had been proved reducible in the forty years following Birkhoff's paper was not even close to being sufficient for a proof of the Four-Color Conjecture.

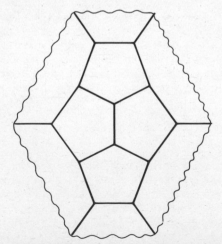

Figure 10. Birkhoff's reducible "diamond" configuration.

In other words, these configurations did not even come close to forming an unavoidable set. Only a few mathematicians developed unavoidable sets of configurations; and they seemed to have little hope that their work would lead to an unavoidable set of reducible configurations. In fact, the primary use of reducible configurations during the first half of this century was to raise the so-called Birkhoff number of Franklin from 22 to 36 (i.e., to show that every map with fewer than 36 countries was four colorable). This was the best result prior to 1950.

Heinrich Heesch of the University of Hanover, who began his work on the Four-Color Conjecture in 1936, seems to have been the first mathematician (after Kempe) who publicly stated a belief that the Four-Color Conjecture could be proved by finding an unavoidable set of reducible configurations. In 1950 he conjectured that not only could such a set be found but that the configurations in the set would be of certain restricted sizes and about ten thousand in number. At that time it seemed extremely difficult to actually produce such a set (and to prove that each of its members was reducible). However, with the advent of high-speed digital computers, an attack on these problems became technically possible. The pessimism of earlier researchers, which appeared justified by the difficulties of hand computation, had to be reevaluated in the light of machines of ever-increasing speed and power. Heesch formalized the known methods of proving configurations reducible and observed that at least one of them (a straightforward generalization of the method used by Kempe) was, in principle, a sufficiently mechanical procedure to be done by computer.

Heesch's student Karl Dürre then wrote a computer program using this procedure to prove configurations reducible. Whenever such a program succeeds in proving a configuration reducible, the configuration is certainly reducible. However, a negative result shows only that the particular method of proving reducibility is not sufficient to prove the configuration reducible; it might be possible to prove it reducible by other methods. In some cases, when Dürre's program failed (to prove a configuration reducible) Heesch succeeded: he was able to show the configurations reducible by using data generated by the program and further calculations to implement a stronger technique which was, in principle, described by Birkhoff.

Dual Graphs and Reduction Obstacles

Heesch described reducible configurations in a somewhat more convenient way than did his predecessors. He began by recasting the original map into what mathematicians call a "dual" form. To do this, mark the capital in each country and then, whenever two countries are neighbors, join their capitals by a road across the common border (see Figure 11). Now delete everything except the capitals (called *vertices*) and the roads (called *arcs* or *edges*) you have added; what remains is known as the *dual graph* of the original map. It

Figure 11. The dual of Birkhoff's diamond.

is possible, and often convenient, to redraw the dual so that all of the arcs are actually segments of straight lines. The edges of a graph divide the plane into regions which are usually called *faces*. If we begin with a normal map—which is all we ever consider here—all of these faces are triangles, because faces in the dual graph correspond to vertices of the original map, and in a normal map each vertex joins precisely three edges. In this case, the whole dual graph is called a *triangulation*. The number of edges that end at a particular vertex (in the dual graph) is called the *degree* of that vertex and is equal to the number of neighbors of the country (in the original map) that is represented by that vertex. A path of edges that starts and ends at the same vertex and does not cross itself separates the graph into two parts: its interior and its exterior. Such a path is called a *circuit*. Figure 11 shows the dual of the configuration of Figure 10, with the countries of Figure 10 lightly indicated; note that the ring of six surrounding countries in the original map is replaced by a circuit of six vertices and six edges in the dual configuration.

In the vocabulary of dual graphs, a configuration is a part of a triangulation consisting of a set of vertices plus all of the edges joining them. The boundary circuit is called the *ring* of the configuration. The configuration of Figure 11 (which is drawn in dual form) is called a six-ring configuration since its ring contains six vertices; it corresponds exactly to the ring of six countries surrounding the original configuration.

While testing configurations for reducibility, Heesch observed a number of distinctive phenomena that provided clues to the likelihood of successful reduction. For instance, there were certain conditions involving the neighbors of vertices of the configuration under which no reducible configurations had ever been found. No reducible configuration had ever been found that

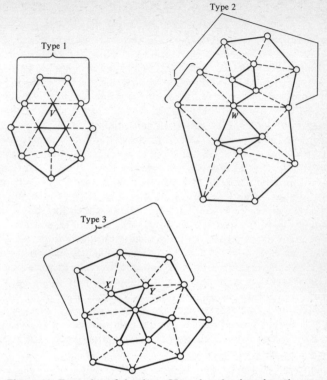

Figure 12. Examples of the three Heesch reduction obstacles.

contained, for example, at least two vertices, a vertex adjacent to four ring vertices, and no smaller configuration that was reducible. While no proof is known that reducible configurations with these "reduction obstacles" could not exist, it seemed prudent to assume that if one wanted reducible configurations one should avoid such configurations. Heesch found three major reduction obstacles (see Figure 12), including the one described above, that were easily describable. No configuration containing one of them has yet been proved reducible.

Discharging

With Heesch's work the theory of reducible configurations seemed extremely well developed. While certain improvements in the attacks on reducibility have since been made, all of the ideas on reducibility that were needed for the proof of the Four-Color Theorem were understood in the late

1960's. Comparable progress had not been made in finding unavoidable sets of configurations. Heesch introduced a method that was analogous to moving charge in an electrical network to find an unavoidable set of configurations (not all reducible), but he had not treated the idea of unavoidability with the same enthusiasm as that of reducibility. The "method of discharging" that first appeared in rather rudimentary form in the work of Heesch has been crucial in all later work on unavoidable sets. In a much more sophisticated form it became the central element in the proof of the Four-Color Theorem, so we will explain it in some detail.

A triangulation that represents a minimal five-chromatic map, by the correct part of Kempe's work, cannot have any vertices with *fewer* than five neighbors. Thus, in what follows, we will, for convenience, use the word triangulation to mean those triangulations with no vertices of degree less than five. It follows from Kempe's work that if we assign the number $6 - k$ to every vertex of degree k (i.e., with k neighbors) then the sum of the assigned numbers (which we shall call charges) is exactly 12 (see Figure 13 for an example). (This somewhat surprising result depends both on the fact that the graph is drawn on a plane and that it is a triangulation.) The particular sum of 12 is not very important. What is extremely important in what follows is that for every planar triangulation this charge sum is positive. It is also important to notice that since vertices of degree k are assigned charge $6 - k$, vertices of degree greater than six (such vertices will be called *major* vertices) are assigned negative charge and only vertices of degree five are given positive charge. (Recall that by the convention mentioned above, we are not considering vertices of smaller degree.)

Now suppose that the charges in such a triangulation are moved around without losing or gaining charge in the entire system. In particular, positive charge is moved from some of the positively charged (degree-five) vertices

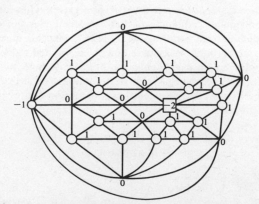

Figure 13. A small triangulation with the associated charges at the vertices.

to some of the negatively charged (major) vertices. While it is certainly not possible to change the sum of the charges by these operations, the vertices having positive charge may change; for example, some degree-five vertices may lose all positive charge (become discharged), while some major vertices may gain so much that they end up with positive charge (become overcharged).

The purpose of this "discharging" of positive vertices is to find a precise procedure describing exactly how to move charge in such a way as to insure that every vertex of positive charge remaining in the resulting distribution must belong to a reducible configuration. Then, since every triangulation must have vertices of positive charge, the configurations signalled by this procedure must be unavoidable. So if all these configurations are also reducible, then the Four-Color Conjecture is proved. (A simple example of a discharging procedure, as it is commonly called, is contained in the box on p. 170.) Of course, if not all of the resulting configurations are reducible, no real progress has been made, since one would be in a position no better than that of Kempe. In fact, one may consider Kempe's unavoidable set as that resulting from the procedure of moving no charges at all.

The Problem as of 1970

In 1970 Haken noticed certain methods of improving discharging procedures and began to hope that such improvements might lead to a proof of the Four-Color Conjecture. However the difficulties still appeared formidable.

First of all, it was conjectured that very large configurations (with rings of neighbors containing as many as eighteen vertices) would be included in any unavoidable set of reducible configurations. Although testing configurations of small ring size (say up to eleven) for reducibility was reasonably simple on a computer, the computer time involved increased by a factor of four for every unit increase in ring size. To make things worse, the computer storage requirements increased just as quickly. When Dürre's program was applied to a particularly difficult fourteen-ring configuration it took twenty-six hours to prove that the configuration did not satisfy just the most mechanical definition of reducibility. Even if the average time required for examining fourteen-ring configurations was only 25 minutes, the factor of four to the fourth power in passing from fourteen- to eighteen-rings would imply that the average eighteen-ring configurations would require over 100 hours of time and much more storage than was available on any existing computer. It was known that experts like Heesch and Jean Mayer, a professor of French literature at Université Paul Valery in Montpellier, France, could often prove configurations reducible by elegant methods that were much shorter than existing computer programs, but even they seldom essayed configurations of very large ring size. The possibility remained that some of their ingenious

A Discharging Procedure

The key to the proof of the Four-Color Conjecture is the redistribution of "charges" among the vertices of the graph in such a way as to locate reducible configurations in the vicinity of positively charged vertices. We begin by assigning to each vertex a charge equal to 6 less the degree of the vertex (see Figure 13). Thus degree-five vertices begin with charge of $+1$. Since the triangulations involved in the proof of the Four-Color Conjecture have no vertices of degree less than five, those of degree five are the only ones with positive initial charge.

In order to obtain a simple example of a discharging procedure, transfer $1/5$ unit of charge from every vertex of degree five to each of its major neighbors, that is, to those neighbors of degree seven or greater. In the new charge distribution the inevitable presence of a vertex of positive charge implies that the triangulation must contain either the configuration consisting of two degree-five vertices joined by an edge or the configuration consisting of a degree-six vertex joined by an edge to a degree-five vertex:

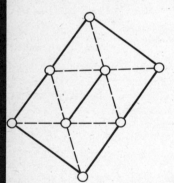 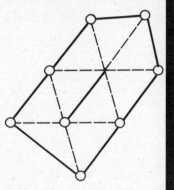

This may be verified by examining all possibilities. A vertex of degree five has positive charge (under the distribution resulting from the transfer) only if not all of its neighbors are major, i.e., if it has a neighbor of degree five or six. A vertex of degree six cannot have positive charge since, not being major, it never receives any. A vertex of degree seven can have positive charge only if it has at least six neighbors of degree five; certainly at least two of these are adjacent. No vertex of degree greater than seven can have positive charge, since the charge supplied to it at the rate of $1/5$ unit per degree-five neighbor cannot overcome its original negative charge. These two configurations (in addition to the configurations consisting of single vertices of degrees two, three, and four) form an unavoidable set. They are not reducible, however, for each of the two configurations contains reduction obstacles of Type 1. This particular discharging provides a simple proof of an unavoidability result which was obtained by P. Wernicke of the University of Göttingen in 1904. If, instead of using $\frac{1}{5}$ in the example above, one used $\frac{1}{4}$, then a proof of a result slightly stronger than a result of Franklin (1922) would be obtained. If $\frac{1}{3}$ is used, an even better unavoidable set is obtained:

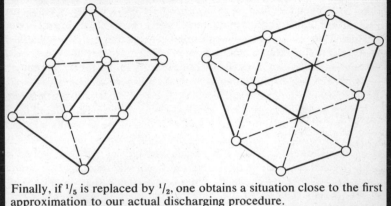

Finally, if $\frac{1}{5}$ is replaced by $\frac{1}{2}$, one obtains a situation close to the first approximation to our actual discharging procedure.

methods might be included in a computer program to speed up the proof of reducibility of those configurations that were reducible.

A second difficulty was that no one knew for sure exactly how many reducible configurations would be needed to form an unavoidable set. It seemed likely that the number would be in the thousands but no reasonable upper bound had been established. In terms of computation time, the numbers seemed just on the border of the possible. Suppose, for example, that to show an eighteen-ring configuration reducible on a computer with enough storage were to take 100 hours. If there were a thousand eighteen-ring configurations in the unavoidable set, the time to prove them reducible would be 100,000 computer hours or over eleven years on an extremely large computer. For all practical purposes, if the set had been this large the proof would have at least required waiting for computers much faster than those currently available.

Even if the theorem could be proved by finding an unavoidable set of reducible configurations, the proof would not satisfy those who demanded mathematical elegance. There would certainly be no hope that a human being could personally check the reducibility of all of the configurations in the unavoidable set. On the other hand, by 1970 many experts on the Four-Color Conjecture had become very pessimistic about the possibility of a moderately short proof. Since the problem is so easily stated, a very large number of mathematicians, amateur and professional, had tried to solve it. Some approaches offered quite reasonable (albeit unsuccessful) bases for attacks on the problem. Although these approaches often led to results of great importance for other areas of pure and applied mathematics, they have never come close to solving the Four-Color Problem.

Two types of standard "proofs" have appeared. The more usual type, often by amateurs, has either been based on a misunderstanding of the problem (for example, leading to a re-proof of the DeMorgan result mentioned earlier) or else contained an immediately recognizable error. The second type usually contains quite sophisticated ideas and is extremely difficult to analyze. Frank Bernhart, whose father Arthur Bernhart of the University of Oklahoma made very significant contributions to the theory of reducibility, has become expert at finding flaws in such arguments and has found an error in every one he has examined. At present, further proofs of moderate length have been announced but none have been presented in detail for expert scrutiny. It is certainly conceivable that one of these proofs is correct, but it is also reasonable to consider the possibility that no correct proof will ever be found except one based on an unavoidable set of reducible configurations. In that case it seems likely that any proof will require a computation that cannot be checked by hand. Thus it is possible that the Four-Color Theorem is an easily stated theorem that requires an argument quite different from those that have been previously used in mathematics, an argument that is not checkable by a human being alone.

Edward F. Moore of the University of Wisconsin, who created the map in Figure 6, developed powerful techniques for constructing maps that do not contain any small reducible configurations. The map in Figure 6, for instance, has no reducible configuration of ring size less than twelve. (It can be colored with four colors and does contain, as indicated, a reducible twelve-ring configuration.) While Moore's example shows that any unavoidable set of reducible configurations must contain at least one configuration of ring size twelve or greater, it appears very likely that thirteen-ring configurations are also necessary and a considerable computational effort will be required even for the best possible solution to the problem.

Geographically Good Configurations

When we began our work on the problem in 1972 we felt certain that the techniques we had available at that time would not lead to a nonmachine proof. We were even quite doubtful that they could lead to any proof at all before much more powerful computers were developed. Our first step in attacking the problem of finding an unavoidable set of reducible configurations was to determine whether there was any hope of finding such a set with configurations of ring size sufficiently small that the computer time required for the reductions could be expected to be within reason. By the very nature of this question it was clear that we should not begin by examining the reducibility of all configurations considered, otherwise the time spent in making the estimate would exceed the expected time for the entire task.

Here the idea of reduction obstacles (see Figure 12) proved extremely useful. It is very easy to determine if a given configuration contains a reduction obstacle; and configurations without reduction obstacles, on the basis of known data, have a very good chance of being reducible. If there was a reasonable unavoidable set of configurations free of reduction obstacles, we felt that there would have to be an unavoidable set of roughly the same size containing only reducible configurations.

We decided, therefore, to study first certain kinds of discharging procedures to determine the types of sets of obstacle-free configurations that might arise. To gain understanding of what was needed even for this study, we limited this restricted problem to what are called "geographically good" configurations—those that avoided the first two of the three reduction obstacles of Heesch (see Figure 12). Geographically good configurations can be characterized very easily: no vertex inside the configuration can have more than three neighbors on the ring of the configuration, and if a vertex has precisely three such neighbors then these lie in consecutive order on the ring.

Dialogue with a Computer

In the fall of 1972 we wrote a computer program which would carry out the particular type of discharging procedure that seemed most reasonable to us and which would give, as output, the configurations that resulted from the most important situations. Although a computer program could not be expected to proceed quite as cleverly as a human being, the immense speed of the computer made it possible to accept certain inefficiencies. In any event, the program was written in such a way that its output could be easily checked by hand.

The first runs of the computer program in late 1972 gave us much valuable information. First, it seemed that, with the original type of procedure, geographically good configurations of reasonable size (ring size at most sixteen) would be found close to most vertices of ultimately positive charge. Second, the same configurations occurred sufficiently frequently that it might be expected that the list of different configurations might be reasonable in length. Third, as the procedure was originally organized, the computer output would be too large to handle; similar cases repeated the same argument too frequently. Fourth, there were clearly some flaws in both the type of procedure and in the details, since there were some vertices of ultimately positive charge in whose neighborhoods no geographically good configurations could be guaranteed. Fifth, the program obtained a tremendous amount of information in a few hours of computer time, so the idea of experimenting frequently was going to be feasible.

The program and the basic ideas of the discharging procedure had to be modified to overcome the problems indicated by the first runs. Since the basic program structure could be preserved, these changes were not too difficult to accomplish and a month later we made a second set of runs. Now that the gross problems indicated by the first runs were largely corrected, the program enabled us to pinpoint more subtle problems and the need for changes of detail. After some study, we found solutions for these problems and again modified the computer program.

This man-machine dialogue continued for another six months until it appeared that we had a feasible method of obtaining an unavoidable set of geographically good configurations. At this point we decided to prove formally that our method would provide a finite unavoidable set of geographically good configurations. To do this we were forced to put aside the experimental approach and describe the total procedure. It was necessary to prove that all cases had really been covered and that those cases that were not handled by the computer program really were as simple as they appeared. Much to our surprise this task proved extremely difficult and took over a year.

The problem stemmed from the necessity in pure mathematics to formulate general definitions of terms and to prove abstract statements about these terms. Special cases had to be examined in detail, often requiring rather complicated analyses, even if they seemed unlikely to arise in practice. The end result was a lengthy proof that an unavoidable set of geographically good configurations did exist, together with a procedure for constructing such a set with precise (but much larger than desirable) bounds on the sizes of the configurations in the set. The procedure itself was very important to us since we intended to use it in a possible attack on the Four-Color Conjecture. Soon thereafter, Walter Stromquist, at that time a graduate student at Harvard and a major contributor to reduction theory, gave a proof of the existence of unavoidable sets of geographically good configurations in a rather elegant way. But since Stromquist's proof did not provide a method of actually constructing the configurations in the set, it appeared unlikely that it would be immediately applicable to the Four-Color Conjecture itself.

Experiments and Modifications

In the fall of 1974, after completing the proof that our procedure would work for geographically good configurations, we discovered that we still had rather little knowledge of exactly how complicated it would be to actually carry out this procedure. To find out more we decided to try it out on a restricted problem, namely, triangulations that contained no pairs of adjacent degree-five vertices. Of course, this was a strong restriction, but the corresponding set of unavoidable geographically good configurations was quite small (47 configurations) and needed no configurations of ring size larger than sixteen. We tried to determine how much more complicated the general problem would be and decided that it might be fifty times as bad (this turned out to be a bit optimistic) and that there was good reason to proceed.

In early 1975 we modified the experimental program to yield obstacle-free configurations and forced it to search for arguments that employed configurations of small ring size. The resulting runs pointed out the need for new improvements in the procedure, but also yielded a very pleasant surprise: replacing geographically good configurations by obstacle-free ones did not seem to more than double the size of the unavoidable set.

At this point the program, which had by now absorbed our ideas and improvements for two years, began to surprise us. At the beginning we would check its arguments by hand so we could always predict the course it would follow in any situation; but now it suddenly started to act like a chess-playing machine. It would work out compound strategies based on all the tricks it had been "taught" and often these approaches were far more clever than those we would have tried. Thus it began to teach us things about how to proceed that we never expected. In a sense it had surpassed its creators in some aspects of the "intellectual" as well as the mechanical parts of the task.

Reducibility Programs

By the summer of 1975 it became clear that there was a good chance of mounting a successful attack on the Four-Color Conjecture. It seemed reasonably certain that we could find an unavoidable set of configurations that were all obstacle-free and likely to be reducible. Although it seemed very likely that such a set would contain some irreducible configurations, it seemed that there was a good chance that some small change of the procedure could be found to replace them with reducible configurations. Now, for the first time, we would need to test configurations for reducibility. Since we expected a set of configurations of ring size up to seventeen, it appeared that some sophisticated shortcut guessing would be needed in order to show configurations reducible. Since our methods, which would necessarily be somewhat restrictive, would not be certain to detect reducible configurations, we were also concerned about the probability of actually determining if a configuration was reducible by any of the known approaches.

We decided to begin writing an efficient program for testing the most mechanical form of reducibility. To do this we employed the assembler language for the IBM 360 computer at the University of Illinois. In late 1974 we were joined by John Koch, who was then a graduate student in computer science and is presently at Wilkes College. He decided to write a dissertation on reducibility for configurations of small ring size. (Frank Allaire, then of the University of Calgary, and Edward Swart, then at the University of Rhodesia, were then doing somewhat parallel work of which we were unaware.) By the fall of 1975, Koch had written programs to check the most mechanical definition of reducibility for configurations of ring size up to eleven and had begun his more general investigations. In the last half of 1975, programs for checking reducibility of twelve-, thirteen-, and fourteen-ring configurations were written by appropriately modifying Koch's work on eleven-ring configurations. These programs were improved to make use of the more general reduction procedure of Birkhoff and then we were almost ready to attack the main problem directly.

Discharging Procedure

Meanwhile, the work on the discharging procedure had gotten to the point where the changes needed to make improvements were structural changes rather than technical adjustments. Since each such change would have required a major modification in the program, we decided that the program should be discarded and that the final form of the discharging procedure could best be implemented by hand. Doing this would provide greater flexibility and would enable the procedure to be modified "locally" whenever it seemed desirable. In December 1975 we discovered that one of the rules that had been used to define the discharging procedures was too rigid. Relaxation of this rule resulted in a discharging procedure that was considerably more efficient. It now seemed possible that one might find an unavoidable set of reducible configurations of smaller ring size than would have been needed by the previous procedures. This meant that the required computer time might be less than previous estimates.

Soon after the discovery of the improved procedure we received a communication from Mayer pointing out that if the problem of "isolated five-vertices" were treated as a special problem rather than as a particular case of the general procedure, the unavoidable set could be greatly improved. (Mayer needed only fourteen configurations of ring size up to fourteen instead of our forty-seven configurations with ring sizes up to sixteen.) This led us to apply our new discharging procedure to this special case. Our general procedure for attacking the Four-Color Conjecture could not quite match the efficiency of Mayer's special procedure for this case: it yielded twenty-eight configurations of ring size up to thirteen. However, it appeared that the new procedure might cut the size of the resulting unavoidable set in half and limit the ring sizes of the configurations to fifteen or even fourteen.

Working Out the Proof

In January 1976 we began the construction of an unavoidable set of reducible configurations by means of our new discharging procedure. The final version of the discharging procedure had one further advantage for insuring the reducibility of the final configurations: the procedure was essentially self-modifying. It began by a first approximation to the final procedure. We considered each possible instance in which a major vertex was forced to become positive, and in each such case the neighborhood of the positive vertex was examined to find an obstacle-free configuration. If none was found, the neighborhood was called "critical," which means that the discharging procedure would have to be modified to avoid this problem. But even when an obstacle-free configuration was found, we could not yet guarantee a reducible configuration. The new reduction programs were used to try to find some obstacle-free configuration that was reducible. If none was found, the neighborhood was also called critical. In classifying neighborhoods as critical, we did not distinguish between configurations with obstacles and those we could not show reducible by our programs.

This method of developing an unavoidable set of reducible configurations was only possible by another dialogue with the computer. To determine which neighborhoods were critical it was necessary to check for reducibility quickly, both in terms of computer time and in terms of real time. We were very fortunate in this dialogue, for it was seldom necessary to wait more than a few days for results, even though a considerable amount of computer time was often needed. Since this extensive man-machine interaction was absolutely essential for our success, we should explain the circumstances that made it all possible.

Although our arrangement for computer usage seemed quite natural to us at the time, we have since discovered that we were indeed extremely fortunate to be working at the University of Illinois, where a combination of a large computing establishment and an enlightened policy toward research use of computers gave us an opportunity that seemed unavailable at almost any other university or research establishment. When we approached the Computer Center and the University Research Board to ask for over a thousand hours of computer time, we could give no guarantee that the work would result in a proof of the Four-Color Conjecture. (We had no external support, although we had made applications; such support seems to depend on unanimity among the referees that an approach will succeed, and certainly no such unanimity is likely to exist with respect to a problem of this nature.) We were told by the Computer Center that since the University's computers were not fully utilized at all times by classwork and ordinary research, we could be included in a small group of computer users who were allowed to share the surplus computer time. We realize in retrospect that this policy was extraordinary and quite courageous; at many institutions the policy is to leave computers idle rather than take a chance that a project which uses a very large amount of time will be unsuccessful and will subject those who approve the request to bureaucratic difficulties. In any event, this policy provided as much time as we could use without disturbing the day-to-day flow of computer work at the University, and was essential to our success.

From January 1976 until June 1976 we worked to define the last details of the discharging procedure and simultaneously to create the unavoidable set of reducible configurations which it produced. Over a thousand hours of time on three computers was used and it was possible to do the reductions quickly enough (in real time) to keep pace with the development of the final discharging procedure (which was done by hand).

The discharging procedure involved about 500 special discharging situations (resulting from critical neighborhoods) that modified the first approximation of January 1976. It required analysis of about ten thousand neighborhoods of vertices of positive charge by hand and analysis of reducibility of over two thousand configurations by machine. While not all of this material became part of the final proof, a considerable part, including the proof of reducibility of about 1500 configurations, is essential. A person could carefully check the part of the discharging procedure that did not involve reducibility computations in a month or two, but it does not seem possible to check the reducibility computations themselves by hand. Indeed, the referees of the paper resulting from our work used our complete notes to check the discharging procedure, but they resorted to an independent computer program to check the correctness of the reducibility computations.

The Nature of Proof: Limits and Opportunities

The fundamental reason that the unavoidable set argument worked whereas other approaches to the Four-Color Conjecture did not is that all other approaches need somewhat stronger theoretical tools to make their methods apply. While these might be possible to create, there is no guarantee that they are actually possible; and if they are, there is no obvious way to go about finding them.

On the other hand, many mathematicians have believed that an unavoidable set of reducible configurations might exist, but that a smallest such set was beyond the bounds of reasonable computation. This attitude appears justified when the problem is considered with respect to the tools available prior to 1960. After 1960, with the advent of faster computers, there were still strong reasons to believe that the computations would be infeasibly large, but there were certainly no theoretical difficulties to overcome other than the choice of a method for obtaining an unavoidable set. Thus by 1970 it became a problem of discovering whether efficient use of known techniques and technical (as opposed to theoretical) improvements would enable one to find an unavoidable set of reducible configurations.

Most mathematicians who were educated prior to the development of fast computers tend not to think of the computer as a routine tool to be used in conjunction with other older and more theoretical tools in advancing mathematical knowledge. Thus they intuitively feel that if an argument contains parts that are not verifiable by hand calculations it is on rather insecure ground. There is a tendency to feel that verification of computer results by independent computer programs is not as certain to be correct as independent hand checking of the proof of theorems proved in the standard way.

This point of view is reasonable for those theorems whose proofs are of moderate length and highly theoretical. When proofs are long and highly computational, it may be argued that even when hand checking is possible, the probability of human error is considerably higher than that of machine error; moreover, if the computations are sufficiently routine, the validity of programs themselves is easier to verify than the correctness of hand computations.

In any event, even if the Four-Color Theorem turns out to have a simpler proof, mathematicians might be well advised to consider more carefully other problems that might have solutions of this new type, requiring computation or analysis of a type not possible for humans alone. There is every reason to believe that there are a large number of such problems. After all, the argument that almost all known proofs are reasonably short can be answered by the argument that if one only employs tools which will yield short proofs that is all one is likely to get.

One might well ask whether this work has any practical value. The answer seems to be that the value is surely greater to mathematics than to cartography. The example of the Four-Color Theorem may help to clarify the possibilities and the limitations of the methods of pure mathematics and those of computation. It may be that a problem cannot be solved by either of these alone but can be solved by a combination of the two methods. There is a certain parallel in the early history of science. From the time of Plato until the late Middle Ages, mathematical methods were regarded as so superior to experimental methods that experimental physics was not considered (socially) acceptable among serious scientists. This severely handicapped the development of certain branches of physics. For instance, the laws of free fall of bodies under the influence of gravity were incorrectly stated by Aristotle, who tried to derive them theoretically, and the error was not corrected for about two thousand years until the simple observations and experiments of Galileo clarified the matter and initiated a rapid progress of mechanical dynamics. As soon as the importance of experimentation was recognized (and somewhat stronger limitations were recognized as applying to purely mathematical methods), a very fruitful development of physics was achieved through a combination of the two methods. Thus the fact that our result hints at somewhat more stringent limitations of purely mathematical methods than some mathematicians would like to see should be interpreted not as a negative result but rather as an indication of a direction for progress.

It may be argued that there is nothing of greater practical importance than obtaining an approximately correct idea of the powers and limitations of one's methods, since here misjudgments may have the most severe negative consequences. We hope that our result yields some progress in this direction and that this justifies the great effort in human resources which has been made in the attack on the Four-Color Problem in the years since 1852.

Suggestions for Further Reading

General

Biggs, N.L., Lloyd, E.K. and Wilson, Robin J. *Graph Theory* 1736–1936. Clarendon Pr, Oxford, 1976.

> Contains a history of the Four-Color Problem, along with other classic problems in graph theory. A pleasure to read both for the professional mathematician and for the curious beginner. A first-rate bibliography of older papers is included.

Technical

Appel, Kenneth and Haken, Wolfgang. Every planar map is four colorable, part I: discharging. *Illinois J. of Mathematics* **21** (1977) 429–490.

Appel, Kenneth, Haken, Wolfgang, and Koch, John. Every planar map is four colorable, part II: reducibility. *Illinois J. of Mathematics* **21** (1977) 491–567.

> The reader interested in the full details of the proof will find them here, along with over 460 pages of microfiched checklists.

Harary, Frank. *Graph Theory*. Addison-Wesley, Reading, 1969.

> For the reader who wishes to see more of the general subject of graph theory; contains an excellent bibliography.

Ore, Oystein. *The Four Color Problem*. Academic Pr, New York, 1967.

> Highly recommended survey of the status of various approaches to the problem in the mid-1960's. It also has a voluminous bibliography.

Saaty, Thomas L. Thirteen colorful variations on Guthrie's four color conjecture. *American Mathematical Monthly* **79** (1972) 2–43.

Part Three

Combinatorial Scheduling Theory

Ronald L. Graham

Things had not been going at all well in the assembly section of the Acme Bicycle Company. For the past six months, the section had consistently failed to meet its quota and heads were beginning to roll. As newly appointed foreman of the assembly section, you have been brought in to remedy this sad state of affairs. You realize that this is your big chance to catch the eye of upper management, so the first day on the job you roll up your sleeves and begin finding out everything you can about what goes on in the section.

The first thing you learn is that the overall process of assembling a bicycle is usually broken up into a number of specific smaller jobs:

FP —Frame preparation which includes installation of the front fork and fenders.

FW—Mounting and aligning front wheel.

BW—Mounting and aligning back wheel.

DE —Attaching the derailleur to the frame.

GC —Installing the gear cluster.

CW—Attaching the chain wheel to the crank.

CR —Attaching the crank and chain wheel to the frame.

RP —Mounting right pedal and toe clip.

LP —Mounting left pedal and toe clip.

FA —Final attachments which includes mounting and adjusting handlebars, seat, brakes, etc.

You also learn that your recently departed predecessor had collected reams of data on how long each of these various jobs takes a trained assembler to

The Acme Bicycle

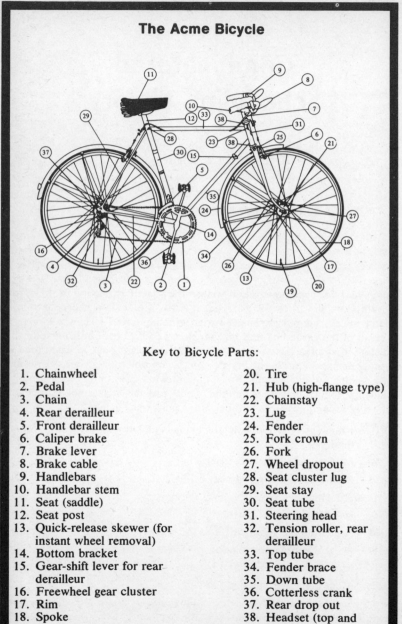

Key to Bicycle Parts:

 1. Chainwheel
 2. Pedal
 3. Chain
 4. Rear derailleur
 5. Front derailleur
 6. Caliper brake
 7. Brake lever
 8. Brake cable
 9. Handlebars
10. Handlebar stem
11. Seat (saddle)
12. Seat post
13. Quick-release skewer (for instant wheel removal)
14. Bottom bracket
15. Gear-shift lever for rear derailleur
16. Freewheel gear cluster
17. Rim
18. Spoke
19. Valve

20. Tire
21. Hub (high-flange type)
22. Chainstay
23. Lug
24. Fender
25. Fork crown
26. Fork
27. Wheel dropout
28. Seat cluster lug
29. Seat stay
30. Seat tube
31. Steering head
32. Tension roller, rear derailleur
33. Top tube
34. Fender brace
35. Down tube
36. Cotterless crank
37. Rear drop out
38. Headset (top and bottom)

perform, which he had conveniently summarized in the following table:

Job:	FP	FW	BW	DE	GC	CW	CR	RP	LP	FA
Time:	7	7	7	2	3	2	2	8	8	18

Because of space and equipment constraints in the shop, the 20 assemblers in the section are usually paired up into 10 teams of 2 people each, with each team assembling one bicycle at a time. You make a quick calculation: One bicycle requires altogether 64 minutes of total assembly time, so a team of two *should* manage this in 32 minutes. This means that in an eight-hour day, each team could assemble 15 bicycles and with all 10 teams doing this, your quota of 150 bicycles per day can be met. You can already taste your next promotion.

Your enthusiasm dwindles considerably, however, when you realize that bicycles can't be put together in a random order. Certain jobs must be done before certain others. For example, it is extremely awkward to mount the front fork to the frame of a bicycle if you have first already attached the handlebars to the fork! Similarly, the crank must be mounted on the frame before the pedals can be attached. After lengthy discussions with several of the experienced assemblers, you prepare the following chart showing which jobs must precede which others during assembly.

This job	must be preceded by	These jobs
FA		FP, FW, BW, GC, DE
BW		GC, DE
GC, CW		DE
LP, RP		CR, CW, GC
CR		CW

In addition to these mechanical constraints on the work schedule, there are also two rules (known locally as "busy" rules) which management requires to be observed during working hours:

Rule 1: No assembler can be idle if there is some job he or she can be doing.

Rule 2: Once an assembler starts a job, he must continue working on the job until it is completed.

The customary order of assembling bicycles at Acme Bicycle has always been the one shown in the schedule in Figure 1. The schedule shows the activity of each assembler of the team beginning at time zero and progressing to the time of completed assembly, called the *finishing time*, some 34 minutes later. Although this schedule obeys all the required order-of-assembly constraints given above, it allows each team to complete only slightly over 14 bicycles per day. Thus the total output of the section is just over 140 bicycles per day, well under the quota of 150.

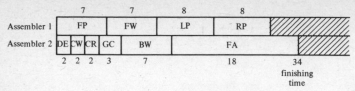

Figure 1. Standard assembly schedule.

After wasting numerous pieces of paper trying out various alternative schedules with no success, you decide, in haste, to furnish all the assemblers with rented all-electric tools. This decreases the times of *each* of the jobs by exactly one minute, so the total time required for all the individual jobs is only 54 minutes. With a little luck, you hope it will be possible to push the output per team to nearly 18 bicycles per day. However, at the end of the first week using the rented tools, you notice that production has now gone *down* to less than 14 bicycles per day. This seems hard to understand so you start jotting down a few possible schedules using the new reduced job times. Surprisingly, you find the best you can do is the one shown in Figure 2. All schedules obeying Rules 1 and 2 take at least 35 minutes for each assembled bicycle!

So you return the rented electrical tools and in desperation decide on a brute force approach: you hire 10 extra assemblers and decree that from now on, each of the 10 teams will consist of *three* assemblers working together to put the miserable bicycles together. You realize that you have increased labor costs by 50%, but you are determined to meet the quota or else.

It only takes you two days this time to decide that something is seriously wrong. Production has now dropped off to less than 13 bicycles per day for each 3-man team! Reluctantly, you again outline various schedules the teams might use. A typical one is shown in Figure 3. Curiously enough, you discover that every possible schedule for a 3-man team obeying Rules 1 and 2 requires 37 minutes for each assembled bicycle. You spend the next few days wandering around the halls and muttering to yourself "Where did I go wrong?" Your termination notice arrives at the end of the week.

This parable actually has quite serious implications for many real scheduling situations. Indeed, some of the earliest motivation for studying these types of scheduling anomalies arose from work on the design of com-

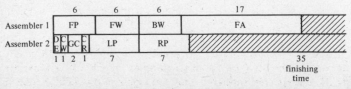

Figure 2. Best schedule with reduced job times.

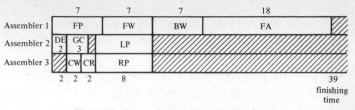

Figure 3. Best schedule for a 3-man team.

puter programs for antiballistic missile defense systems. There it was discovered (fortunately, by simulation) that decreasing job times and increasing computer resources could result in *more* time being used to finish the whole job. (An interesting discussion of this well-known "too-many-cooks-spoil-the-broth" effect, sometimes called "Brooks's Law", can be found in the book "The Mythical Man-Month" by Frederick Brooks.) Furthermore, it was found that adding more computing power to the system was no guarantee that the assigned jobs would all be completed on time. In a missile defense system, this was clearly a cause for some concern.

One might well ask just where it was that our hypothetical foreman at Acme Bicycle did go wrong. It will turn out that he was a victim of Rules 1 and 2 (and a little bad luck). The short-sighted greediness they demand resulted, as it often does, in an overall loss in performance of the system as a whole. In each case, assemblers were forced (by Rule 1) to start working on jobs which they couldn't interrupt (by Rule 2) when a more urgent job eventually came up.

Generally speaking, scheduling problems arise in almost all areas of human activity. Typically such problems can range from the allocation of manpower and resources in a highly complex project (such as the Viking Mars mission which required coordination of the simultaneous activities of more than 20,000 people or the instantaneous control of a modern electronic telephone network) to something as (relatively) simple as the preparation of a 7-course French meal. During the past few years, a number of scheduling models have been rather thoroughly studied in order to understand just why the unpredictable behavior like that in our bicycle example occurred, how bad such effects can ever be and how they can be avoided or minimized. What we will do in this article is to describe what has been recently learned about some of these problems and to point to the exciting new directions that researchers are beginning to pursue. Frequently we will examine a particular problem from several different viewpoints, showing how a variety of approaches can furnish us with a powerful arsenal of tools.

A very important aspect of this subject, both from the point of view of understanding specific scheduling procedures as well as for discovering exactly what is true and what is not true, is the use of examples. Indeed, much of the article will be devoted to various examples which often illustrate very vividly the unexpected subtleties that can occur in this field. From this discussion we hope that the reader will gain insight not only into scheduling theory itself, but also into the kind of productive interaction which can (and often does) occur between mathematics (in this case, combinatorics), computer science (in this case, the theory of algorithms), and problems from the real world.

A Mathematical Model

In order to discuss our scheduling problems in a somewhat more precise way, we need to isolate the essential concepts occurring in the bicycle example. We do this by describing an abstract model of the scheduling situation. The model will consist of a system of m identical processors (before, the assemblers) denoted by $P_1, P_2, \ldots, P_m$, and a collection of tasks $A, B, C, \ldots, T, \ldots$ (before, the various jobs FP, BW, ...) which are to be performed by the processors according to certain rules. Any processor is equally capable of performing any of the tasks although at no time can a processor be performing more than one task. Each task T has associated with it a positive number $t(T)$ known as its processing time (before, these were the job times). Once a processor starts performing (or executing) a task T, it is required to continue execution until the completion of T, altogether using $t(T)$ units of time (this was Rule 2 before).

Between various pairs of tasks, say A and B, there may exist precedence constraints (before, these were the "order-of-assembly" constraints), which we indicate by writing $A \rightarrow B$ or by saying that A is a predecessor of B and that B is a successor of A. What this signifies is that in any schedule, task A must always be completed before task B can be started. Of course, we must assume that there are no cycles in the precedence constraints (for example, $A \rightarrow B$, $B \rightarrow C$, and $C \rightarrow A$) since this would make it obviously impossible to finish (or even start) these tasks. If at any time a processor can find no available task to execute, it becomes idle. However, a processor is not allowed to be idle if there is some task it could be performing (this was Rule 1 before).

The basic scheduling problem is to determine how the finishing time depends on the processing times and precedence constraints of the tasks, the number of processors, and the strategies used for constructing the schedules. In particular, we would like a method to determine schedules that have the earliest possible finishing time.

Performance Guarantees

The approach we will focus on for measuring the performance of our scheduling procedures will be that of "worst-case analysis." What we try to determine in this approach is the *worst possible* behavior that can occur for any system of tasks satisfying the particular constraints under consideration.

As a simple example of this approach, we might want to know just how much the finishing time can increase because of a *decrease* in the processing times of the individual tasks. The answer is given by the following result (due to the author) which dates back to 1966 and is one of the earliest results of this type in the literature.

Let f denote the finishing time of a schedule for m processors using times $t(A), t(B), \ldots$ and let f' denote the finishing time of a schedule using processing times $t'(A), t'(B), \ldots$. If $t'(T) \leq t(T)$ for all tasks T, then

$$\frac{f'}{f} \leq 2 - \frac{1}{m}.$$

For example, if there are two processors, $m = 2$; so $f'/f \leq 3/2$. This means that there can be an increase of at most 50% in finishing time if processing times are decreased.

This bound is a performance guarantee. It asserts that no matter how complicated and exotic a set of tasks with its precedence constraints and processing times happens to be, and no matter how cleverly or stupidly the schedules are chosen, as long as $t'(T) \leq t(T)$ for all tasks T (and this allows for the possibility that $t'(T) = t(T)$ for all T) then it is always true that the ratio of finishing times f'/f is never greater than $2 - (1/m)$. Furthermore, it turns out that this bound of $2 - (1/m)$ cannot be improved upon. What we mean by this is that we can always find examples for which $f'/f = 2 - (1/m)$. We give such an example for $m = 6$. In this example the times $t(T)$ are as follows:

Task T:	A	B	C	D	E	F	G	H	I	J	K	L	M
Time $t(T)$:	7	4	5	6	5	3	7	6	5	5	4	3	12

There are no precedence constraints and $t(T) = t'(T)$ for all tasks T. In Figures 4 and 5 we show the worst and best possible schedules. The corresponding finishing times are $f' = 22$ and $f = 12$, respectively. The ratio f'/f is $22/12$ which is exactly equal to $2 - (1/m)$ when $m = 6$.

In the box on p. 192-193 we give a brief sketch showing how a result of this type can be proved. With such a result, it is possible to know in ad-

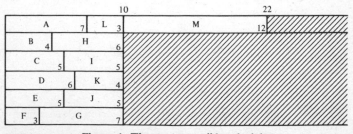

Figure 4. The worst possible schedule.

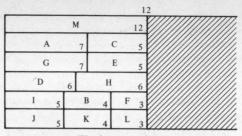

Figure 5. The best possible schedule.

vance the possible variation in f and so to build in a suitable safety factor (of 2, for example) to allow for the potential increase in finishing time. This could be extremely valuable in situations such as the previously mentioned ABM defense system application.

Critical Path Scheduling

One of the most common methods in use for constructing schedules involves what is known as "critical path" scheduling. This forms the basis for the construction of so-called PERT networks which have found widespread use in project management. The basic idea is that any time a processor has finished a task, it should try to choose the "most urgent" task as the next task to start working on. By "most urgent" we mean that task which heads the chain of unexecuted tasks which has the greatest sum of processing

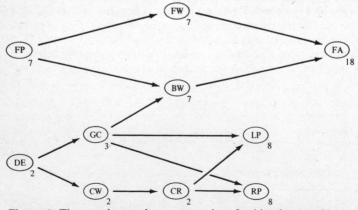

Figure 6. Times and precedence constraints for bicycle assembly at Acme Bicycle.

times. This "longest" chain is termed a "critical path" since its tasks are the most likely to be the bottlenecks in the overall completion of the whole set of tasks. In critical path (CP) scheduling, tasks which head the current critical paths are always chosen as the next tasks to execute.

As an example let us look again at our bicycle assembly example. We can combine the information on task times and precedence constraints in Figure 6. At the start the critical paths are FP → FW → FA and FP → BW → FA, both with length 32. This, in a CP schedule, FP is started first. Once FP is started, the new critical path becomes DE → GC → BW → FA which has length 30. Hence, in a CP schedule, the other processor (assembler) will start on DE. If we continue this scheduling strategy, we will finally end up with the schedule shown in Figure 7. Note that this

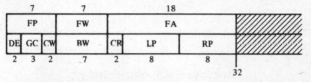

Figure 7. A critical path schedule for bicycle assembly.

schedule has a finishing time of 32 which is clearly the best we could hope for. If only our hypothetical foreman had known about CP scheduling.

Of course, this example is really too small to show the power of CP scheduling. The extensive literature on this technique attests to its wide usefulness in a variety of applications. Nevertheless, from our point of view of worst-case analysis, it should be noted that CP scheduling can perform very poorly on some examples. In fact, not only is there no guarantee that CP scheduling will be close to optimal, but it can happen that this procedure will result in the worst schedule possible!

Here is an example with four processors. The tasks, processing times and precedence constraints are shown in Figure 8, as are a CP schedule with finishing time $f_{CP} = 23$ and an optimal schedule with finishing time $f_{OPT} = 14$. Note that the ratio of finishing times f_{CP}/f_{OPT} is equal to $^{23}/_{14}$, only slightly less than $2 - (^1/_4)$ which (according to our performance guarantee result) is the *worst* possible value of the ratio for any two finishing times for four processors.

Scheduling Independent Tasks

Critical path scheduling gives a rational and, in practice, useful method for constructing schedules. We did find, however, that it is possible for a CP schedule to be very bad. In fact, there is no guarantee that CP scheduling won't produce the worst possible schedule, giving a finishing time ratio (compared to least possible finishing time) of $2 - (^1/_m)$. There are special situations, though, where CP scheduling will never perform this poorly. The

Performance Guarantee: A Proof

In order to give the mathematically-minded reader a feeling for how the bound of $2 - (1/m)$ on the ratio f'/f is obtained, we outline a proof. Consider a schedule which uses the execution times $t'(T) \leq t(T)$ and has finishing time f':

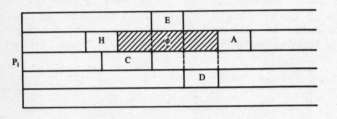

In particular, let us look at an interval of time during which some processor, say P_i, is idle; such intervals are denoted above by the symbol $\emptyset$. From the figure we see that P_i became idle after task H was completed and did not start executing tasks again until task A was begun. The only reason that A wasn't started immediately after the completion of H was that some predecessor of A, say D, was not yet completed at that time. Now either D was already being executed when H was completed or D hadn't yet been started. If it hadn't yet been started, then again this must be caused by one of D's predecessors, say E, not being finished then. By continuing this argument, we conclude that there must be a chain of tasks $\ldots C \rightarrow E \rightarrow D \rightarrow A \ldots$ which are being executed at all times that P_i is idle.

The same ideas show in fact that there must be some chain $\mathscr{C}$ of tasks with the property that some task of $\mathscr{C}$ is being executed whenever *any* processor is idle. Let $t'(\mathscr{C})$ denote the sum of the execution times of all the tasks in the chain $\mathscr{C}$ and let $t'(\emptyset)$ denote the sum of all the idle times in the schedule. It follows from what we have just said that

$$t'(\emptyset) \leq (m-1)\, t'(\mathscr{C}) \tag{1}$$

since the only time a processor can be idle is when some task in $\mathscr{C}$ is being executed, and then we must have at most $m-1$ processors actually being idle.

Of course

$$f \geq t(\mathscr{C}) \tag{2}$$

since the tasks of $\mathscr{C}$ form a chain and consequently must be executed one at a time in any schedule. Thus, if t^* denotes the sum of all the decreased execution times and $\bar{t}$ denotes the sum of all the original execution times, we see that

$$f' = \frac{1}{m}(t^* + t'(\varnothing))$$ (since between times 0 and f' each processor is either executing a task or is idle)

$$\leq \frac{1}{m}(t^* + (m-1)t'(\mathscr{C}))$$ (by (1))

$$\leq \frac{1}{m}(t^* + (m-1)t(\mathscr{C}))$$ (since $t'(T) \leq t(T)$ for all T)

$$\leq \frac{1}{m}(t^* + (m-1)f)$$ (by (2))

$$\leq \frac{1}{m}(\bar{t} + (m-1)f)$$ (since $t^* \leq \bar{t}$)

$$\leq \frac{1}{m}(mf + (m-1)f)$$ (since $\bar{t}$ cannot exceed mf)

$$= (2 - \frac{1}{m})f.$$

In other words,

$$\frac{f'}{f} \leq 2 - \frac{1}{m}$$

which is the desired bound.

most common of these, which we now take up, is the situation where there are no precedence constraints among the tasks.

Imagine that you are the supervisor of a typing pool in a large company. Part of your job each morning is to assign the various papers submitted the night before to particular typists in the pool. Because of differences in type-writers, internal page number references, etc., each paper must be typed by a single typist. How should you assign papers so as to minimize the time required to complete all the papers?

As we saw in the example in Figure 8, a bad CP schedule can take almost twice as much time as a good schedule. But if CP scheduling is used for a situation similar to the typing pool, then the following inequality always holds:

$$\frac{f_{CP}}{f_{OPT}} \leq \frac{4}{3} - \frac{1}{3m}.$$

Thus, the CP schedule for independent tasks (no precedence constraints) always finishes within 33 1/3% of the optimum finishing time. Furthermore, as in the case of the $2 - (1/m)$ bound, it is not hard to find examples which actu-

Task:	A	B	C	D	E	G	F	H	I	J	K	L
Time:	1	1	1	1	10	10	10	3	3	3	3	10

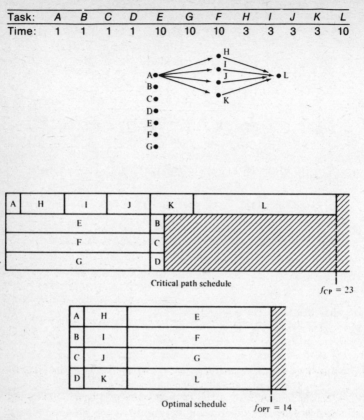

Figure 8. A critical path schedule does not always yield the best results. In this example, the critical path schedule for four processors is nearly twice as long as the optimal schedule.

ally achieve the bound $(^4/_3) - (^1/_{3m})$ showing that it is not possible in general to improve the bound. In Figure 9 we give such an example for $m = 5$. A quick calculation shows that the ratio f_{CP}/f_{OPT} is equal to $^{19}/_{15} = (^4/_3) - (^1/_{15})$ which is just the bound $(^4/_3) - (^1/_{3m})$ with $m = 5$.

NP-complete Problems

At this point one might well ask, "Why should I settle for a schedule which is not best possible? Why not use a method that will always generate the schedule with the least possible finishing time?" A laudable ambition, to be sure. After all, any particular set of tasks you might be faced with is a *finite*

Task:	A	B	C	D	E	F	G	H	I	J	K
Time:	9	9	8	8	7	7	6	6	5	5	5

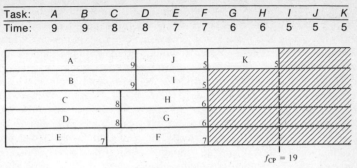

Critical path schedule

$f_{CP} = 19$

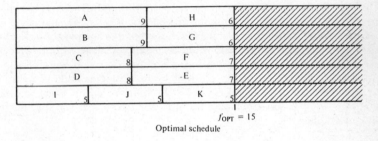

$f_{OPT} = 15$

Optimal schedule

Figure 9. A critical path schedule and an optimal schedule for five processors. Critical path scheduling without precedence constraints, such as in a typing pool, produces results no worse than 4/3 of the optimal schedule.

set so you could always just examine *all* possible schedules and choose the best one.

The trouble with this brute force approach is that the number of possible schedules grows so explosively that there is no hope of looking at even a small fraction of them when the number of tasks is large. If we start with n tasks, then the number of different schedules with 2 processors is 2 multiplied by itself n times, or 2^n. Even for relatively small numbers n, 2^n is a very large number. For example, when $n = 70$, even if we could check 1,000,000 schedules each second, it would require more than 300,000 centuries to check all 2^{70} possible schedules. What we really need is a procedure (or, as computer scientists call it, an algorithm) which will not blow up so drastically as the number of tasks increases.

Unfortunately, this goal is very likely to remain beyond our reach for all time. This gloomy prospect is the result of the fundamental work of Stephen

Cook of the University of Toronto, who in 1972 introduced the concept of "NP-complete" problems. This class of problems is now known to contain literally hundreds of different problems notorious for their computational intractability. NP-complete problems, which occur in areas such as computer science, mathematics, operations research and economics, have two important properties: First, if any particular NP-complete problem had an efficient solution procedure then all of them would. Second, all methods currently known for finding general solutions for any of the NP-complete problems can always blow up exponentially in a manner similar to the behavior of 2^n. Mathematicians strongly suspect (but have not yet proved) that our inability to discover an efficient solution procedure is inherent in the nature of NP-complete problems: they believe that no such procedure can exist. We illustrate in the box on p. 197 one of the more well-known examples of an NP-complete problem, the so-called Traveling Salesman Problem.

As the reader may by now suspect, scheduling problems are, in general, NP-complete. In fact, even without precedence constraints and using only two processors, scheduling is still an NP-complete problem. This is the basis for the pessimistic outlook we took at the beginning of the section. One of its effects has been to redirect much of the earlier effort of trying to find good methods for determining *exact* solutions to the problems to the more fruitful direction of determining good *approximate* solutions easily. What we will do next is to look at the scheduling problem for two-processor systems from this viewpoint.

Getting Closer to the Best Schedule

Critical path scheduling for tasks having no precedence constraints is, as we have seen, always guaranteed to finish within a factor of $4/3$ of the shortest possible finishing time. But to determine a CP schedule we must arrange the processing times of the n tasks in decreasing order. This sorting takes time: for n tasks it can be done in an amount of time which grows like $n \log_2 n$, which is only slightly faster than linear growth in n.

Suppose now that we are willing to do more work, but we want an answer which we know would be very close to the best possible. One approach for doing this is to choose, for some integer k, the $2k$ longest tasks and construct the best possible two-processor schedule for them; then schedule the remaining tasks arbitrarily. If we denote the finishing time of this schedule by f_k then

$$\frac{f_k}{f_{\text{OPT}}} \leq 1 + \frac{1}{2k+2}.$$

For a set of n tasks, the whole procedure can be done in at most $n \log_2 n + 2^{2k}$ operations, where the term $n \log_2 n$ comes from choosing and

The Traveling Salesman Problem

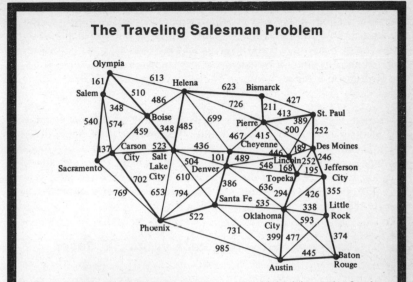

The problem of finding the shortest route which visits each of a given collection of "cities," finally returning to the city from which it began, is traditionally called the Traveling Salesman Problem. Such problems occur in a variety of contexts, e.g., collecting the money from coin telephones, periodic servicing of a dispersed set of vending machines, security guard inspections of locations in a factory, delivery routes for a product to different stores in a city, etc.. In the figure below we show the state capitals of the states west of the Mississippi with approximate road mileages between some of them. The route shown in bold has a total length of 8119 miles. However, the shortest route has length only 8117 miles. Can you find it?

sorting the $2k$ longest tasks and the term 2^{2k} comes from examining all possible schedules on two processors. Since k is a fixed number, the growth of this function as n increases is still moderate.

For example, choosing $k = 3$, we can guarantee that $f_3/f_{OPT} \leq {}^9/_8$ with an amount of work proportional to at most $n \log_2 n + 64$. More generally, any desired degree of accuracy can be guaranteed provided we are willing to pay for it. Unfortunately the price we may have to pay can increase very rapidly. For instance, to guarantee a value of f_k within 2% of the optimum could take time proportional to $n \log_2 n + 2^{49}$, which is sufficient to use up more than a few computer budgets.

This behavior should not be too surprising. After all, if an exponential amount of time seems to be required to find an optimal solution, we might well expect the cost of approximate solutions to behave similarly as their guaranteed accuracy increased. What is surprising is that this exponential increase in cost can be avoided. Oscar Ibarra of the University of Minnesota and Chul Kim of the University of Maryland have very recently developed an algorithm for producing schedules with finishing times f_k which are guaranteed to satisfy

$$\frac{f_k}{f_{\text{OPT}}} \leq 1 + \frac{1}{k}$$

and which requires at most only kn^2 steps. The function kn^2 is an example of a "polynomial" in k and n. When k and n become large, the value of kn^2 is much smaller than the exponential function 2^k. The ideas underlying this procedure involve a clever combination of dynamic programming and "rounding" and are beyond the scope of this article. However, this type of approximation may well be able to guarantee very nearly optimal results using a reasonable amount of computer time.

Finding the Best Schedule in Special Cases

Even though the goal of finding efficient techniques for constructing optimal schedules appears, in general, to be permanently out of reach, there are several interesting special cases of the scheduling problems for which this is actually possible. In this section we would like to discuss two of these cases and briefly describe the philosophy behind the various approaches in order to help understand why they work.

Much of the complexity in our scheduling problems comes from the complicated structure the precedence constraints can have and the complex number-theoretic properties the processing times can have. In our first special case, we restrict both of these factors: we assume that all processing times are equal to 1, and that the precedence constraints are tree-like. What this means is that every task T has at most one successor, i.e., there is at most one arrow leaving T. We show an example of tree-like precedence constraints in Figure 10.

It turns out with these particular restrictions critical path schedules are always optimal. This result, proved in 1961 by T.C. Hu of the University of California, was one of the first results in the field. To construct critical path schedules in this case we simply assign to each task T the length $L(T)$ of the longest chain headed by T. This number $L(T)$ is also known as the *level* of T. (We have assigned levels to tasks in Figure 10.) Any time a processor is available, we always choose a task which has the highest level possible. Figure 10 also shows a CP schedule on 3 processors for the tasks displayed in that Figure.

The second case we examine will allow the complexity of arbitrary (not necessarily tree-like) precedence constraints; but we will still require all processing times to be 1 and, in addition, we only allow *two* processors.

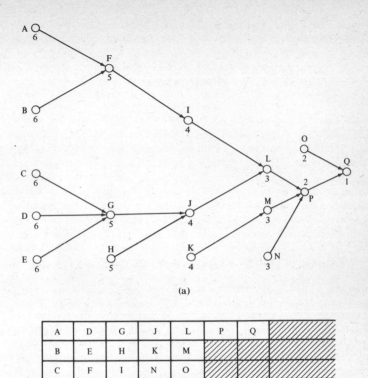

(a)

(b)

Figure 10. For tasks of equal length connected with tree-like precedence constraints (a), CP schedules are optimal (b).

There are now known three essentially different methods for producing optimal schedules in this situation, each somewhat complex; the box on p. 201-203 is devoted to these methods.

From these examples we hope the reader can get a feeling for the variety of techniques which can be successfully applied to special scheduling problems. One might hope that extensions of these ideas would lead to similarly successful algorithms for related questions, e.g., for the problem of three processors with all execution times equal to 1. At present, however, this tantalizing question remains completely unanswered. It should be noted that the "slightly" generalized problem having two processors with processing times of either 1 or 2 has recently been shown to be NP-complete. As we have previously mentioned, this provides strong theoretical evidence that in this case no efficient algorithms can exist which are guaranteed to produce optimal schedules.

Bin Packing

One of the most interesting types of scheduling problems is one which turns the usual question around. Instead of fixing the number of processors and trying to finish as soon as possible, we can ask how few processors can be used and still complete all the tasks by a given deadline d. (For the time being, we shall restrict ourselves to the case where there are no precedence constraints between the tasks.) This new scheduling problem commonly goes under the name of *bin packing*. In the standard statement of this problem we are given a set of items $I_1, I_2, \ldots$ with item I_k having weight w_k. The object is to pack all the items in a minimum number of identical containers, called *bins*, so that no bin contains items having a total weight greater than some fixed number W which is called the *capacity* of the bin. (In the scheduling analogy, the bins are processors: we try to pack jobs into as few processors as possible, subject to the condition that no processor can be assigned jobs whose total time exceeds the processor's capacity, namely, the deadline by which all jobs must be completed.)

The bin packing problem occurs in numerous practical situations and in a variety of guises. For example: a plumber may be trying to minimize the number of standard-length pipes needed from which some desired list of pipe lengths can be cut; a paper producer must furnish customers with various quantities of paper rolls of assorted widths which he forms from rolls of a standard width by "slicing," and he would like to minimize the number of standard rolls he must use; or a television network would like to schedule its advertisers' commercials of varying lengths into a minimum number of standard length station breaks. Another easily pictured example is that of a person confronting a standard postage stamp machine with an armful of assorted letters in one hand and a handful of quarters in the other.

In general, bin packing problems can be extremely difficult. At present, the only known methods for producing optimal packings, i.e., those using the minimum number of bins, involve examining essentially all possible packings and then choosing the best one. For very small problems this may be feasible but as soon as the problem size becomes moderately large, these techniques are hopelessly inadequate. See, for example, the problem in Figure 11. The approach commonly taken to circumvent these difficulties is to design efficient heuristic procedures which may not always produce optimal packings but which are guaranteed to find packings which are reasonably close to optimal. We have already seen examples of this philosophy in some of the earlier scheduling problems. Some of the deepest results in the theory

Two-Processor Algorithms

There is now a variety of methods for determining an optimal schedule for two processors dealing with tasks of identical length. Here are three of them:

The FKN algorithm (after M. Fujii, T. Kasami, and K. Ninomiya (1969)).

The idea behind this method is that we can execute two tasks A and B during the same time interval only if they are *incomparable*, i.e., if neither task must be finished before the other is started. First create a diagram in which we connect two tasks if and only if they are incomparable. (This is called the incomparability graph of the set of tasks.)

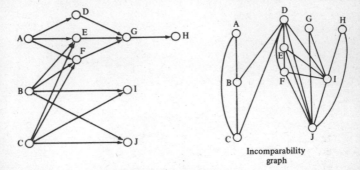

Incomparability
graph

Then find the largest number of lines between tasks in the diagram so that no two go to a common task. (In our example, the lines between A–B, C–D, E–F, G–I, and H–J form such a collection.) These pairs designate the tasks to execute simultaneously; the other tasks we execute one at a time. (Once in a while we may have to do a mild amount of interchanging in order to actually produce a valid schedule. For example, if we have the set of four tasks P, Q, R, and S with precedence constraints $P \rightarrow Q$ and $R \rightarrow S$, then the FKN algorithm can choose the pairs P–S and R–Q to execute simultaneously, which is impossible. However, we *can* exchange R and S and execute P–R and S–Q with no trouble.) Since general methods are known for finding the largest number of disjoint lines in a graph, requiring at most $n^{5/2}$ steps for graphs with n points, the FKN method does indeed satisfy our requirements of efficiency and optimality.

The CG algorithm (after Edward Coffman of the University of California and Ronald Graham (1972)).

The approach here is much in the spirit of the level algorithm discussed on p. 198. The idea is that we are going to assign numbers to the tasks which will depend on the numbers assigned to *all* the successors of a task, and not just on a single successor as was the case for the level algorithm. In order to apply this algorithm we must first remove all extraneous precedence constraints. In other words, if $A \to B$ and $B \to C$, then we do not include $A \to C$ since this is implied automatically. (Removing these extraneous edges is called forming the "transitive reduction" of the set of precedence constraints; it can be done in at most $n^{2 \cdot 81}$ operations for a set of n tasks.) The CG algorithm proceeds by assigning the number 1 to some task having no successor and thereafter, for each task which has *all* its successors numbered, forming the *decreasing* sequence of its successors' numbers and choosing that task which has the smallest sequence in so-called "dictionary" order. (For example, $(5, 3, 2)$ is earlier in "dictionary" order than $(6, 1)$, $(5, 4, 2)$ and $(5, 3, 2, 1)$.) After all tasks have been numbered, using the numbers from 1 to n, the schedule is then formed by trying to execute tasks in order of *decreasing* numbers beginning with the task having number n.

In the figure below we give a CG numbering with some of the successor number sequences also shown.

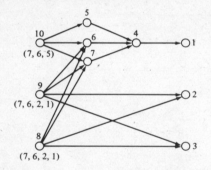

Note that after D, E, F, G, H, I, and J have been numbered, then among the three candidates for label 8, B and C with successor sequences $(7, 6, 2, 1)$ are chosen over A which has the larger successor sequence $(7, 6, 5)$. This results in A getting the number 10 and hence, being executed first. Basically, what the CG algorithm does is to give tasks which either head long chains *or* have many successors larger numbers and consequently it tends to place them earlier in the schedule. Note that if we just use simple CP scheduling, executing tasks only on the basis of the longest chain they head, there is no reason why B and C couldn't be executed first, resulting in the schedule shown below.

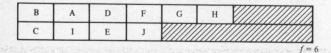

The GJ algorithm (after Michael R. Garey and David S. Johnson of Bell Laboratories (1975)).

In order to apply this algorithm we must put *in* all arrows implied by the precedence constraints. (This is called forming the "transitive closure" of the set of precedence constraints and like the transitive reduction, can also be done for a set of n tasks in at most $n^{2.81}$ operations.) Suppose that we would like to try to finish all tasks by some deadline d. The GJ algorithm computes a set of deadlines (which are always $\leq d$) which would have to be met if in fact all tasks could be executed by time d. This is done by first assigning all tasks with no successors a deadline of d. Thereafter, for any task T which has all its successors assigned deadlines, we do the following: For each deadline d' assigned to a successor of T, determine the number $N_{d'}$ of successors of T having modified deadline d' or less. Set the deadline of T to be the smallest value achieved by any of the quantities $d' - [\,^1\!/_2 N_{d'}\,]$ (where $[x]$ denotes the smallest integer which is greater than or equal to x).

It is easy to see that if T has $N_{d'}$ successors which all have deadlines which are $\leq d'$, then T must be finished by time $d' - [\,^1\!/_2 N_{d'}\,]$ if all deadlines are to be met. After all deadlines have been assigned, the GJ schedule is formed by choosing the tasks in order of earliest deadlines first. The theorem which Garey and Johnson prove is that if there is any schedule which finishes by time d then this method will produce one. It follows from this that a GJ schedule has the minimum possible finishing time (and it doesn't matter what d we started with in forming it!). In the figure below we assign deadlines to the tasks using the value $d = 5$.

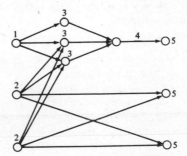

As is apparent, the earliest deadline (1) belongs to task A, which as we saw, must be executed first in any optimal schedule.

Finally, here is the optimal schedule produced by each of these three algorithms:

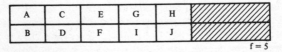

A	C	E	G	H	
B	D	F	I	J	

$f = 5$

1415926535	5820974944	8979323846	5923078164	2643383279
8214808651	4811174502	3282306647	8410270193	0938446095
4428810975	4564856692	6659334461	3460348610	2847564823
7245870066	7892590360	0631558817	0113305305	4881520920
3305727036	0744623799	5759591953	6274956735	0921861173
9833673362	6094370277	4406566430	0539217176	8602139494
0005681271	1468440901	4526356082	2249534301	7785771342
4201995611	5187072113	2129021960	4999999837	8640344181
5024459455	7101000313	3469083026	7838752886	4252230825
5982534904	8903894223	2875546873	2858849455	1159562863
0628620899	5028841971	8628034825	6939937510	3421170679
8521105559	5058223172	6446229489	5359408128	5493038196
4543266482	3786783165	1339360726	2712019091	0249141273
4882046652	9628292540	1384146951	9171536436	9415116094
1885752724	8193261179	8912279381	3105118548	8301194912
2931767523	6395224737	8467481846	1907021798	7669405132
4654958537	7577896091	1050792279	7363717872	6892589235
2978049951	5981362977	0597317328	4771309960	1609631859
5875332083	3344685035	8142061717	2619311881	7669147303
9550031194	8823537875	6252505467	9375195778	4157424218

Figure 11. Can the 100 weights shown be packed into 10 bins of capacity 50,000,000,000? Even with the use of all the computing power in the world currently available, there seems to be no hope of knowing the answer to this question. This example, while not particularly realistic (at present), illustrates the enormous difficulties in solving even relatively small bin packing problems.

of algorithms have arisen in connection with this approach to bin packing problems. We now turn to several of these.

To begin, let us arbitrarily arrange the weights of the items we are to pack into a list $L = (w_1, w_2, \dots)$; no confusion will arise by identifying an item with its weight. One obvious way to pack the weights of L, called the *first-fit* packing of L, is to pack the weights in the order in which they occur in L, filling each bin as much as possible before proceeding to subsequent bins. More precisely, when it is w_k's turn to be packed, it is placed into the bin B_i having the *least* index i in which w_k can validly be packed, i.e., so that the new total of the weights in B_i still does not exceed W. (Of course, we assume at the outset that no w_k exceeds W since otherwise no packing at all is possible.)

We let $FF(L)$ denote the number of bins required when the first-fit packing algorithm is applied to the list L and, similarly, we let $OPT(L)$ denote the number of bins required in an optimal packing of L. The natural question of performance guarantee is this: How much larger than $OPT(L)$ can $FF(L)$ ever be? The answer, due to Jeffrey Ullman of Princeton University, is quite interesting. Ullman showed in 1973 that for any list of weights L,

$$FF(L) \leq \frac{17}{10} OPT(L) + 2.$$

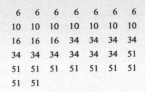

List of weights

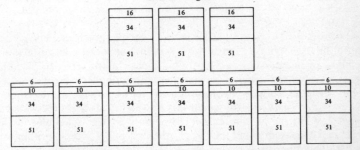

Optimal packing in 10 bins of size $W = 101$

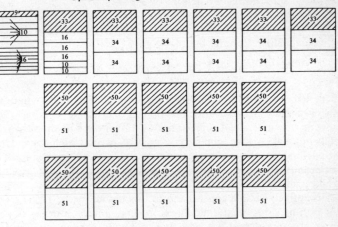

First-fit packing requires 17 bins of size $W = 101$.

Figure 12. The given weights can be packed, optimally, in 10 bins of capacity $W = 101$. But the first-fit algorithm—packing the weights in order of their appearance on the list—requires 17 bins of this same size. The ratio $^{17}/_{10}$ is the worst that can ever happen with the first-fit algorithm.

He also showed that the unexpected fraction $^{17}/_{10}$ cannot be replaced by any smaller number. In Figure 12 we give an example of a list L which has $FF(L) = 17$ and $OPT(L) = 10$, so that $FF(L) = ^{17}/_{10}OPT(L)$. It is suspected that, in fact, Ullman's inequality always holds with the term $+2$ removed. However, it is not known if in this case equality can ever hold when OPT(L) becomes very large.

The appearance of the rather mysterious value $^{17}/_{10}$ turns out ultimately to depend on properties of so-called "Egyptian" fractions, that is, fractions which have numerator 1. Such fractions were treated extensively in one of the earliest known mathematical manuscripts, the famous Rhind papyrus of Aahmes (c. 1690 B.C.); at that time it was common to express fractional quantities as sums of "Egyptian" or unit fractions. For example, $^{25}/_{28}$ would be written as $(^1/_2) + (^1/_4) + (^1/_7)$.

The reason FF(L) was so large in the example given in Figure 12 was because all the large weights were left until last to be packed. As in the scheduling of independent tasks, it would make more sense to place large weights near the beginning of the list and let the smaller weights be used at the end for filling up gaps. More precisely, let us arrange the weights of L into a new *decreasing* list in which larger weights always precede smaller weights (exactly as in our earlier formation of CP schedules with no precedence constraints) and then apply the first-fit packing algorithm to this new list. This packing is called the *first-fit decreasing* packing of L; the number of bins it requires we denote by FFD (L).

The behavior of first-fit decreasing packings turns out to be surprisingly good. Corresponding to the $^{17}/_{10}$ bound for FF(L), we have in this case the following inequality: For any list of weights L,

$$\text{FFD}(L) \leq \frac{11}{9} \text{ OPT}(L) + 4 \cdot$$

Thus, for lists of weights requiring a large number of bins (where the $+4$ becomes relatively insignificant), the first-fit decreasing packing is guaranteed never to use more than approximately 22% more bins than does the theoretically optimal packing, compared to a 70% waste which could occur in a first-fit packing which uses a particularly unlucky choice for its list. The class of examples given in Figure 13 shows that the coefficient $^{11}/_9$ cannot be replaced by any smaller number.

The deceptively simple appearance of this bound for FFD(L) gives little evidence of the substantial difficulties one encounters in trying to prove it. The only proof known at present, due to David Johnson at Bell Laboratories, runs over 75 pages in length. In contrast to the situation for $^{17}/_{10}$, no one currently understands the appearance of the fraction $^{11}/_9$.

One reason for the complex behavior of various bin packing algorithms is the existence of certain counterintuitive anomalies. For example, suppose the list of weights we are required to pack is

442	252	127	106	37	10	10
252	252	127	106	37	10	9
252	252	127	85	12	10	9
252	127	106	84	12	10	
252	127	106	46	12	10	

51	27	26	23	23
51	27	26	23	23
51	27	26	23	23
51	27	26	23	23
51	27	26	23	23
51	27	26	23	23

List of weights

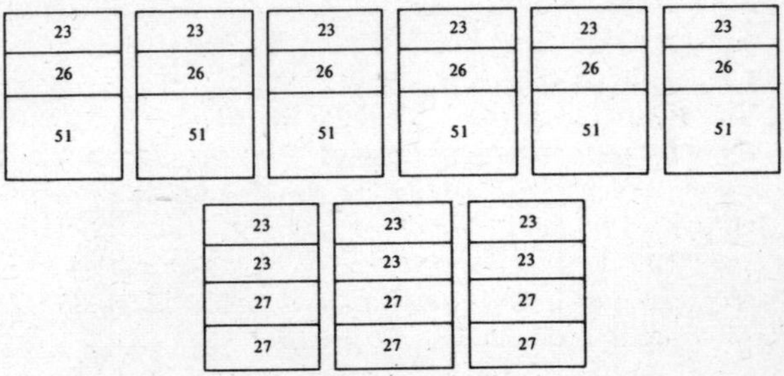

Optimal packing in bins of capacity W = 100

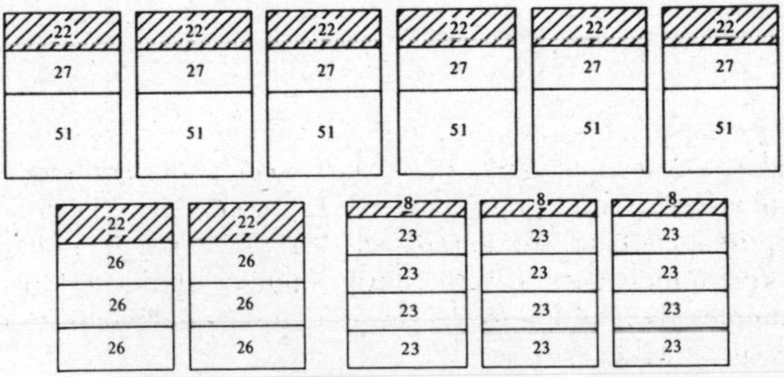

First-fit decreasing algorithm requires 11 bins.

Figure 13. First-fit packing of the listed weights arranged in decreasing order requires 11 bins of capacity W — 100 instead of the optimal 9. By repeating each weight *n* times, we can create as large an example as desired which has this $^{11}/_9$ ratio.

and suppose the bin capacity W is 524. In Figure 14 we show a first-fit decreasing packing of L; seven bins are required. However, if the weight 46 is removed from the list, a first-fit decreasing packing of the diminished list L' now requires *eight* bins (see Figure 14). Not surprisingly, this kind of behavior can cause serious difficulties in a straightforward approach to bin packing problems.

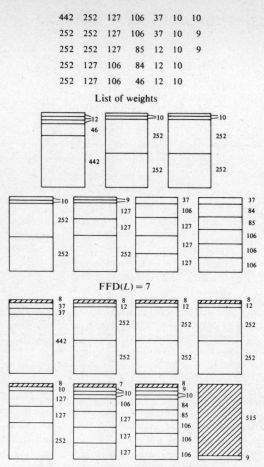

442 252 127 106 37 10 10
252 252 127 106 37 10 9
252 252 127 85 12 10 9
252 127 106 84 12 10
252 127 106 46 12 10

List of weights

FFD(L) = 7

FFD (L') = 8 Where L' = L −{46}.

Figure 14. Decreasing first-fit packing of the listed weights requires 7 bins of size 524, but if one weight is removed from this list, then this same algorithm requires 8 bins!

Pattern Arrangements

One natural extension of the bin packing problem just discussed is the so-called two-dimensional bin packing problem. In this we are given a list of plane regions of different sizes and shapes which we are to pack without overlapping into as few as possible "standard" regions. Common examples

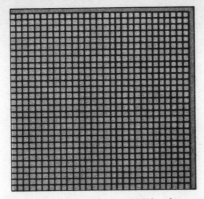

Figure 15. The obvious packing of $(100,000)^2$ unit squares into a large square whose side is 100,000.1 leaves slightly over 20,000 square units uncovered. But no unit square can fit in the uncovered region.

are the placement of sewing patterns onto pieces of material, or the arrangement of patterns for shoe components on irregular pieces of leather. It might be thought that the main source of difficulty here is due primarily to the irregularities in the shapes of the pieces involved. However, even for pieces with very regular shapes, it is usually far from obvious what the best packings are. Indeed, such questions will probably never be completely answered.

As an example, let us consider the following simple geometrical packing problem: How many nonoverlapping unit squares can be placed inside a large square of side s? Of course, if s is equal to some integer N then it is easy to see that the right answer is N^2. But what if s is *not* an integer, e.g., $s = N + \frac{1}{10}$. What should we do in this case? Certainly one option is to fill in an N by N subsquare with N^2 unit squares in the obvious way (see Figure 15) and surrender the uncovered area (nearly s/5 square units) as unavoidable waste. But is this really the best which can be done? Quite surprisingly the answer is *no*. It has been shown very recently by Paul Erdös of the Hungarian Academy of Science, Hugh Montgomery of the University of Michigan, and the author, that as s becomes large, there actually exist packings for any s by s square which leave an uncovered area of at most $s^{(3-\sqrt{3})/2} \approx s^{0.634}\dots$ square units; this is significantly less than the s/5 units of uncovered area left by the obvious packing when s becomes very large. For example, when $s = 100,000.1$ the conventional packing (as in Figure 15) fits in $100,000^2 = 10^{10}$ unit squares, leaving slightly over 20,000 ($= 100,000/5$) square units uncovered. However, it is possible by clever packing to fit in more than 6000 extra squares (see Figure 16). The figure $s^{0.634}$ is probably not the best possible ultimate bound for large values of s; it seems to be extremely difficult to decide what the correct order of growth for the unavoid-

Figure 16. By placing the squares into a different arrangement, it is possible to fit more than $(100,000)^2 + 6000$ unit squares into a large square whose side is 100,000.1. The area left uncovered in the new arrangement is less than in the standard arrangement. (The illustration only suggests how this rearrangement may be done. The actual pattern cannot be drawn with as few squares as appear above.)

ably uncovered area really is, although $\sqrt{s} = s^{0.5}$ looks like a likely candidate.

The particular scheduling model we have examined in this article, while quite simple and basic, still possesses enough structure to exhibit the unpredictable behavior so typical of more complex (and more realistic) situations. What happens is that processors start relatively unimportant tasks (since unnecessary idleness is not permitted) and having started, cannot stop until these tasks are completed, thereby delaying tasks which had since become more urgent. The fact that such behavior is not uncommon in real world scheduling situations testifies to the reasonableness of the model's assumptions.

The problem of deciding the proper order in which the tasks should be chosen so as to minimize the overall finishing time is extremely difficult. In most cases of realistic complexity, we must be satisfied with obtaining solutions which (we hope) are reasonably close to the optimum. Fortunately, it is often possible to find efficient procedures for closely approximating optimal solutions; this is frequently the most useful approach for tackling scheduling problems; in fact, for NP-complete problems it is the only general method that offers any hope at present.

Naturally, many extensions of our basic model are possible. They can include, for example, interruption of tasks before completion, introduction of various resource requirements for the tasks, random arrival of tasks, nonidentical processors and different performance measures, to name a few. By subjecting these extended models to the type of analysis we have described, researchers today are rapidly gaining insight into the very difficult problems of scheduling.

Suggestions for Further Reading

General

Graham, Ronald L. and Garey, Michael R. The limits to computation. 1978 *Yearbook of Science and the Future*. Encyclopaedia Britannica, 1977, pp. 170–185.

Knuth, Donald E. Mathematics and computer science: coping with finiteness. *Science* **194** (December 17, 1976) 1235–1242.

Kolata, Gina Bari. Analysis of algorithms: coping with hard problems. *Science* **186** (1974) 520–521.

Steen, Lynn Arthur. Computational unsolvability. *Science News*, **109** (1976) 298–301.

Technical

Garey, Michael R., Graham, Ronald L., and Johnson, D.S. Performance guarantees for scheduling algorithms. *Operations Research* 26(1978) 3–21.

Garey, Michael R. and Johnson, David S. *Computers and Intractability: A Guide to the Theory of NP-Completeness*. Freeman, San Francisco, 1978.

Graham, Ronald L. Bounds on the performance of scheduling algorithms. In Coffman, E. G., *Computer and Jobshop Scheduling Theory*. Wiley, New York, 1976.

Statistical Analysis of Experimental Data

David S. Moore

In 1976, the food color additive most widely used in the United States, Red 2, was banned as a possible carcinogen. The Food and Drug Administration based its action on statistical analysis of an experiment which involved feeding different levels of Red 2 to five different groups of rats: the number of malignant tumors discovered was significantly higher in rats fed high levels of Red 2 than in those fed lower levels.

At about the same time, the Veterans Administration was conducting a study of treatments for duodenal ulcers. Whenever surgery was necessary, it began with an abdominal exploration to determine which of four possible surgical treatments were tenable for the particular patient. That done, the surgeon opened an envelope prepared in advance by a statistician that contained instructions specifying which of the tenable operations should actually be carried out. By following the patterns set forth by the statisticians, the surgeons were able to insure meaningful results from the study, results that might otherwise have been clouded by unanticipated extraneous factors.

As these examples illustrate, the statistical design of experiments and statistical analysis of experimental data are essential tools of the scientist in many disciplines. Statistical designs for gathering data by sampling large populations are even more prominent, as they provide the foundation for public opinion polls and for government, economic, and social statistics. Use of statistics as a guide to decision making is a pervasive feature of contemporary science and government.

Statistics (like Caesar's Gaul) is divided into three parts: collecting data, organizing and summarizing data, and drawing conclusions from data. There is a close relationship among these parts of the subject. In particular, methods for drawing conclusions (called statistical inference) are based on a description of the process by which the data were collected. This description is usually stated in mathematical terms, thereby making possible a mathematical study of statistical inference. A new and fertile field for the application of mathematics, statistical inference is almost entirely a creation of the twentieth century. Yet it is now very widely used in science and public policy.

We will focus on one area of statistics, inference from experimental data, and will concentrate on statistical procedures that rest on only one of the many mathematical fields applied in modern statistics, namely the behavior of random paths, especially their boundary-crossing properties. The study of such paths actually predates statistical applications by many years. One earlier problem motivating the mathematical analysis of random paths was the irregular ("Brownian") motion of fine particles (such as dust motes in a sunbeam) under frequent collisions with air molecules. The interplay of this mathematical theory with statistical ideas has created both new mathematics and new tools for gathering and analyzing experimental data. We begin our study by examining the ways in which experimental data are gathered.

Classical Design of Experiments

Experimentation is active data collection, in which treatments are applied to experimental material and their effects observed. An experiment is usually a much more efficient tool for gaining knowledge than is passive observation of nature, for the experimenter can create specific conditions of interest to him. For centuries the physical sciences have engaged in experimentation with little call for the aid of statistics. But the physical scientist is commonly blessed with a controlled laboratory environment and with data that show little variation. Not so the agricultural or medical researcher, who must deal with variable units (plants and people) that are inevitably subject to the effects of many factors other than those the experimenter wishes to study. Statistical design of experiments first arose in such settings and is still most applied outside the older physical sciences.

The pioneering work was done in the 1920's by the British geneticist and statistician Ronald A. Fisher (1890–1962), who was involved in the design of agricultural field trials. The general patterns or strategies for arranging experiments that grow from Fisher's work are today termed classical design of experiments. These designs can be studied in their own right and applied with suitable variations to many fields of scientific work. Since Fisher's basic ideas also clarified the way in which mathematics could be applied to the collection and analysis of data, they are a fitting introduction to our subject.

The first requisite of a design for data collection is that it yield data which can be meaningfully interpreted. A medical trial in which patients are simply given the treatment under study will rarely yield valid data, for the effect of the treatment will be obscured by other effects such as the natural course of the disease and the surprisingly strong "placebo effect" due to merely

Ronald A. Fisher

Sir Ronald A. Fisher is among the most eminent scientists of the century in two fields, genetics and statistics. As a statistician, he developed experimental design and contributed profound theoretical ideas on measuring the information in data, condensing data without loss of information, and estimating unknown parameters in a model. Fisher held strong views on the philosophy of statistical inference, views which he expressed frequently and pugnaciously. Statisticians remain deeply divided in their approach to inference, though all admire Fisher's technical work.

receiving attention and treatment. The experimenter ought therefore to divide the patients into two groups, so that those receiving the treatment can be directly compared with a control group who receive instead a placebo or an alternative treatment (see box on p. 216). Several treatments can be compared by dividing the patients into several groups. (This is the pattern of the Veterans Administration ulcer study: four surgical procedures and a control treatment not involving surgery were each applied to a group of patients.)

Such comparative experiments have been common in some fields since the nineteenth century. But even a comparison study will be biased unless the groups of experimental units are equivalent at the beginning of the exper-

Experimental Design

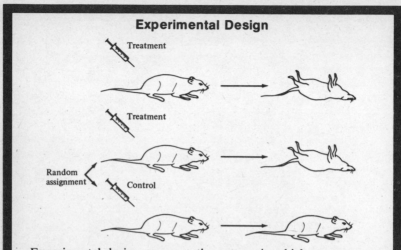

Experimental design concerns the pattern in which treatments are applied to the experimental material. Most laboratory experiments in the physical sciences are simple in design (top), though the treatment applied may be very complex. However, when the observations are affected by factors other than the treatment, such experiments cannot isolate the effect of the treatment. The simplest statistical design (bottom) remedies this by using randomization to create two groups. These groups are treated identically, save that only one receives the treatment that the experiment is designed to study. Conclusions are based on the comparison of the difference between the groups observed at the end of the experiment with the difference expected due to random assignment alone.

iment. One corn variety may be favored by experimental plots of higher fertility, or one surgical procedure discriminated against by the unconscious tendency of the surgeon to use it on the most seriously ill patients. What is more, it was not clear how to interpret the results of comparative experiments. If Variety A of corn has a higher yield than Variety B in an agronomic experiment, is this because Variety A is truly superior, or merely the cumulative effect of minor variations between plots in such factors as fertility, soil type and weed content?

Fisher realized that if experimental units (plots of ground, patients, etc.) are assigned to groups *at random*, as by a coin toss, then equivalent groups will result, at least on the average. His experimental designs therefore involved both comparison and randomization. Practitioners were somewhat slow to accept this innovation, preferring to use their own judgment to set up matching groups of plots or patients. But Fisher won his case as awareness spread of the ease with which bias can infiltrate judgment. Randomization avoids unconscious experimenter bias, which is endemic in medical trials and other experiments on human subjects. Moreover, random selection shows no favoritism with regard to any factor, even those whose importance is unknown to the experimenter. Agricultural researchers, who had long struggled to create groups of field plots which matched simultaneously in all the many factors affecting crop growth, were the first to appreciate the "all-at-one-blow" character of randomization as a device for eliminating bias.

Randomization has a second benefit, less obvious but more important than the elimination of bias: it makes the results of experimentation directly accessible to mathematical study. Random assignment of experimental subjects produces the same behavior as do such other random phenomena as games of chance: although each individual outcome is unpredictable, there is a stable pattern of outcomes from many repetitions. These patterns are studied by a branch of mathematics, the theory of probability. The existence of a probabilistic description for the data collection process is the foundation of statistical inference. Fisher did not originate the use of probability in statistical inference, but he greatly advanced it, and his insistence on randomization created data to which probability theory could be validly applied.

Fisher's approach to inference via probability was to evaluate the significance of the observed variation by comparing it to the inherent variation due to uncontrolled factors. Suppose that Variety A and Variety B of corn are each grown on a number of plots that were randomly assigned to the two varieties. The yield varies from plot to plot, but overall the yield from A, say, exceeds that from B. The variation among plots planted with the *same* variety allows us to estimate the inherent variability in yield due to all the other factors affecting plant growth. Then the amount by which A exceeds B on the average is compared with this inherent (or "within-variety") variability. Because plots were randomly assigned to A and B, the theory of probability allows us to calculate how likely A's observed superiority would be if only the chance effect of the different plots assigned the two varieties were operating. If A's average yield exceeds B's by an amount so large (relative to the within-variety variability) that it would occur by chance only (say) once in 100 experiments in the long run, this is strong evidence that something more than chance is favoring Variety A. This something more, if the experiment was carefully conducted, must be the inherently greater productivity of Variety A. (The box on p. 218 outlines the mathematical form of Fisher's technique, commonly called analysis of variance.)

The availability of a formal mathematical analysis frequently makes possible the study of situations too complex for the experimenter's unaided intuition. Armed with a good mathematical model it becomes possible to study simultaneously the effects of many factors and the interactions between them, as well as to choose designs which make efficient use of resources, yielding most precise information about the factors of greatest interest at the expense of less information about other factors. Fisher's own work created

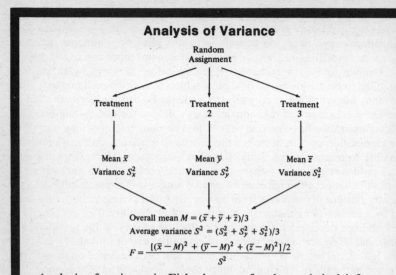

Analysis of Variance

Analysis of variance is Fisher's name for the statistical inference method he developed to analyze data collected by his randomized, comparative experimental designs. In the diagram above, the same number of units is assigned randomly to each of three treatments. Analysis of variance first finds the mean and variance of the responses to each individual treatment. (The mean of responses $x_1, \ldots, x_n$ is the ordinary average $\bar{x} = (x_1 + x_2 + \cdots x_n)/n$. The variance is the sum of the squared distances $(x_i - \bar{x})^2$ of the responses from their mean, divided by the number of responses less one. The variance is a measure of the variability of the responses.) The average of the three treatment means, M, provides the overall mean of the experiment; it sets the average level against which the variation caused by the treatments can be measured. The average variance, S^2, measures variability when the treatment is held fixed; this is the variability "within" groups. Finally, the variance of the treatment means is compared to S^2 to determine whether the average responses to the different treatments differ (from one another) by more than the within-treatment variability would lead us to expect. The ratio of variances used in this comparison was later named the F-statistic in honor of Fisher.

many such possibilities, and other statisticians have continued to extend them. Although the mathematics required by classical design of experiments is not advanced, this cornerstone of statistics establishes the pattern we seek to display in our discussion of some more recent applications of mathematics to statistics. The most valuable advances in statistics are neither mathematical gems isolated from the rude touch of use, nor clever tricks with data which remain suspect because the clarity of mathematical analysis is lacking. We applaud rather an idea which, like randomization in experimental design, both advances practice and brings practice within the scope of theory.

Optional Stopping

Classical statistics assumed that the number of observations to be taken in an experiment was fixed in advance. Indeed, choosing this number was part of the task of designing an experiment or survey. But often the data arrive not in one batch, but sequentially. This is true, for example, in continuing medical trials in which patients are treated as they arrive. In this case it seems unethical as well as unnecessary to continue to assign patients at random to reach a fixed sample size once it becomes clear that one treatment is superior to the other. Even when data do not inherently arrive sequentially, it is very often reasonable to allow the number of observations to depend on the data rather than being fixed in advance. In acceptance sampling of manufactured products, lots of very high or very low quality can be accepted or rejected after inspection of fewer items than are required for a borderline lot. And in scientific work, many a researcher with not quite decisive results has decided to take a few more observations.

Sequential collection of data—that is, stopping at a point dictated by the data already collected—requires that inference from these data proceed quite differently than in classical statistics. To see why, consider the problem of detecting extra-sensory perception (ESP). The standard experiment to detect ESP uses cards marked in five suits (waves, stars, circles, squares, and crosses). The experimenter, using a shuffled deck, concentrates on each card in turn as he sits screened from the subject; the subject must call the suit of each card. If the experiment is carefully conducted, a subject without ESP has probability $1/_5$ of guessing the correct suit. When a fixed number, say n, of trials is made, the situation is exactly analogous to making n tosses of a coin weighted so that the probability of a head is $1/_5$. A quite simple probability calculation (see box on p. 220) tells us how likely it is that the subject will attain any stated proportion of correct calls. If he attains a proportion of correct calls so high that it is very unlikely to occur by merely guessing, we have convincing evidence that he is not merely guessing. (Deciding whether the subject has replaced guessing by ESP or by some other means of gaining an advantage requires very careful investigation of the conduct of the experiment.)

This type of inferential reasoning is possible, here as in classical experimental design, because in both cases a probabilistic description of the data-collection process can be given. A common way of organizing the inference process is to set in advance a specific criterion by deciding how improbable the subject's proportion of correct calls must be to convince us that he is not

Distribution of Probabilities

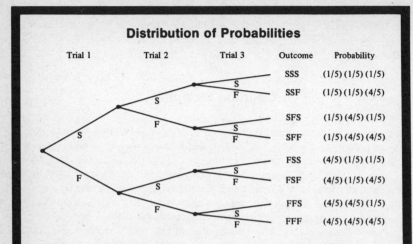

Trial 1	Trial 2	Trial 3	Outcome	Probability
			SSS	(1/5) (1/5) (1/5)
			SSF	(1/5) (1/5) (4/5)
			SFS	(1/5) (4/5) (1/5)
			SFF	(1/5) (4/5) (4/5)
			FSS	(4/5) (1/5) (1/5)
			FSF	(4/5) (1/5) (4/5)
			FFS	(4/5) (4/5) (1/5)
			FFF	(4/5) (4/5) (4/5)

Distribution of probabilities for three calls in the ESP experiment by a subject without ESP can be calculated from the probability of success on a single call and the fact that the correctness of one call does not influence that of any other. The subject has probability $^1/_5$ of succeeding (and therefore probability $^4/_5$ of failing) on any single trial. Two consecutive successes therefore occur with probability $(^1/_5)$ $(^1/_5)$, and two successes followed by a failure with probability $(^1/_5)$ $(^1/_5)$ $(^4/_5)$. The branches of the tree above show all possible outcomes of three trials with success denoted by S and failure by F. To find the probability of exactly two successes in three trials, we add the probabilities of the three outcomes (SSF, SFS, FFS) having two successes. Each has probability $(^1/_5)^2$ $(^4/_5)$, so that (3) $(^1/_5)^2$ $(^4/_5)$ is the probability of two successes in three trials. The collection of probabilities derived in this way is called a binomial probability distribution.

just guessing. Suppose our criterion is that the proportion correct must be so high that it would occur with probability no more than 5% if the subject is guessing. A probability calculation translates this into a specific criterion for any fixed number of trials: the subject must be correct in at least 15 of 50 trials, or in at least 27 of 100 trials, and so on. Such outcomes are said to be statistically significant at the 5% level.

Suppose now that an experimenter, anxious to find evidence of ESP, decides to continue taking observations until for some n the criterion for significance at the 5% level is satisfied. Then, and only then, will he stop—and announce that significant evidence was obtained. This is optional stopping: the experimenter stops at his option rather than after making a prespecified number of observations. That this is fraudulent is clear without mathematics. If the subject is simply guessing, with probability $^1/_5$ of guessing correctly, there is (by definition) a 5% probability that he will satisfy the criterion for significance at the 5% level for any particular n. The probability of eventually satisfying that criterion for some n must be much higher. To discover just how often the experimenter will falsely find ESP using this fraudulent inference procedure requires a mathematical description of the new data-gathering process.

Just as before, a guessing subject can be compared to independent tosses of a coin which has probability $^1/_5$ of landing heads on each toss. But we must now contemplate an indefinitely long sequence of guesses. Any such sequence can be depicted by a graph of the number of correct guesses against the number of guesses made. In Figure 1 such a graph appears as a path which starts at the origin, advances one unit horizontally with each guess, and also steps up one unit if that guess is correct. The process of generating such a path by independent trials, each with the same probability of a correct guess, is called a random walk. In this case, the random walk at each step moves upward with probability $^1/_5$. Just as simple probability theory tells us how likely 27 successes in 100 trials are, more advanced probability theory can compute the chance that a random walk will produce a path with specified properties. Figure 1 also contains (as a boundary on the graph) the sequence of criteria for significance at the 5% level for each fixed number of guesses. The likelihood of fraudulent inference under the optional stopping strategy is the probability that the random walk will somewhere cross that boundary.

A mathematical analysis yields the striking fact that this probability is 1: the path is certain to cross the boundary! In fact, there is a limiting boundary

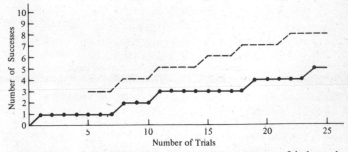

Figure 1. A random walk is generated by a sequence of independent trials, each having the same probability of succeeding. The path of the random walk (solid line) moves forward at each trial, and also moves up if the trial succeeds. The path shown was produced by making 25 trials, each having probability $^1/_5$ of success. (If this experiment were repeated, the pattern of failures and successes would almost certainly be different, so that a different path would result.) The broken line indicates the number of successes required to decide (at the 5% significance level) that the probability of success exceeds $^1/_5$. The particular sequence of trials shown in the solid line would not justify this (incorrect) decision for any n up to 25.

Number of Trials	Criterion for Significance	Limiting Boundary
30	9	9.4
40	12	12.1
50	15	14.7
70	20	19.7
100	27	27.0
200	50	50.3
300	72	72.9
400	94	95.1
500	115	117.1
1000	221	224.9
2000	430	436.0

Figure 2. Comparison between the upper limit of the fluctuations of the random walk for a guessing subject (third column) and the criteria for 5% significance (second column) shows the similar shapes of the two boundaries. As the number of trials increases, the significance boundary gradually falls below the limiting boundary. The law of the iterated logarithm (see box on p. 223) states that if the subject is just guessing the random walk will inevitably reach the significance boundary, if the experiment continues indefinitely, thus giving a false decision.

which (roughly speaking) the random walk is certain to approach infinitely often but cross only finitely often. For large numbers of trials, the significance boundary lies inside this limiting boundary (see Figure 2), so that the experimenter will always be able to claim significance at the 5% level if he is patient enough. Since series of several thousand trials are common in ESP experiments, an ignorant or deceitful experimenter will often have the required patience. The theorem which describes just how widely the random walk will fluctuate is called the law of the iterated logarithm (see box on p. 223), from the form of the limiting boundary. It was discovered in 1924 by the Soviet mathematician Alexander Khintchine. Many forms of this law have been found for more general random paths, and these continue to be a focus of research in probability.

Sequential Analysis

The discussion of optional stopping makes clear what was perhaps not so clear to the scientist tempted to take a few more observations: it is incorrect to allow the data to decide how many observations to take and then to ignore that part of the data-collection strategy when the data are analyzed. But instead of crying "foul" and trying to rule out sequential collection of data, the response of the mathematical statistician is to provide procedures which are appropriate in this setting. This is done by the very same mathematical anal-

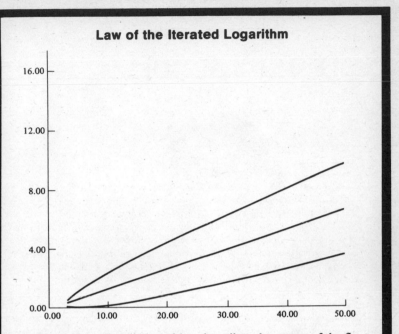

Law of the Iterated Logarithm

The law of the iterated logarithm describes the extent of the fluctuations of a random walk. For the ESP experiment with a guessing subject, the walk drifts upward at an average rate of $0.2n$ units in n steps, since each trial has probability 0.2 of succeeding. This drift is represented by the straight (middle) line on the graph. Fluctuations about this mean drift are expressed in terms of the standard deviation of the vertical position after n trials, which in this case is $0.4\sqrt{n}$. The law of the iterated logarithm (LIL) boundaries lie $\sqrt{2\log\log n}$ standard deviations on either side of the mean. The boundaries drawn above are therefore given by the formula $0.2n \pm (0.4\sqrt{n})\sqrt{2\log\log n}$. The law of the iterated logarithm states, precisely, that a boundary of the form $0.2n \pm \lambda\,(0.4\sqrt{n})\sqrt{2\log\log n}$ is certain to be crossed infinitely often if $\lambda < 1$ and only finitely often if $\lambda > 1$.

ysis that showed the illegitimate consequences of using an inference procedure not designed for sequential data collection.

The analysis of sequential data was first studied systematically during the Second World War by the Statistical Research Group at Columbia University and by a similar wartime advisory group in England. Since munitions are destroyed in testing, the potential for saving observations in quality control and acceptance sampling was then particularly attractive. The leader in the mathematical study was Abraham Wald (1902–1950), one of the many

Abraham Wald

Abraham Wald entered statistics late, and worked in the field for only a dozen years before his death in a plane crash in 1950. Despite this short career, he was a prolific contributor to statistical theory. In two cases, he created new subjects within statistics. One of these is sequential analysis, the other his general decision theory. Wald showed that the idea of seeking optimal methods for making decisions could be used to give a general framework for statistical problems, and he carried out a profound study of which procedures are acceptable for use in a particular problem. Wald's decision-oriented approach to inference did not find favor with Fisher.

European-born scientists who came to the United States in the exodus of the 1930's.

In classical statistics, an inference procedure consists of a recipe for drawing conclusions from the data. The recipe is of course only valid when the data are collected according to the design assumed by the recipe, but the data influence only the final conclusion. Not so in sequential analysis. Here the data are used to decide when to stop taking observations as well as for drawing substantive conclusions. Wald therefore added to the mathematical description of a statistical procedure the notion of a stopping rule. This is a rule which decides at each stage whether to stop or to take another observation, based only on observations already in hand. Stopping rules are the conceptual basis for sequential statistical analysis. They have also proved to be one of the most fruitful notions in modern probability theory.

Many interesting problems can be formulated simply in terms of finding the best stopping rule in a particular situation, as gamblers seeking a rule for quitting while ahead will recognize. Sequential analysis, however, requires making an inference as well as deciding when to stop. Wald's greatest specific contribution to sequential analysis was a stopping-and-deciding rule for situations calling for a Yes-or-No decision. Acceptance sampling is such a situation. So is deciding whether or not a subject has ESP on the basis of a sequence of attempts to call the suit of unseen cards. In such cases, the inference and stopping rules can be described together by drawing two boundaries on the graph of the number of successes against the number of trials. As long as the random walk remains between these boundaries, more trials are run. If its path crosses the lower boundary (too few successes), we stop and conclude that the subject is guessing. If the upper boundary is crossed (too many successes), we stop and conclude that the subject has found a way to call the correct suit with higher probability than the $1/5$ allowed by mere guessing.

What form should these boundaries take? Wald argued as follows. The probability of a successful call by guessing is $1/5$. Any higher probability represents ESP. Choose a degree of ESP we wish to be confident of detecting, say probability $2/5$ of a correct call. For each n, find the probability that the sequence of right and wrong calls actually observed in the first n trials would occur by guessing; call this probability p_n. Then find the probability of the same event in the case of the stated degree of ESP (success probability $2/5$ on each call), and call this probability q_n. The ratio q_n/p_n will be small (p_n exceeds q_n) when the observed outcome is characteristic of a subject who is guessing, and it will be large (q_n exceeds p_n) when the first n calls are more probable assuming him to have ESP. It is therefore reasonable to propose to stop and decide that the subject has ESP as soon as q_n/p_n rises above a fixed level, and to stop and decide that he is guessing as soon as q_n/p_n falls below a fixed lower level. A calculation shows that in terms of the random walk this procedure corresponds to using decision boundaries which are parallel straight lines. These lines are shown in Figure 3.

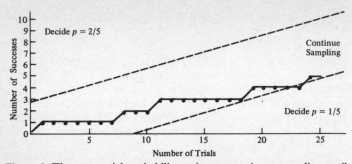

Figure 3. The sequential probability ratio test continues sampling until the random walk crosses one of two parallel straight line boundaries. The boundaries are determined by choosing two values of the probability p of a success on a single trial (here $p = \frac{1}{5}$ and $p = \frac{2}{5}$) and the two probabilities of false decisions about these values (here probability 5% of deciding $p = \frac{2}{5}$ when in fact $p = \frac{1}{5}$, and probability 10% of deciding $p = \frac{1}{5}$ when in fact $p = \frac{2}{5}$). The path of the random walk (solid line) shown here is that of Figure 1, and was generated by trials for which $p = \frac{1}{5}$. The path crosses the lower boundary after 22 observations, at which point the SPRT stops sampling and correctly decides that $p = \frac{1}{5}$. Requiring lower probabilities of error would give boundaries farther apart and increase the average number of observations required to reach a decision.

Wald went on to specify how particular boundaries should be chosen. As in nonsequential inference, he asked that there be a stated probability (say 5%) of declaring that the subject is not guessing when in fact he is. This is the level of significance. Now we ought also to be concerned about the other kind of error, deciding that the subject is guessing when in fact he has ESP. (Jerzy Neyman and Egon Pearson in England had introduced consideration of this second kind of error in nonsequential problems in the 1930's, an innovation opposed by Fisher.) Having agreed that we wish to be fairly confident of detecting a correct-call (ESP) probability of $\frac{2}{5}$, let us require that the probability be (say) 10% of incorrectly declaring the subject to be guessing when he actually has this level of ESP. Wald showed that choosing two alternatives ($\frac{1}{5}$ for guessing and $\frac{2}{5}$ for ESP) and the two corresponding error probabilities (5% and 10%) produces specific decision boundaries, and he gave a recipe for finding them. Figure 3 shows the result for the choices made here. Wald named this procedure the sequential probability ratio test (SPRT).

The SPRT appeals to common sense. Since the boundaries are parallel straight lines, it is easy to use. But it was characteristic of Wald to aim not for simplicity but for optimality. He made great advances toward the goal of constructing statistical procedures which are not simply reasonable, but are "best" according to some specific criterion. Here again he continued in a direction initiated by Neyman and Pearson. In the case of a Yes-or-No decision problem, a possible criterion is to reach a decision using (on the average) as few observations as possible for specified probabilities of the two kinds of error. It was immediately clear that in practice the SPRT often saved half the observations required by a fixed sample size procedure with the same error probabilities, a fact of such importance in munitions testing that the SPRT was for some time a military secret. In 1948, Wald and his Columbia University colleague Jacob Wolfowitz were able to prove that the average number of observations needed for a SPRT to reach a decision is less than that of any other procedure (sequential or not) having the same or smaller probabilities of error. Remarkably, Wald had achieved both optimality and simplicity.

An SPRT can be constructed whenever we must decide between two alternatives based on a sequence of independent observations. It always amounts to studying a random walk with two parallel boundaries, if we allow the term "random walk" to describe somewhat more general random paths. In the ESP example, each observation is a call which is either right or wrong. If we record a right call as "1" and a wrong call as "0," the random walk is just a graph of the sum of the first n observations (that's the number of correct calls) against n. When numerical observations which may have values other than 0 or 1 are taken sequentially, the graph of the sum of the first n observations against n is the random path used in the SPRT. These generalized random walks are mathematically more complicated, but Wald's work covered all such cases.

Later work in sequential analysis has extended Wald's pioneering explorations in many directions. While Wald showed that his SPRT is certain to stop eventually, and that on the average it stops sooner than any other procedure with the same error probabilities, a regular user of the SPRT will find that the test occasionally requires very many observations to reach a decision. So when observations are expensive, sequential tests having a definite upper limit to the sample size are frequently used in place of the SPRT; one of these modified tests is described in the box on p. 228. Another type of problem occurs in many acceptance sampling applications where it is physically more convenient to take items in batches than individually; sequential procedures allowing this option are widely used in industry.

A more difficult problem concerns the number of options under study. The SPRT is designed only for a Yes-or-No decision problem, whereas many problems involve deciding among more than two alternatives. Many statisticians are actively developing sequential procedures for these more general decision problems. In each case, the goal is to emulate Wald by finding (via mathematical analysis) ways of implementing practically appealing innovations. The mathematical problems which must be faced involve boundary-crossing behavior for quite general boundaries and quite general random paths in the plane and in spaces of higher dimension. These are exceedingly difficult problems. There is much unfinished business in both the mathematics and the application of sequential analysis.

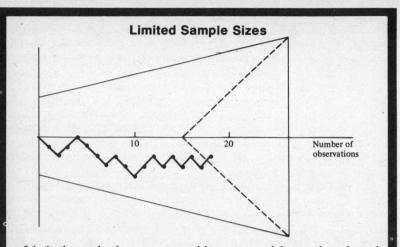

Limited Sample Sizes

Limited sample sizes are ensured by a sequential test whose bound-aries are as shown above. This test is useful whenever the large sample sizes occasionally occurring with the SPRT cannot be tol-erated. Suppose that two medical treatments are compared by giving both (at different times and in random order) to the same patients. Each patient's case yields a preference for one or the other treatment. The path which moves up one unit if Treatment A is preferred and down one if Treatment B is preferred is a random walk. We wish to test the hypothesis that the treatments are equally effective against the alternative that one is better than the other. The latter decision is made if the random walk crosses the upper or lower boundary, the former if it crosses the end boundary. The end boundary is an upper limit on the number of patients who will be tested. The end boundary can be contracted (dotted lines), since once the dotted lines are crossed, the path must eventually cross the end boundary. Note that this test is designed, unlike the SPRT, to detect a two-sided alterna-tive: "equal" is tested against "either better or worse."

In the figure, the end boundary allows no more than 25 observa-tions. The random walk crosses the dotted lines on the 18th observa-tion, at which point we stop and decide that there is no difference be-tween the treatments. Since the random walk was generated by tosses of a fair coin, this is the correct decision in this case.

Sequential Assignment of Experimental Subjects

Acceptance sampling and ESP experiments lack the emphasis on comparing different treatments which is the heart of classical experimental design. Comparative experiments utilizing subjects who enter the experiment sequentially are common. In medical experimentation particularly, subjects often arrive sequentially but must be assigned immediately either to a treatment or to a control group. In a well-planned experiment, the assignment is done at random, usually by opening an envelope provided by a statistician that contains the results of a randomization (such as a coin toss). The envelopes' contents are hidden from the experimenter, lest knowledge of the assigned treatment bias the decision to accept the subject as suitable for the study. Such an experiment may be carried out sequentially, using data on previous patients to decide whether or not to continue taking observations. But frequently sequential analysis is complicated by the fact that full results on a subject admitted to the experiment are often not available for long periods of time. It is therefore quite common to assign patients until predetermined fixed sample sizes are reached, making no use of the fact that patients arrive sequentially.

Ethical concerns suggest seeking a procedure which reduces the number of patients assigned to the inferior treatment. In 1969, Marvin Zelen of the State University of New York at Buffalo introduced a simple sequential rule for assigning patients which he named the "play-the-winner" rule after a familiar gambling strategy. The key idea is to assign the first patient by a coin toss and thereafter assign patients to the same treatment as long as a run of successes continues, switching treatments whenever a failure occurs. Since the outcome of the immediately preceding patient's case is often not available, a modification of the play-the-winner idea is required. Each success with a particular treatment generates a future trial of that treatment on a new patient. Each failure generates a future trial with the alternate treatment. The future trials are put into a hat as they are generated, and as each patient enters the experiment, an assignment is drawn at random from the hat. If the time required to observe the outcome of a trial is long, the hat may be empty. In this case the patient is assigned by a coin toss.

The play-the-winner rule aims to assign patients to the superior treatment as often as possible while accumulating evidence on the relative effectiveness of the two treatments. If the delay in observing outcomes is too long, the rule will make assignments mainly by coin toss, and is of little use. But with short delays, the rule is quite effective in reducing the number of patients assigned to the inferior treatment. To make this advantage precise and to find proper inference rules requires a mathematical study. That study can again be based on the behavior of a random path, which advances one unit at each step and also moves up one unit at step n if patient n is assigned to the first treatment. This is not, technically, a random walk (though the paths look similar), because what happens at each step depends on the results of previous steps. A different set of tools from the probabilist's study of random paths is now required.

An instructive contrast to the play-the-winner rule is provided by a sequential assignment scheme having quite a different objective. Experimenters prefer to compare groups of about equal size when fixed group sizes are used. Tossing a coin as each patient appears will in the long run assign equal proportions of patients to treatment and control. But random assignment of the first 10, or the first 30, patients can be quite unbalanced. If the disease is rare and the treatments expensive, the first 30 patients may comprise the entire experiment. In 1971, Bradley Efron of Stanford University offered a "biased coin" design which tends to correct any imbalance in the allocation of subjects. The plan is as follows:

If more patients have previously been assigned to control than to treatment, the present patient is assigned to treatment with probability $\frac{2}{3}$.

If equal numbers have been assigned to control and to treatment, the present patient is assigned to treatment with probability $\frac{1}{2}$.

If more patients have been assigned to treatment than to control, the present patient is assigned to control with probability $\frac{2}{3}$.

The effectiveness of the biased coin design in achieving balanced groups can be studied by the mathematics of random walks, just as in the quite dissimilar optional stopping and SPRT examples (see box on p. 231). Such study shows that, for example, the biased coin design with 10 subjects creates groups more unbalanced than 6 patients in one and 4 in the other only 9% of the time. Groups more unbalanced than 6-to-4 occur 34% of the time when subjects are assigned by coin-toss randomization.

Unlike the play-the-winner rule, the biased coin design is of limited practical interest. It is simpler to achieve balance, if that is the goal, by randomly assigning patients in pairs, one to each treatment. Yet the contrast between the two allocation schemes (see Figure 4) illustrates the variety of sequential experiments and the possibility of fitting the design to the desires of the experimenter. Sequential allocation can be combined with a sequential inference procedure (a "stopping-and-deciding" rule) or used merely to assign a fixed number of patients. Even in the latter case, the effect on inference must be examined, since the randomization is not that used in classical experimental design. Fortunately, classical statistical procedures are not seriously perturbed by sequential allocation of a fixed number of subjects. This of course contrasts with the serious distortion arising when classical procedures are used with optional stopping.

Inference Without Error

In statistics as in other fields, mathematics is not merely the servant of practice. It has in addition a life of its own. Mathematicians study random walks, for example, as objects interesting in themselves. Such "pure mathematics" provides the tools needed to analyze practical problems, and may even suggest possibilities which a user of statistics could not have imagined.

An example of this consciousness-expanding function of mathematical research is found in studies related to the law of the iterated logarithm (LIL)

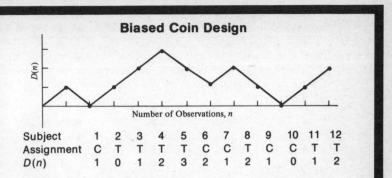

Subject	1	2	3	4	5	6	7	8	9	10	11	12
Assignment	C	T	T	T	T	C	C	T	C	C	T	T
$D(n)$	1	0	1	2	3	2	1	2	1	0	1	2

The biased coin design can be studied through a random walk by concentrating on the absolute difference $D(n)$ between the number of subjects among the first n who were assigned to treatment and the number assigned to control. The figure above is a typical path of $D(n)$ against n, obtained by using biased coin randomization to assign 12 patients. At each step, $D(n)$ moves up one unit with probability $1/3$ and down the probability $2/3$ if the previous value is greater than 0. When the previous value is 0, $D(n)$ is 1. This path is a random walk because $D(n)$ is the sum of n independent observations, each having the value 1 or -1. But there is a complication: whenever the path hits the horizontal axis it bounces up one unit at the next step. This is called a random walk with reflecting barrier at 0. The mathematics of random walks can be applied to show that when n is large the probability that $D(n)$ is no greater than 2 is $7/8$ for even n and $3/4$ for odd n. (Notice that $D(n)$ must be odd for n odd and even for n even.) These probabilities can also be computed for specific numbers of patients. For example, when $n = 10$, $D(n)$ is 0 with probability 0.53 and 2 with probability 0.38.

initiated in 1967 by Donald Darling of the University of Michigan and Herbert Robbins of Columbia University. Return for simplicity to the ESP example, where a subject guessing the suit of a sequence of cards produces a random walk which advances one unit at each step, and with probability $1/5$ also moves up one unit. The path of the random walk drifts upward, since on the average there are $n/5$ successes in n trials. Draw a boundary above the line of this average drift, and denote by P_m the probability that the walk crosses this boundary at the mth or any later step. The law of the iterated logarithm (see box on p. 223) gives the boundary which divides curves for which P_m is 1 for every m from those for which P_m approaches 0 as m increases; the former curves lie below the LIL boundary and are crossed infinitely often, whereas the latter curves lie above the LIL boundary and are crossed only finitely often. The LIL says nothing about how fast P_m

Treatment	Fair Coin 1	Fair Coin 2	Biased Coin 1	Biased Coin 2	Play-the-Winner 1	Play-the-Winner 2
1		F		F		F
2	S		S		S	
3	F		F		F	
4	S		S			S
5	S		S			F
6		S	S		S	
7		F		F	S	
8	S		S		S	
9	F			F	F	
10	S			F		F
11		F	F		S	
12	S		S		S	
13		F		F	S	
14		F		F	S	
15	S		S		S	
16	S			S	S	
17		F		F	S	
18		S	S		S	
19	S		S		S	
20	S			S	S	
21		F		F	F	
22		S	S			F
23	S			F	S	
24		F	S		S	
25	S		S		S	
26	S			F	F	
27	S		S			F
28	F			S	S	
29		F		S	S	
30	S		S		S	

Figure 4. Allocation of 30 patients between two medical treatments by three methods, as simulated using a table of random digits. The methods are coin toss randomization (left), the biased coin design (middle), and play-the-winner without observation delays (right). In this simulation it was assumed that Treatment 1 has probability 0.8 of being successful, and Treatment 2 probability 0.4. The same sequence of random digits was used to simulate success (S) or failure (F) under Treatment 1 in all three cases. Another set simulated the outcomes of Treatment 2, while a third sequence was used to make all random assignments. Coin toss randomization produced a somewhat unbalanced assignment (18 to 12). The biased coin design achieved equal group sizes, while play-the-winner assigned 24 patients to the superior Treatment 1. Inference is based on comparing the proportions of successes for the two treatments. These are 83% vs. 25% for the coin toss sample, 87% vs. 33% for the biased coin, and 83% vs. 17% for the play-the-winner. The superiority of Treatment 1 is clear in all cases.

approaches 0, either for the LIL boundary itself or for boundaries lying above it. Such questions of rates of convergence are attractive to mathematicians.

Darling and Robbins studied this problem, obtaining a wealth of results. Since the LIL boundary is the lowest boundary for which P_m approaches 0 at all, we suspect that approach to be slow. Darling and Robbins showed that in this case P_m is, for all m, less than a constant multiple of $1/\log\log m$. (That "$\log\log m$" is the logarithm of the logarithm of m, the "iterated logarithm" in the LIL. It grows, and its reciprocal therefore shrinks, very slowly indeed: the iterated logarithm of 10,000 is 2.2 and that of 100,000,000 is only 2.9.) This rate of convergence result, like the LIL itself, is true for many random paths other than random walks. In particular, it applies to the generalized random walks found in sequential analysis. Darling and Robbins gave more exact results for particular situations, and found much faster rates of convergence of P_m to 0 for boundaries only slightly above the LIL boundary.

This detailed study of boundary crossing probabilities is impressive "pure mathematics." Yet Darling and Robbins at once suggested an ingenious statistical application of their work. Suppose that in the ESP experiment we adopt the following criterion for deciding that the subject is not just guessing. Choose an m large enough to make P_m less than a fixed small number, say q. Information on the rate at which P_m approaches 0 allows us to do this. We will decide that the subject is not guessing if the path his responses produce crosses the boundary at some point beyond m. As long as the path remains below the boundary, we continue to take more observations.

The characteristics of this inference procedure are in striking contrast to those of more conventional statistical procedures, sequential or not. Statistical procedures generally control the probabilities of error; for example, the specific SPRT of Figure 3 has probability 5% of falsely deciding in favor of ESP and probability 10% of failing to detect a specific degree of ESP. These error probabilities can be made small by taking many observations, but there always remains a positive chance of both kinds of error. Not so in the Darling–Robbins sequential procedure. If the subject is guessing, the probability of crossing the boundary beyond m is no greater than q, which can be chosen as small as we wish. This is the probability of an error of the first kind. The second kind of error, failing to detect true ESP, will *never* occur. For if the probability of a correct call on each card exceeds $^1/_5$, the random walk drifts upward more rapidly, and the upper limit to its fluctuations given by the LIL moves upward correspondingly (see Figure 5). Hence the random walk of responses is now certain to cross the boundary infinitely often. Here is a recipe for a sequential procedure having one error probability as small as desired and the other error probability equal to 0.

This is not the proverbial free lunch. The usual sequential procedures, such as the SPRT, are sure to eventually stop and make a decision. The Darling–Robbins procedure is sure to stop sampling eventually if the subject is not guessing. But if he is guessing, we have probability $1 - q$ of continuing

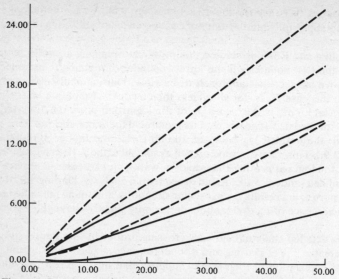

Figure 5. Increasing the probability of success on each trial changes the behavior of a random walk. The solid lines here represent the upward drift ($0.2n$ in n steps) and LIL boundaries for a sequence of trials having success probability $1/_5$. The dashed lines represent the drift and LIL boundaries if the success probability is increased to $2/_5$. The law of the iterated logarithm implies that the random walk in the second case is sure to cross the upper solid boundary. This is the basis of the Darling–Robbins "inference without error" procedure.

forever. This is the price of never making an error of the second kind. In applications such as acceptance sampling, this price is too high. It is always too high if observations have any cost to us. But suppose that someone else claims that a subject has ESP, and regularly publishes the outcomes of new card-calling trials by that subject. These observations cost us nothing. If the subject lacks ESP, we are content to watch the trial results as long as they continue, forever if necessary. We are secure in the knowledge that if the subject has ESP, however weak, we will eventually be convinced; if he does not, the Darling–Robbins procedure will give a false signal with probability no greater than q.

Searching for Optimality

A procedure for solving a statistical problem is "optimal" if it is, by some specific criterion, the best possible procedure for that problem. Suppose that in the case of independent ESP trials by a subject with unknown probability

p of guessing correctly we wish to distinguish $p = \frac{1}{5}$ from $p = \frac{2}{5}$. A possible criterion is to minimize the average sample size needed to make a decision when either $p = \frac{1}{5}$ or $p = \frac{2}{5}$ is true, subject to the condition that the probabilities of making the wrong decision must not exceed 5% when $p = \frac{1}{5}$ and 10% when $p = \frac{2}{5}$. The Wald–Wolfowitz theorem, which we mentioned earlier, states that in this situation the SPRT of Figure 3 is optimal.

Any claim of optimality must be judged relative to the formulation of the problem and to the criterion used. If, as is common in acceptance sampling, we are concerned that the procedure never require very large numbers of observations, the criterion of simply minimizing average sample size is no longer appropriate and the "optimality" of the SPRT is no longer directly relevant. The variability of the sample size is now also important. Moreover, the optimality of the SPRT applies to a simplified formulation of the ESP problem which considers only $p = \frac{1}{5}$ and $p = \frac{2}{5}$. The SPRT of Figure 3 need not have minimum sample size for detecting levels of ESP between guessing ($p = \frac{1}{5}$) and $p = \frac{2}{5}$. So the Wald–Wolfowitz theorem does not say that the SPRT is the "best possible" procedure in the actual conditions of use.

This fact does not, however, reduce that theorem to empty mathematics. Knowledge of the Wald–Wolfowitz result suggests that SPRT can be modified to obtain procedures which reduce the variability of the sample size without increasing the average sample size too much. How heavy a price in average sample size such procedures pay for their other advantages can be measured by comparison with the SPRT's smallest possible average sample size. Thus the optimality result establishes the SPRT as a standard and a point of departure in sequential analysis.

Searching for optimality is a reflection within statistics of the spirit of mathematics. Specific criteria for optimality are almost always oversimplified, as critics of this approach are quick to point out. Yet, as in the case of the SPRT, this mathematical spirit has yielded great practical benefits. Here then is a wide class of challenges for statistical theory. We must first formulate criteria which are realistic measures of performance, then strive to find procedures that are optimal by these criteria. Whether or not this search succeeds, there remains the work of comparing the performance of competing procedures which may offer other advantages to offset lack of optimality. In situations such as clinical trials, where no one criterion summarizes all of the statistical, practical, and ethical considerations present, assessment by several criteria is the only way to insure an informed choice.

Optimality problems in sequential analysis require the choice of both a stopping rule and a decision rule. It can be shown that in many circumstances such a problem can be reduced to two stages: first choose an "optimal stopping rule," then for each possible number of observations study the statistical decision problem as if the number of observations taken before stopping had been fixed in advance. Optimality problems involving only stopping rules are called optimal stopping problems. Such a problem is formally stated as follows: we have a sequence of observations $y_1, y_2, \ldots$, and must give a rule which decides when to stop observing. If we stop after n observations, we pay a penalty x_n which is a known function of $y_1, \ldots, y_n$. Our goal is to give a stopping rule which minimizes the average penalty we will pay if we use the same procedure repeatedly.

Since the sequence of observations $y_1, y_2, \ldots$ can be represented by a graph of the sums $y_1 + y_2 + \ldots + y_n$ against n, many optimal stopping problems can be visualized in terms of choosing boundaries on such a graph. For example, Darling and Robbins showed that many boundaries close to that given by the law of the iterated logarithm have the properties needed to implement their "inference without error" scheme. A reasonable criterion for the "best" such boundary among all those giving a false declaration of ESP with probability q or less is the one which the random walk crosses most quickly (on the average) when the subject truly has ESP. For all such boundaries are certain to detect true ESP, and will cry wolf only with probability q, leaving the number of observations required to detect true ESP as the "penalty." Finding a "best possible" boundary is a harder mathematical problem than studying the characteristics of a particular boundary. In this example, no optimal boundary has been found.

Certain optimal stopping problems on which mathematicians have made much progress are those known as "secretary problems." In the simplest such problem, there is a pool of N applicants for a secretarial position, who rank from 1 (best) to N (worst). These applicants approach the employer in random order, so that the best (or any other rank) is equally likely to be the first, second, or last applicant interviewed. From each interview, the employer can rank that secretary relative to those already interviewed. This is the only information available; the overall ranking of the candidates cannot be known until all N have been interviewed. But the employer must either hire or reject each applicant at the end of the interview, and cannot recall earlier applicants. We must find a procedure that is most likely to select the best secretary. In this optimal stopping problem, the observation y_n is the relative rank of the nth secretary interviewed among the first n interviewees. It is a whole number between 1 and n. The penalty x_n for stopping at step n is zero if the nth applicant has overall rank 1, and one if this applicant is not the best in the pool. That is, we are penalized equally for selecting any but the top applicant.

This is a finite problem. Since N steps will exhaust the pool of applicants, there are only finitely many possible stopping rules. In principle (but not in practice when N is large) we could find the optimal stopping rule by studying each possible rule individually. In a finite problem it is at least clear that there *is* a best possible procedure. When an unlimited sequence of observations can be taken, it sometimes happens that there is an infinite sequence of stopping rules whose average penalties approach the smallest possible value (or even grow more negative without any bound), but no optimal rule. Such possibilities make infinite problems more difficult.

There is a simple way of actually constructing the optimal stopping rule in the basic secretary problem, called the method of backward induction. Suppose first that we are required to interview all N applicants. This forces us to hire the last, with probability $1/_N$ that the best applicant happened to be the last interviewed. Now move back one step, and suppose that we are allowed to stop either after N-1 or after N interviews. If the $(N$-1)st candidate does not rank first among the first N-1, stopping there has *no* chance of choosing the top secretary, since candidate N-1 is not as good as an earlier candidate. So we should go on, accepting the probability $1/_N$ that candidate N will be better than all others. But if candidate N-1 *is* the best of those interviewed so far, the situation changes. The remaining candidate may be still better than any yet seen, but it is not likely: the probability of this happening is only $1/_N$. That leaves probability $1 - (1/_N)$ that candidate N-1 is the best of all. So if N, the number of candidates, is at least 3, we have a higher probability of success by hiring candidate N-1. The backward induction procedure continues by next allowing us to stop after N-2, N-1, or N interviews and comparing the chance of success after N-2 interviews with that for continuing. In this step-by-step way we can construct the optimal stopping rule for any N. It has this form: allow a certain number of candidates to pass, then hire the first who is best among those already interviewed (see Figure 6). When N is large, the optimal rule passes by about the first 37% of the candidates, and chooses the best candidate about 37% of the time. This is the best we can hope to do under the given circumstances. The optimal rule depends on knowledge of N, the total number of candidates. Fortunately, rules which pass approximately 37% of the candidates have nearly optimal success rates, so that near-optimal rules can be found when N is only roughly known.

The secretary problem is a potential model for economic decision problems such as accepting or rejecting successive job offers or successive bids on a piece of property. As stated, the model is unrealistic for these purposes, since it assumes that no information other than the relative ranks of the successive candidates is available. Economic problems usually include information about the "market" which can be used to assess the excellence of a candidate. Such prior information creates a quite different optimal stopping problem, one for which the best rules do not always let the first candidates go by to build up information, as does the optimal rule in the secretary problem. Another modification of the secretary problem, which is more realistic in some situations, is to replace the "all or nothing" goal of choosing the best (first ranked) secretary by the goal of keeping the rank of the candidate hired as low as possible. This is not so major a modification. The optimal rule has been found; as might be guessed, it reduces the chance of being left with the Nth candidate. What is surprising is that the optimal rule is able to choose (on the average) candidates with very low ranks. When there are $N = 100$ candidates, the average rank of the candidate chosen by the optimal rule in many repetitions is 3.6; when $N = 1000$, it is 3.83. Choosing

Figure 6. The optimal stopping rule for the secretary problem with $N = 4$ applicants passes over the first applicant, then chooses the first (if any) who is better than all preceding candidates. The table below shows all possible orders in which the candidates can appear for interviews:

1 2 3 **4**	*2 1 3 4	*3 1 2 4	*4 1 2 3
1 2 **4** 3	*2 1 4 3	*3 1 4 2	*4 1 3 2
1 3 2 **4**	*2 3 1 4	3 2 1 4	4 2 1 3
1 3 4 **2**	*2 3 4 1	3 2 4 1	4 2 3 1
1 4 2 3	*2 4 1 3	*3 4 1 2	4 3 1 2
1 4 3 **2**	*2 4 3 1	3 4 2 1	4 3 2 1

We assume that Candidate 1 is the best, Candidate 4 the worst; boldface type indicates which applicant is selected by the optimal stopping rule for each possible order of interview. An asterisk marks each order for which the optimal rule chooses the best applicant; this occurs in 11 of 24 cases. Since the 24 orders are equally likely to occur, the optimal rule succeeds with probability $^{11}/_{24}$. This is a substantial improvement over simply guessing among the candidates, which succeeds with probability $^1/_4$.

(on the average) one of the top 4 of 1000 candidates is excellent performance.

Mathematical study of finite optimal stopping problems has produced optimal rules for many models. Infinite problems and sequential decision problems are less well understood, though progress continues. It is reasonable to expect that the work of mathematicians will eventually allow optimal decision rules to join sequential analysis and statistical design of experiments as useful aids to practical decision making.

Suggestions for Further Reading

General

Cochran, William G. Early development of techniques in comparative experimentation. In D.B. Owen, *On the History of Statistics and Probability*. Marcel Dekker, New York, 1976, pp. 3–25.

A relatively nontechnical historical account which places Fisher's work in its context of nineteenth and early twentieth century methods for agricultural field trials.

Tanur, Judith M., et al. (Eds.). *Statistics, A Guide to the Unknown*. Holden-Day, San Francisco, 1972.

A collection of excellent essays on applications of statistics. The best nontechnical book available on statistics. Unfortunately, complete avoidance of mathematics leaves the essays very light on inference.

Wallis, W. Allen and Roberts, Harry V. *The Nature of Statistics*. Free Pr, Glencoe, 1962.

Although old, this is better than most recent expositions emphasizing concepts over techniques. Most other popular expositions take the "how to lie" approach rather than looking at statistics as a science.

Technical

Feller, William. *An Introduction to Probability Theory and Its Applications*, Vol. 1, Third Edition. Wiley, New York, 1968.

This classic text contains an account of random walks and the law of the iterated logarithm. It is a beautiful book often read for pleasure by nonspecialists with college-level preparation in mathematics.

Gilbert, John P. and Mosteller, Frederick. Recognizing the maximum of a sequence. *Journal of the American Statistical Association* 61 (1966) 35–73.

Varieties of the secretary problem, with more exposition than journal editors usually allow.

What is a Computation?

Martin Davis

On numerous occasions during the Second World War, members of the German high command had reason to believe that the allies knew the contents of some of their most secret communications. Naturally, the Nazi leadership was most eager to locate and eliminate this dangerous leak. They were convinced that the problem was one of treachery. The one thing they did not suspect was the simple truth: the British were able to systematically decipher their secret codes. These codes were based on a special machine, the "Enigma," which the German experts were convinced produced coded messages that were entirely secure. In fact, a young English mathematician, Alan Turing, had designed a special machine for the purpose of decoding messages enciphered using the Enigma. This is not the appropriate place to speculate on the extent to which the course of history might have been different without Turing's ingenious device, but it can hardly be doubted that it played an extremely important role.

In this essay we will discuss some work which Alan Turing did a few years before the Second World War whose consequences are still being developed. What Turing did around 1936 was to give a cogent and complete logical analysis of the notion of "computation." Thus it was that although people have been computing for centuries, it has only been since 1936 that we have possessed a satisfactory answer to the question: "What is a computation?" Turing's analysis provided the framework for important mathematical investigations in a number of directions, and we shall survey a few of them.

Turing's analysis of the computation process led to the conclusion that it should be possible to construct "universal" computers which could be programmed to carry out any possible computation. The existence of a logical analysis of the computation process also made it possible to show that

Alan M. Turing

Alan M. Turing was born in 1912, the second son in an upper class English family. After a precocious childhood, he had a distinguished career as a student at Cambridge University. It was shortly after graduation that Turing published his revolutionary work on computability. Turing's involvement in the deciphering of German secret codes during the Second World War has only recently become public knowledge. His work has included important contributions to mathematical logic and other branches of mathematics. He was one of the first to write about the possibility of computer intelligence and his writings on the subject are still regarded as fundamental. His death of cyanide poisoning in June 1954 was officially adjudged suicide.

certain mathematical problems are incapable of computational solution, that they are, as one says, *unsolvable*. Turing himself gave some simple examples of unsolvable problems. Later investigators found that many mathematical problems for which computational solutions had been sought unsuccessfully for many years were, in fact, unsolvable. Turing's logical proof of the existence of "universal" computers was prophetic of the modern all-purpose digital computer and played a key role in the thinking of such pioneers in the development of modern computers as John von Neumann. (Likely these ideas also played a role in Turing's seeing how to translate his cryptographic work on the German codes into a working machine.) Along with the development of modern computers has come a new branch of applied mathematics: *theory of computation*, the application of mathematics to the theoretical understanding of computation. Not surprisingly, Turing's analysis of computation has played a pivotal role in this development.

Although Turing's work on giving a precise explication of the notion of computation was fundamental because of the cogency and completeness of his analysis, it should be stated that various other mathematicians were independently working on this problem at about the same time, and that a number of their formulations have turned out to be logically equivalent to that of Turing. In fact the specific formulation we will use is closest to one originally due to the American mathematician Emil Post.

The Turing—Post Language

Turing based his precise definition of computation on an analysis of what a human being actually does when he computes. Such a person is following a set of rules which must be carried out in a completely mechanical manner. Ingenuity may well be involved in setting up these rules so that a computation may be carried out efficiently, but once the rules are laid down, they must be carried out in a mercilessly exact way. If we watch a human being calculating something (whether he is carrying out a long division, performing an algebraic manipulation, or doing a calculus problem), we observe symbols being written, say on a piece of paper, and the behavior of the person doing the calculating changing as he notes various specific symbols appearing as results of computation steps.

The problem which Turing faced and solved was this: how can one extract from this process what is essential and eliminate what is irrelevant? Of course some things are clearly irrelevant; obviously it does not matter whether our calculator is or is not drinking coffee as he works, whether he is using pencil or pen, or whether his paper is lined, unlined, or quadruled. Turing's method was to introduce a series of restrictions on the calculator's behavior, each of which could clearly be seen to be inessential. However, when he was done all that was left were a few very simple basic steps performed over and over again many times.

We shall trace Turing's argument. In the first place, he argued that we can restrict the calculator to write on a linear medium, that is, on a tape, rather

Emil L. Post

Emil L. Post was born in Poland in 1897, but arrived in New York City at the age of seven, and lived there for the remainder of his life. His life was plagued by tragic problems: he lost his left arm while still a child and was troubled as an adult by recurring episodes of a disabling mental illness. While still an undergraduate at City College he worked out a generalization of the differential calculus which later turned out to be of practical importance. His doctoral dissertation at Columbia University initiated the modern metamathematical method in logic. His researches while a postdoctoral fellow at Princeton in the early 1920's anticipated later work of Gödel and Turing, but remained unpublished until much later, partly because of the lack of a receptive atmosphere for such work at the time, and partly because Post never completed the definitive development he was seeking. His work in computability theory included the independent discovery of Turing's analysis of the computation process, various important unsolvability results, and the first investigations into degrees of unsolvability (which provide a classification of unsolvable problems). He died quite unexpectedly in 1954 while under medical care.

than on a two-dimensional sheet of paper. Instead of paper tape (such as is used in an adding machine) we can, if we prefer, think of magnetic tape as used in a tape recorder. (Of course, in this latter case, the symbols occur as magnetic signals rather than as marks on paper, but conceptually this makes no difference whatsoever.) It is easy to convince oneself that the use of a two-dimensional sheet of paper plays no essential role in the computational process and that we really are not giving up any computational power by restricting ourselves to a linear tape. Thus the "two-dimensional" multiplication:

$$\begin{array}{r} 26 \\ \times\ 32 \\ \hline 52 \\ 780 \\ \hline 832 \end{array}$$

can be written on a "tape" as follows:

$$26 \times 32 = 52 + 780 = 832.$$

We suppose that the linear tape is marked off into individual squares and that only one symbol can occupy a square. Again, this is a matter of convenience and involves no particular limitations. So, our multiplication example might look like this:

| 2 | 6 | × | 3 | 2 | = | 5 | 2 | + | 7 | 8 | 0 | = | 8 | 3 | 2 |

The next restriction we impose (here we are actually going a bit further than Turing did) is that the only symbols which may appear on our tape are 0 and 1. Here we are merely making use of the familiar fact that all information can be "coded" in terms of two symbols. It is this fact, for example, which furnishes the basis for Morse code in which the letters of the alphabet are represented as strings of "dots" and "dashes." Another example is binary arithmetic which forms the basis of modern digital computation.

Our next restriction has to do with the number of different symbols our calculator can take note of (or as we shall say, "scan") in a single observation. How many different symbols can a human calculator actually take in at one time? Certainly no one will be able to take in at a glance the distinction between two very long strings of zeros and ones which differ only at one place somewhere in the middle. One can take in at a glance, perhaps, five, six, seven, or eight symbols. Turing's restriction was more drastic. He assumed that in fact one can take in only a single symbol at a glance. To see that this places no essential restriction on what our calculator can accomplish, it suffices to realize that whatever he does as a result of scanning a group of, say, five symbols can always be broken up into separate operations performed viewing the symbols one at a time.

What kinds of things can the calculator actually do? He can replace a 0 by a 1 or a 1 by a 0 on the square he is scanning at any particular moment, or he can decide to shift his attention to another square. Turing assumed that this shifting of attention is restricted to a square which is the immediate neighbor, either on the left or on the right, of the square previously scanned. Again, this is obviously no essential restriction: if one wants to shift one's attention to a square three to the right, one simply shifts one to the right three successive times. Also the calculator may observe the symbol in the square being scanned and make a decision accordingly. And presumably this decision should take the form: "Which instruction shall I carry out next?" Finally, the calculator may halt, signifying the end of the computation.

To summarize: any computation can be thought of as being carried out by a human calculator, working with strings of zeros and ones written on a linear tape, who executes instructions of the form:

> Write the symbol 1
> Write the symbol 0
> Move one square to the right
> Move one square to the left
> Observe the symbol currently scanned and choose the next step accordingly
> Stop

The procedure which our calculator is carrying out then takes the form of a list of instructions of these kinds. As in modern computing practice, it is convenient to think of these kinds of instructions as constituting a special *programming language*. A list of such instructions written in this language is then called a *program*.

We are now ready to introduce the Turing–Post Programming Language. In this language there are seven kinds of instructions:

> PRINT 1
> PRINT 0
> GO RIGHT
> GO LEFT
> GO TO STEP i IF 1 IS SCANNED
> GO TO STEP i IF 0 IS SCANNED
> STOP

A Turing–Post program is then a list of instructions, each of which is of one of these seven kinds. Of course in an actual program the letter i in a step of either the fifth or sixth kind must be replaced by a definite (positive whole) number.

In order that a particular Turing–Post program begin to calculate, it must have some "input" data. That is, the program must begin scanning at a specific square of a tape already containing a sequence of zeros and ones. The

1. PRINT 0
2. GO LEFT
3. GO TO STEP 2 IF 1 IS SCANNED
4. PRINT 1
5. GO RIGHT
6. GO TO STEP 5 IF 1 IS SCANNED
7. PRINT 1
8. GO RIGHT
9. GO TO STEP 1 IF 1 IS SCANNED
10. STOP

Figure 1. Doubling Program

symbol 0 functions as a "blank"; although the entire tape is infinite, there are never more than a finite number of ones that appear on it in the course of a computation. (A reader who is disturbed by the notion of an infinite tape can replace it for our purposes by a finite tape to which blank squares—that is, squares filled with zeros—are attached to the left or the right whenever necessary.)

Figure 1 exhibits a Turing–Post program consisting of ten instructions which we will use repeatedly for illustrative purposes. The presence of the "GO TO" instruction makes it possible for the same instruction to be executed over and over again in the course of a single computation. This can be seen in some detail in Figure 2 which shows the successive steps in one particular computation by the program of Figure 1. The computation is completely determined by the initial arrangement of symbols on the tape together with a specification of which square is initially scanned. In Figure 2 this latter information is given by an upward arrow (↑) below the scanned square. (Of course only a finite number of symbols from the tape can actually be explicitly exhibited; in Figure 3, we exhibit six adjacent symbols, and assume that all squares not explicitly shown are blank, that is contain the symbol 0.) Such combined information, consisting of the symbols on the tape (pictorially represented by showing a finite number of consecutive squares, the remainder of which are presumed to be blank) and the identity of the scanned square (designated by an arrow just below it) is called a *tape configuration.*

Figure 2 gives a list of such tape configurations, with the initial configuration at the top, each of which is transformed by an appropriate step of the program (from Figure 1) into the configuration shown below it. The program steps are listed alongside the tape configurations. The computation begins by executing the first step (which in our case results in replacing the 1 on the scanned square by 0) and continues through the successive steps of the program, except as "GO TO" instructions cause the computation to return to earlier instructions. Ultimately, Step 9 is executed with the tape configuration as shown at the bottom of Figure 2. Since 0 is being scanned, the computation continues to Step 10 and then halts.

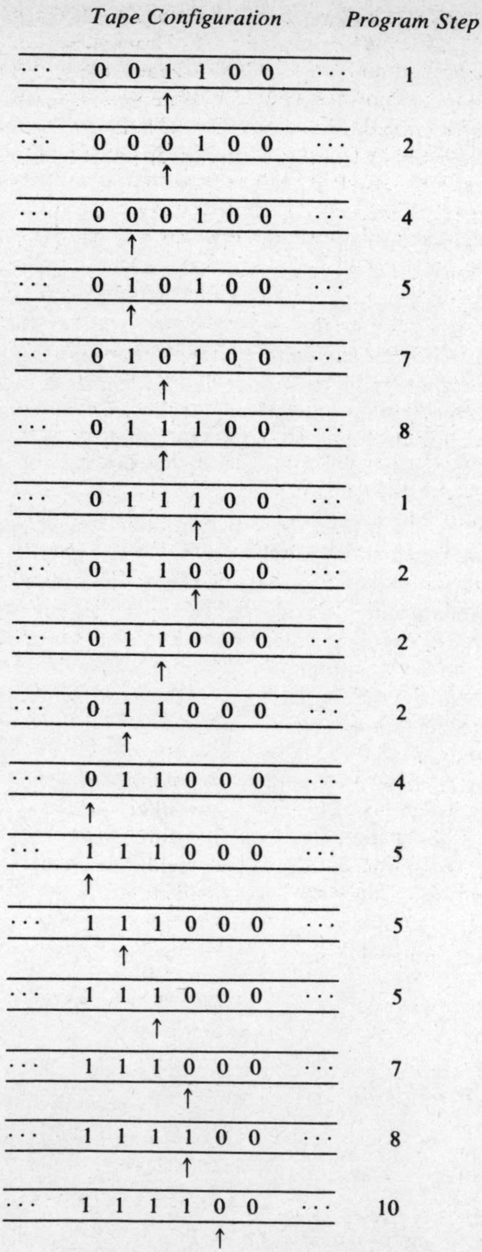

Tape Configuration	Program Step
· · · 0 0 1 1 0 0 · · · ↑	1
· · · 0 0 0 1 0 0 · · · ↑	2
· · · 0 0 0 1 0 0 · · · ↑	4
· · · 0 1 0 1 0 0 · · · ↑	5
· · · 0 1 0 1 0 0 · · · ↑	7
· · · 0 1 1 1 0 0 · · · ↑	8
· · · 0 1 1 1 0 0 · · · ↑	1
· · · 0 1 1 0 0 0 · · · ↑	2
· · · 0 1 1 0 0 0 · · · ↑	2
· · · 0 1 1 0 0 0 · · · ↑	2
· · · 0 1 1 0 0 0 · · · ↑	4
· · · 1 1 1 0 0 0 · · · ↑	5
· · · 1 1 1 0 0 0 · · · ↑	5
· · · 1 1 1 0 0 0 · · · ↑	5
· · · 1 1 1 0 0 0 · · · ↑	7
· · · 1 1 1 1 0 0 · · · ↑	8
· · · 1 1 1 1 0 0 · · · ↑	10

Figure 2. Steps in a Computation by Doubling Program

Given an alphabet of three symbols a, b, c, and three equations

$$ba - abc$$
$$bc = cba$$
$$ac = ca$$

we can obtain other equations by substitution:

bac $= abcc$
bac $=$ **bc**$a = c$**ba**$a = ca$**b**$ca = cab$**c**$a = a$**c**$bca = \ldots$
$\qquad\qquad or = cab$**c**$a = cab$**a**$c = \ldots$
$\qquad\qquad or = ca$**b**$ca = cacbaa = \ldots$

(The expressions in boldface type are the symbols about to be replaced.) In this context can be raised questions such as: "Can we deduce from the three equations listed above that $bacabca = acbca$?" The word problem defined by the three equations is the general question: to determine of an arbitrary given equation between two words, whether or not it can be deduced from the three given equations.

Figure 3. A Word Problem

The computation shown in Figure 2 begins with two ones on the tape and ends with four. It is because this happens generally that we call the program in Figure 1 a "doubling program." To put it precisely: beginning with a tape configuration the nonblank portion of which consists of a row of ones with the scanned square containing the leftmost of the ones, the doubling program will eventually halt with a block of twice as many ones on the tape as were there to begin with. It is by no means obvious at a glance (even to an experienced computer programmer) that our doubling program really behaves in the manner just stated. Readers who want to understand how the doubling program works may examine the "flow chart" in the box on p. 250.

The fact that this doubling program is so short and accomplishes such a simple task should not be permitted to obscure the point of Turing's analysis of the computation process: we have reason to be confident that any computation whatsoever can be carried out by a suitable Turing–Post program.

As we have seen, once a STOP instruction is executed, computation comes to a halt. If, however, no STOP instruction is ever encountered in the course of a computation, the computation will (in principle, of course) continue forever. The question "When can we say that a computation will eventually halt?" will play a crucial role later in our discussion. To see how this can be answered in a simple example, consider the following three-step Turing–Post program:

1. GO RIGHT
2. GO TO STEP 1 IF 0 IS SCANNED
3. STOP

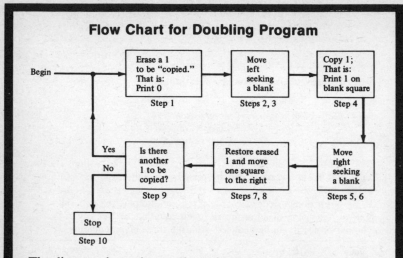

Flow Chart for Doubling Program

The diagram above shows schematically how the doubling program listed in Figure 1 actually works. The underlying idea is to double the number of ones by simply copying them one at a time. Each 1 to be copied is (temporarily) replaced by a 0 which acts as a place marker (Step 1). Next the computation moves left over all the ones (which as the computation progresses will include newly printed ones) seeking the first unused (i.e., blank) square (Steps 2, 3—which will be repeated over and over again until a blank square is encountered). The 1 is now copied (Step 4). Next the computation returns rightward until it encounters the 0 which takes the place of the 1 which has just been copied (Steps 5,6—which again are repeated). The copied 1 is restored (Step 7). The computation moves one square to the right seeking another 1 to copy (Step 8). If there is another 1 to be copied the computation goes back to Step 1; otherwise it advances to Step 10 and halts (Steps 9, 10).

This program will halt as soon as motion to the right reaches a square containing the symbol 1. For once that happens the program will move on to Step 3 and halt. That being the case, suppose we begin with a tape on which there are no ones to the right of the initially scanned square. (For example, the entire tape could be blank or there could be some ones but all to the left of the initially scanned square.) In this case, the first two steps will be carried out over and over again forever, since a 1 will never be encountered. After step 2 is performed, step 1 will be performed again. This makes it clear that a computation from a Turing–Post program *need not actually ever halt*. In the case of this simple three-step program it is very easy to tell from the initial tape configuration whether the computation will eventually halt or continue forever: to repeat, if there is a 1 to the right of the initially scanned square the computation will eventually halt; whereas if there are only blanks to the right of the initially scanned square the computation will continue forever. We shall see later that the question of predicting whether a particular Turing–Post program will eventually halt contains surprising subtleites.

Codes for Turing – Post Programs

All of the dramatic consequences of Turing's analysis of the computation process proceed from Turing's realization that it is possible to encode a Turing–Post program by a string of zeros and ones. Since such a string can itself be placed on the tape being used by another (or even the same) Turing–Post program, this leads to the possibility of thinking of Turing–Post programs as being capable of performing computations on other Turing–Post programs.

There are many ways by which Turing–Post programs can be encoded by strings of zeros and ones. We shall describe one such way. We first represent each Turing–Post instruction by an appropriate sequence of zeros and ones according to the following code:

Code	Instruction
000	PRINT 0
001	PRINT 1
010	GO LEFT
011	GO RIGHT
$10\underbrace{10\ldots01}_{i}$	GO TO STEP i IF 0 IS SCANNED
$11\underbrace{01\ldots10}_{i}$	GO TO STEP i IF 1 IS SCANNED
100	STOP

This table gives the representation of each Turing–Post instruction by a string of zeros and ones. For example the code for the instruction

GO TO STEP 3 IF 0 IS SCANNED

is: 1010001. To represent an entire program, we simply write down in order the representation of each individual instruction and then place an additional 1 at the very beginning and 111 at the very end as punctuation marks.

For example here is the code for the doubling program shown in Figure 1:

100001011011000101111011111000101111010100111

To make this clear, here is the breakdown of this code:

Begin	Step	Step	Step	Step	Step	Step	Step	Step	Step	Step	End
	1	2	3	4	5	6	7	8	9	10	
1	000	010	110110	001	011	110111110	001	011	11010	100	111

It is important to notice that the code of a Turing–Post program can be deciphered in a unique, direct, and straightforward way, yielding the program of which it is the code. First remove the initial 1 and the final 111 which are just punctuation marks. Then, proceeding from left to right, mark off the first group of 3 digits. If this group of 3 digits is 000, 001, 010, 011, or 100 the corresponding instructoin is: PRINT 0, PRINT 1, GO LEFT, GO RIGHT, or STOP, respectively. Otherwise the group of 3 digits is 101 or 110, and the first instruction is a "GO TO." The code will then have one of the forms:

$$10\underbrace{100\ldots01}_{i} \qquad 11\underbrace{011\ldots10}_{i}$$

corresponding to

GO TO STEP i IF 0 IS SCANNED

and

GO TO STEP i IF 1 IS SCANNED

respectively. Having obtained the first instruction, cross out its code and continue the process, still proceeding from left to right. Readers who wish to test their understanding of this process may try to decode the string:

$$10100011010011000001010101010111$$

The Universal Program

We are now ready to see how Turing's analysis of the computation process together with the method for coding Turing–Post programs we have just introduced leads to a conclusion that at first sight seems quite astonishing. Namely, there exists a single (appropriately constructed) Turing–Post program which can compute anything whatever that is computable. Such a program U (for "universal") can be induced to simulate the behavior of any given Turing–Post program P by simply placing *code* (P), the string of zeros and ones which represents P, on a tape and permitting U to operate on it. More precisely, the non-blank portion of the tape is to consist of *code* (P) followed by an input string v on which P can work. (For clarity, we employ capital letters to stand for particular Turing–Post programs and lowercase

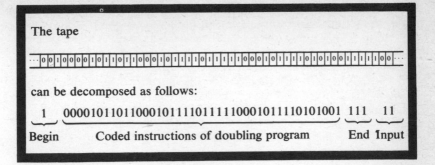

The tape

can be decomposed as follows:

$\underbrace{1}_{\text{Begin}}$ $\underbrace{0000101101100010111101111110001011110101001}_{\text{Coded instructions of doubling program}}$ $\underbrace{111}_{\text{End}}$ $\underbrace{11}_{\text{Input}}$

letters to stand for strings of zeros and ones.) For example, the string shown in the box above signifies that U should simulate the behavior of the doubling program when 11 is the input. Thus, at the end of the computation by U, the tape should look just like the final tape in Figure 2.

Now, a universal Turing–Post program U is supposed to perform in this way not only for our doubling program, but for *every* Turing–Post program. Let us be precise: U is to begin its computation presented with a tape whose nonblank portion consists of *code* (P) for some Turing–Post program P (initially scanning the first symbol, necessarily 1, of this code) followed by a string v. U is then supposed to compute exactly the same result as the program P would get when starting with the string v as the nonblank part of the tape (scanning the initial symbol of v). Such a program U can then be used to simulate any desired Turing–Post program P by simply placing the string *code* (P) on the tape.

What reason do we have for believing that there is such a program U? To help convince ourselves, let us begin by thinking how a human calculator could do what U is supposed to do. Faced with the tape contents on which U is supposed to work, such a person could begin by scanning this string of zeros and ones, from left to right, searching for the first place that 3 consecutive ones appear. This triple 111 marks the end of *code* (P) and the beginning of the input string. Our human calculator can then write *code* (P) on one sheet of paper and the input string on another. As already explained, he can decode the string *code* (P) and obtain the actual Turing–Post program P. Finally, he can "play machine," carrying out the instruction of P, applied to the given input string in a robotlike fashion. If and when the computation comes to a halt, our calculator can report the final tape contents as output. This shows that a human calculator can do what we would like U to do. But now, invoking Turing's analysis of the computation process, we are led to believe that there must be a Turing–Post program which can carry out the process we have just described, a universal Turing–Post program.

The evidence we have given for the existence of such a program is rather unsatisfactory because it depends on Turing's analysis of the computation process. It certainly is not a mathematical proof. But in fact, if one is willing to do some tedious but not very difficult work, one can circumvent the need to refer to Turing's analysis at all and can, in fact, write out in detail an explicit universal Turing–Post program. This was done in fact by Turing himself (in a slightly different, but entirely equivalent context) in his fundamental 1936 paper. And subsequently, it has been redone many times. The success of the construction of the universal program is in itself evidence for the correctness of Turing's analysis. It is not appropriate here to carry out the construction of a universal program in detail; we hope, merely, that the reader is convinced that such a program exists. (Experienced computer programmers will have no difficulty in writing their own universal program if they wish to do so.)

We have conceived of Turing–Post programs as consisting of lists of written instructions. But clearly, given any particular Turing–Post program P, it would be possible to build a machine that would actually carry out the instructions of P in sequence. In particular, this can be done for our universal program U. The machine we get in this way would be an example of an all-purpose or universal computing machine. The code for a particular program P placed on its tape could then be thought of as a "program" for doing the computation which P does. Thus, Turing's analysis leads us, in a very straightforward manner, to the concept of an all-purpose computer which can be programmed to carry out any computation whatever.

The Halting Problem

We are now in a position to demonstrate a truly astonishing result: we are able to state a simple problem, the so-called *halting problem*, for which we can prove that no computational solution exists.

The halting problem for a particular Turing–Post program is the problem of distinguishing between initial tape configurations which lead to the program's eventually halting and initial tape configurations which lead the program to compute forever. We saw above that certain input strings may cause a particular program to run forever, due to an infinite loop caused by the "GO TO" instruction. It would surely be desirable to have a method for determining in advance which input data leads the program to halt and which does not. This is the halting problem: given a particular Turing–Post program, can we computationally test a given tape configuration to see whether or not the program will eventually halt when begun with that tape configuration.

The answer is no. *There is no computaton procedure for testing a given tape expression to determine whether or not the universal program U will eventually halt when begun with that tape configuration.* The fact that there is no such procedure for the universal program shows of course that there can't be such procedures for Turing–Post programs in general, since the universal program is itself a Turing–Post program. Before we see how this un-solvability theorem can be proved, it is worthwhile to reflect on how exciting and remarkable it is that it should be possible to prove such a result. Here is a problem which is easy to state and easy to understand which we *know* can-not be solved. Note that we are not saying simply that we don't know how to solve the problem, or that the solution is difficult. We are saying: *there is no solution.*

Readers may be reminded of the fact that the classical problems of angle trisection and circle squaring also turned out to have no solution. This is a good analogy, but with a very significant difference: the impossibility proofs for angle trisection and circle squaring are for constructions using specific in-struments (straightedge and compass); using more powerful instruments, there is no difficulty with either of these geometric construction problems. Matters are quite different with the halting problem; here what we will show is that there is no solution using any methods available to human beings.

The proof of the unsolvability of the halting problem is remarkably simple. It uses the method known as indirect proof or *reductio ad absurdum*. That is, we suppose that what is stated in italics above is false, that in fact, we possess a computing procedure which, given an initial tape configuration will enable us to determine whether or not the universal program will eventually halt when started in that configuration. Then we show that this supposition is impossible; this is done in the box on p. 256.

Other Unsolvable Problems

In the 1920's the great German mathematician David Hilbert pointed to a certain problem as the fundamental problem of the newly developing field of mathematical logic. This problem, which we may call the decision problem for elementary logic, can be explained as follows: a finite list of statements called *premises* is given together with an additional statement called the *conclusion*. The logical structure of the statements is to be explicitly exhibited in terms of "not," "and," "or," "implies," "for all," and "there exists." Hilbert wanted a computing procedure for testing whether or not the conclusion can be deduced using the rules of logic from the premises. Hil-bert regarded this problem as especially important because he expected that its solution would lead to a purely mechanical technique for settling the truth or falsity of the most diverse mathematical statements. (Such statements could be taken as the conclusion, and an appropriate list of axioms as the premises to which the supposed computing procedure could be applied.) Thus the very existence of an unsolvable mathematical problem (in particu-lar, the halting problem) immediately suggested that Hilbert's decision problem for elementary logic was itself unsolvable. This conclusion turned out to be correct, as was carefully shown by Turing and, quite indepen-dently, by the American logician Alonzo Church. Turing represented the

Unsolvability of Halting Problem

Suppose we possess a computing procedure which solves the halting problem for the universal program U. Then we can imagine more complicated procedures of which this supposed procedure is a part. Specifically, we consider the following procedure which begins with a string v of zeros and ones:

1. Try to decode v as the code for a Post–Turing program, i.e., try to find P with $code\ (P) = v$. If there is no such P, go back to the beginning of Step 1; otherwise go on to Step 2.

2. Make a copy of v and place it to the right of v getting a longer string which we can write as vv (or equivalently as $code\ (P)v$ since $code\ (P) = v$).

3. Use our (pretended) halting problem procedure to find out whether or not the universal program U will eventually halt if it begins with this string vv as the nonblank portion of the tape, scanning the leftmost symbol. If U will eventually halt, go back to the beginning of Step 3; otherwise stop.

This proposed procedure would eventually stop if, first, $v = code(P)$ for some Turing–Post program P (so we will leave Step 1 and go on to Step 2), and, second, if also U will never halt if it begins to scan the leftmost symbol of vv. Since U beginning with $code\ (P)v$ simulates the behavior of P beginning with v, we conclude that our supposed procedure applied to the string v will eventually stop if and only if $v = code(P)$ where P is a computing procedure that will never stop beginning with v on its tape.

By Turing's analysis, there should be a Turing–Post program P_0 which carries out this very procedure. That is, P_0 will eventually halt beginning with the input v if and only if U will never halt beginning with the input vv. Now let $v_0 = code(P_0)$. Does U eventually halt beginning with the input v_0v_0? By what we have just said, P_0 *eventually halts beginning with the input v_0 if and only if U will never halt beginning with the input v_0v_0.* But, as we will show, this contradicts our explanation of how U works as a universal program. Since $v_0 = code(P_0)$, U will act, given the input v_0v_0, to simulate the behavior of P_0 when begun on input v_0. So *U will eventually halt beginning with the input v_0v_0 if and only if P_0 will eventually halt beginning with the input v_0.* But this contradicts the previous italicized statement. The only way out of this contradiction is to conclude that what we were pretending is untenable. In other words, the halting problem for U is not solvable.

theory of Turing–Post programs in logical terms and showed that a solution to the decision problem for elementary logic would lead to a solution of the halting problem. (This connection between logic and programs was redis-covered many years later and now forms the basis for certain investigations into the problem of proving the correctness of computer programs.)

The unsolvability of the decision problem for elementary logic was impor-tant, not only because of the particular importance of this problem, but also because (unlike the halting problem) it was an unsolvable problem that peo-ple had actually tried to solve. A decade went by before another such ex-ample turned up. Early in the century the Norwegian Axel Thue had emphasized the importance of what are now called "word problems." In 1947, Emil Post showed how the unsolvability of the halting problem leads to the existence of an unsolvable word problem. Post's proof is discussed in the box on p. 258. Here we merely explain what a word problem is.

In formulating a word problem one begins with a (finite) collection, called an *alphabet*, of symbols, called *letters*. Any string of letters is called a *word* on the alphabet. A word problem is specified by simply writing down a (fi-nite) list of equations between words. Figure 3 exhibits a word problem specified by a list of 3 equations on the alphabet a, b, c. From the given equations many other equations may be derived by making substitutions in any word of equivalent expressions found in the list of equations. In the ex-ample of Figure 3, we derive the equation $bac = abcc$ by replacing the part ba by abc as permitted by the first given equation.

We have explained how to *specify* the data for a word problem, but we have not yet stated what the problem is. It is simply the problem of deter-mining for two arbitrary given words on the given alphabet, whether one can be transformed into the other by a sequence of substitutions that are le-gitimate using the given equations. We show in the box on p. 258 that we can specify a particular word problem that is unsolvable. In other words, no computational process exists for determining whether or not two words can be transformed into one another using the given equations. Work on unsolvable word problems has turned out to be extremely important, lead-ing to unsolvability results in different parts of mathematics (for example, in group theory and in topology).

Another important problem that eventually turned out to be unsolvable first appeared as the tenth in a famous list of problems given by David Hil-bert in 1900. This problem involves so-called "Diophantine" equations. An equation is called Diophantine when we are only interested in solutions in integers (i.e., whole numbers). It is easy to see that the equation

$$4x - 2y = 3$$

has no solutions in integers (because the left side would have to be even while the right side is odd). On the other hand the equation

$$4x - y = 3$$

An Unsolvable Word Problem

One way to find a word problem that is unsolvable is to invent one whose solution would lead to a solution for the halting problem, which we know to be unsolvable. Specifically, we will show how to use a Turing–Post program P (which we assume consists of n instructions) to construct a word problem in such a way that a solution to the word problem we construct could be used to solve the halting problem for P. Therefore, if we begin with a problem P whose halting problem is unsolvable, we will obtain an unsolvable word problem.

We will use an alphabet consisting of the $n + 4$ symbols:

$$1\ 0\ h\ q_1\ q_2\ \ldots\ q_n\ q_{n+1}.$$

The fact that the ith step of P is about to be carried out and that there is some given tape configuration is coded by a certain word (sometimes called a Post word) in this alphabet. This Post word is constructed by writing down the string of zeros and ones constituting the current nonblank part of the tape, placing an h to its left and right (as punctuation marks) and inserting the symbol q_i (remember that it is the ith instruction which is about to be executed) immediately to the left of the symbol being scanned. For example, with a tape configuration

$$11011$$
$$\uparrow$$

and instruction number 4 about to be executed, the corresponding Post word would be

$$h110q_411h.$$

This correspondence between tape configurations and words makes it possible to translate the steps of a program into equations between words. For example, suppose that the fifth instruction of a certain program is

PRINT 0.

We translate this instruction into the equations

$$q_40 = q_50, \qquad q_41 = q_50,$$

which in turn yield the equation between Post words

$$h110q_411h = h110q_501h$$

corresponding to the next step in the computation. Suppose next that the fifth instruction is

GO RIGHT.

It requires 6 equations to fully translate this instruction, of which two typical ones are

$$q_5 0\ 1 = 0 q_6 1\ , \qquad q_5 1\ h = 1 q_0 0 h\ .$$

In a similar manner each of the instructions of a program can be translated into a list of equations. In particular when the ith instruction is STOP, the corresponding equation will be:

$$q_i = q_{n+1}\ .$$

So the presence of the symbol q_{n+1} in a Post word serves as a signal that the computation has halted. Finally, the four equations

$$q_{n+1}0 = q_{n+1}, \qquad q_{n+1}1 = q_{n+1}$$
$$0q_{n+1} = q_{n+1}, \qquad 1q_{n+1} = q_{n+1}$$

serve to transform any Post word containing q_{n+1} into the word $h q_{n+1} h$. Putting all of the pieces together we see how to obtain a word problem which "translates" any given Turing–Post program.

Now let a Turing–Post program P begin scanning the leftmost symbol of the string v; the corresponding Post word is $h q_1 v h$. Then if P will eventually halt, the equation

$$h q_1 v h = h q_{n+1} h$$

will be derivable from the corresponding equations as we could show by following the computation step by step. If on the other hand P will never halt, it is possible to prove that this same equation will not be derivable. (The idea of the proof is that every time we use one of the equations which translates an instruction, we are either carrying the computation forward, or—in case we substitute from right to left —undoing a step already taken. So, if P never halts, we can never get $h q_1 v h$ equal to any word containing q_{n+1}.) Finally, if we could solve this word problem we could use the solution to test the equation

$$h q_1 v h = h q_{n+1} h$$

and therefore to solve the halting problem for P. If, therefore, we start with a Turing–Post program P which we know has an unsolvable halting problem, we will obtain an unsolvable word problem.

has many (even infinitely many) solutions in integers (e.g., $x = 1$, $y = 1$; $x = 2, y = 5$). The Pythagorean equation

$$x^2 + y^2 = z^2$$

also has infinitely many integer solutions (of which $x = 3$, $y = 4$, $z = 5$ was already known to the ancient Egyptians). Hilbert's tenth problem was to find a computing procedure for testing a Diophantine equation (in any number of unknowns and of any degree) to determine whether or not it has an integer solution.

Since I have been directly involved with this problem and related matters over the past thirty years, my discussion of Hilbert's tenth problem will necessarily have a rather personal character. I first became interested in the problem while I was an undergraduate at City College of New York on reading my teacher Emil Post's remark in one of his papers that the problem "begs for an unsolvability proof." In my doctoral dissertation at Princeton, I proved the unsolvability of a more difficult (and hence easier to prove unsolvable) related problem. At the International Congress of Mathematicians in 1950, I was delighted to learn that Julia Robinson, a young mathematician from California, had been working on the same problem from a different direction: she had been developing ingenious techniques for expressing various complicated mathematical relationships using Diophantine equations. A decade later Hilary Putnam (a philosopher with training in mathematical logic) and I, working together, saw how we could make further progress by combining Julia Robinson's methods with mine. Julia Robinson improved our results still further, and we three then published a joint paper in which we proved that if there were even one Diophantine equation whose solutions satisfy a special condition (involving the relative size of the numbers constituting such a solution), then Hilbert's tenth problem would be unsolvable.

In subsequent years, much of my effort was devoted to seeking such a Diophantine equation (working alone and also with Hilary Putnam), but with no success. Finally such an equation was found in 1970 by the then 22-year old Russian mathematician Yuri Matiyasevich. Matiyasevich's brilliant proof that his equation satisfied the required condition involved surprisingly elementary mathematics. His work not only showed that Hilbert's tenth problem is unsolvable, but has also led to much new and interesting work.

Undecidable Statements

The work of Bertrand Russell and Alfred North Whitehead in their three-volume magnum opus *Principia Mathematica*, completed by 1920, made it clear that all *existing* mathematical proofs could be translated into the specific logical system they had provided. It was assumed without question by most mathematicians that this system would suffice to prove or disprove any statement of ordinary mathematics. Therefore mathematicians were shocked by the discovery in 1931 by Kurt Gödel (then a young Viennese mathematician) that there are statements about the whole numbers which can neither be proved nor disproved in the logical system of *Principia Mathematica* (or similar systems); such statements are called *undecidable*. Turing's work (which was in part inspired by Gödel's) made it possible to understand Gödel's discovery from a different, and indeed a more general, perspective.

Julia B. Robinson

Julia B. Robinson was born in 1919 in St. Louis, Missouri, but has lived most of her life in California. Her education was at the University of California, Berkeley, where she obtained her doctorate in 1948. She has always been especially fascinated by mathematical problems which involve both mathematical logic and the theory of numbers. Her contributions played a key role in the unsolvability proof for Hilbert's tenth problem. In 1975 she was elected to the National Academy of Sciences, the first woman mathematician to be so honored.

Let us write $N(P, v)$ to mean that the Turing–Post program P will *never* halt when begun with v on its tape (as usual, scanning its leftmost symbol). So, for any particular Turing–Post program P and string v, $N(P, v)$ is a perfectly definite statement which is either true (in case P will never halt in the described situation) or false (in case P will eventually halt). When $N(P, v)$ is false, this fact can always be demonstrated by exhibiting the complete sequence of tape configurations produced by P leading to termination. However, when $N(P, v)$ is true no finite sequence of tape configurations will suffice to demonstrate the fact. Of course we may still be able to prove that a particular $N(P, v)$ is true by a logical analysis of P's behavior.

Let us try to be very rigorous about this notion of *proof*. Suppose that certain strings of symbols (possibly paragraphs of English) have been singled out as proofs of particular statements of the form $N(P, v)$. Suppose furthermore that we possess a computing procedure that can test an alleged proof Π that $N(P, v)$ is true and determine whether Π is or is not actually such a proof. Whatever our rules of proof may be, this requirement is surely needed for communication purposes. It must be possible in principle to perform such a test in order that Π should serve its purpose of eliminating doubts concerning the truth of $N(P, v)$. (In practice, published mathematical proofs are in highly condensed form and do not meet this strict requirement. Disputes are resolved by putting in more detail as needed. But it is essential that *in principle* it is always possible to include sufficient detail so that proofs are susceptible to mechanical verification.)

There are two basic requirements which it is natural to demand of our supposed rules of proof:

> *Soundness:* If there is a proof Π that $N(P,v)$ is true, then P will in fact never halt when begun with v on its tape.
> *Completeness:* If P will never halt when begun with v on its tape, then there is a proof Π that $N(P,v)$ is true.

Gödel's theorem asserts that no rules of proof can be both sound and complete! In other words, if a given set of rules of proof is sound, then there will be some true statement $N(P, v)$ which has no proof Π according to the given rules of proof. (Such a true unprovable statement may be called *undecidable* since it will surely not be disprovable.)

To convince ourselves of the truth of Gödel's theorem, suppose we had found rules of proof which were both sound and complete. Suppose "proofs" according to these rules were particular strings of symbols on some specific finite alphabet. We begin by specifying a particular infinite sequence $\Pi_1, \Pi_2, \Pi_3, \ldots$ which includes all finite strings on this alphabet. Namely, let all strings of a given length be put in "alphabetical" order, and let shorter strings always precede longer ones. The sequence $\Pi_1, \Pi_2, \Pi_3, \ldots$ includes all possible proofs, as well as a lot of other things; in particular, it contains a high percentage of total nonsense—strings of symbols combined in completely meaningless ways. But, hidden among the nonsense, are all possible proofs.

Now we show how we can use our supposed rules of proof to solve the halting problem for some given Turing–Post program P. We wish to find out whether or not P will eventually halt when begun on v. We have some friend begin to carry out the instructions of P on input v with the understanding that we will be informed at once if the process halts. Meanwhile we occupy ourselves by generating the sequence $\Pi_1, \Pi_2, \Pi_3, \ldots$ of possible proofs. As each Π_i is generated we use our computing procedure to determine whether or not Π_i is a proof of $N(P, v)$. Now, if P will eventually halt, our friend will discover the fact and will so inform us. And, if P will never halt, since our rules of proof are assumed to be complete, there will be a proof Π_i of

$N(P, v)$ which we will discover. Having obtained this Π_i we will be sure (because the rules are sound) that P will indeed never halt. Thus, we have described a computing procedure (carried out with a little help from a friend) which would solve the halting problem for P. Since, as we well know, P can have an unsolvable halting problem (e.g., P could be the universal program U), we have arrived at a contradiction; this completes the proof of Gödel's theorem.

Of course, Gödel's theorem does not tell us that there is any particular pair P, v for which we will never be able to convince ourselves that $N(P, v)$ is true. It is simply that, for any given sound rules of proof, there will be a pair P, v for which $N(P, v)$ is true, but not provable *using the given rules*. There may well be other sound rules which decide this "undecidable" statement. But these other rules will in turn have their own undecidabilities.

Complexity and Randomness

A computation is generally carried out in order to obtain a desired answer. In our discussion so far, we have pretty much ignored the "answer," contenting ourselves with discussing only the gross distinction between a computation which does at least halt eventually and one which goes on forever. Now we consider the question: how complex need a Turing–Post program be to produce some given output? This straightforward question will lead us to a mathematical theory of randomness and then to a dramatic extension of Gödel's work on undecidability.

We will only consider the case where there are at least 2 ones on the tape when the computation halts. The output is then to be read as consisting of the string of zeros and ones between the leftmost and rightmost ones on the tape, and not counting these extreme ones. Some such convention is necessary because of the infinite string of zeros and ones on the tape. In effect the first and last one serve merely as punctuation marks.

To make matters definite suppose that we wish to obtain as output a string consisting of 1022 ones. When we include the additional ones needed for punctuation, we see that what is required is a computation which on termination leaves a tape consisting of a block of 1024 ones and otherwise blank. One way to do this is simply to write the 1024 ones on the tape initially and do no computing at all. But surely we can do better. We can get a slight improvement by using our faithful doubling program (Figure 1). We need only write 512 ones on the tape and set the doubling program to work. We have already written out the code for the doubling program; it took 39 bits. (A *bit* is simply a zero or a one; the word abbreviates *binary digit*.) So we have a description of a string of 1022 ones which uses $39 + 512 = 551$ bits. But surely we can do better. $1024 = 2^{10}$, so we should be able to get 1024 ones by starting with 1 and applying the doubling program 10 times. In Figure 4

```
 1  PRINT 0
    .
    .
 9.  GO TO STEP 1 IF 1 IS SCANNED
10.  GO RIGHT
11.  GO TO STEP 22 IF 0 IS SCANNED
12.  GO RIGHT
13.  GO TO STEP 12 IF 1 IS SCANNED
14.  GO LEFT
15.  PRINT 0
16.  GO LEFT
17.  GO TO STEP 16 IF 1 IS SCANNED
18.  GO LEFT
19.  GO TO STEP 18 IF 1 IS SCANNED
20.  GO RIGHT
21.  GO TO STEP 1 IF 1 IS SCANNED
22.  STOP
```

Figure 4. A Program for Calculating Powers of 2

we give a 22-step program, the first nine steps of which are identical to the first nine steps of the doubling program, which accomplishes this. Beginning with a tape configuration

$$10\underbrace{11\ldots1}_{n}$$
$\uparrow$

this program will halt with a block of 2^{n+1} ones on the tape.

It is not really important that the reader understand how this program works, but here is a rough account: the program works with two blocks of ones separated by a zero. The effect of Steps 1 through 9 (which is just the doubling program) is to double the number of ones to the left of the 0. Steps 10 through 21 then erase 1 of the ones to the right of the zero and return to Step 1. When all of the ones to the right of the zero have been erased, this will result in a zero being scanned at Step 11 resulting in a transfer to Step 22 and a halt. Thus the number of ones originally to the left of the zero is doubled as many times as there are ones originally to the right of the zero.

The full code for the program of Figure 4 contains 155 bits. To obtain the desired block of 1024 ones we need the input 10111111111. We are thus down to $155 + 11 = 166$ bits, a substantial improvement over 551 bits.

We are now ready for a definition. Let w be any string of bits. Then we say that w has *complexity n* (or equivalently, *information content n*) and write $I(w) = n$ if:

1. There is a program P and string v such that the length of $code(P)$ plus the length of v is n, and P when begun with v will eventually halt with output w (that is with $1w1$) occupying the nonblank part of the tape, and
2. There is no number smaller than n for which this is the case.

If w is the string of 1022 ones, then we have shown that $I(w) \leq 166$. In general, if w is a string of bits of length n, then we can easily show that $I(w) \leq n + 9$. Specifically, let the program P consist of the single instruction: STOP. Since this program does not do anything, if it begins with input $1w1$, it will terminate immediately with $1w1$ still on the tape. Since $Code(P) = 1100111$, it must be the case that $I(w)$ is less than or equal to the length of the string $11001111w1$, that is, less than or equal to $n + 9$. (Naturally, the number 9 is just a technical artifact of our particular formulation and is of no theoretical importance.)

How many strings are there of length n such that, say, $I(w) \leq n - 10$? (We assume $n > 10$; in the interesting cases n is much larger than 10.) Each such w would be associated with a program P and string v such that $Code(P)v$ is a string of bits of length less than or equal to $n - 10$. Since the total number of strings of bits of length i is 2^i, there are only:

$$2 + 4 + \ldots + 2^{n-10}$$

strings of bits of length $\leq n - 10$. This is the sum of a geometric series easily calculated to be $2^{n-9} - 2$. So we conclude: there are fewer than 2^{n-9} strings of bits w of length n such that $I(w) \leq n - 10$.

Since there are 2^n strings of bits of length n, we see that the ratio of the number of strings of length n with complexity $\leq n - 10$ to the total number of strings of length n is no greater than

$$\frac{2^{n-9}}{2^n} = \frac{1}{2^9} = \frac{1}{512} < \frac{1}{500}.$$

This is less than 0.2%. In other words, more than 99.8% of all strings of length n have complexity $> n - 10$. Now the complexity of the string of 1022 ones is, as we know, less than or equal to 166, thus much less than $1022 - 10 = 1012$. Of course, what makes this string so special is that the digit pattern is so regular that a comparatively short computational description is possible. Most strings are irregular or as we may say, *random*.

Thus we are led to an entirely different application of Turing's analysis of computation: a mathematical theory of random strings. This theory was developed around 1965 by Gregory Chaitin, who was at the time an undergraduate at City College of New York (and independently by the world famous A.N. Kolmogorov, a member of the Academy of Sciences of the U.S.S.R.). Chaitin later showed how his ideas could be used to obtain a dramatic extension of Gödel's incompleteness theorem, and it is with this reasoning of Chaitin's that we will conclude this essay.

Let us suppose that we have rules of proof for proving statements of the form $I(w) > n$ where w is a string of bits and n is a positive integer. As before, we assume that we have a computing procedure for testing an alleged proof Π to see whether it really is one. We assume that the rules of proof are sound, so that if Π is a proof of the statement $I(w) > n$, then the complexity of the string w really is greater than n. Furthermore, let us make the very reasonable assumption that we have another computing procedure which, given a proof Π of a statement $I(w) > n$, will furnish us with the specific w and n for which $I(w) > n$ has been proved.

We now describe a new computing procedure we designate as Δ. We begin generating the sequence $\Pi_1, \Pi_2, \Pi_3, \ldots$ of possible proofs as above. For each Π_i we perform our test to determine whether or not Π_i is a proof of a statement of the form $I(w) > n$. If the answer is affirmative we use our second procedure to find the specific w and n. Finally we check to see whether $n > k_0$ where k_0 is some fixed large number. If so, we report w as our answer; otherwise we go on to the next Π_i. By Turing's analysis this entire procedure Δ can be replaced by a Turing–Post program, where the fixed number k_0 is to be chosen at least as large as the length of this program. (The fact that k_0 can be chosen as large as this is not quite obvious; the basic reason is that far fewer than k_0 bits suffice to describe the number k_0.)

Now, a little thought will convince us that this Turing–Post program can never halt: if it did halt we would have a string w for which we had a proof Π_i that $I(w) > n$ where $n > k_0$. On the other hand this very program has length less than or equal to k_0 (and hence less than n) and has computed w, so that $I(w) < n$, in contradiction to the soundness of our proof rules. Conclusion: our rules of proof can yield a proof of no statement of the form $I(w) > n$ for which $n > k_0$. This is Chaitin's form of Gödel's theorem: given a sound set of rules of proof for statements of the form $I(w) > n$, there is a number k_0 such that no such statement is provable using the given rules for any $n > k_0$.

To fully understand the devastating import of this result it is important to realize that there exist rules of proof (presumably sound) for proving statements of the form $I(w) > n$ which include all methods of proof available in ordinary mathematics. (An example is the system obtained by using the ordinary rules of elementary logic applied to a powerful system of axioms, of which the most popular is the so-called Zermelo–Fraenkel axioms for set theory.) We are forced to conclude that there is some definite number k_0, such that it is in principle impossible, by ordinary mathematical methods, to prove that any string of bits has complexity greater than k_0. This is a remarkable limitation on the power of mathematics as we know it.

Although we have discussed a considerable variety of topics, we have touched on only a tiny part of the vast amount of work which Turing's analysis of the computation process has made possible. It has become possible to distinguish not only between solvable and unsolvable problems, but to study an entire spectrum of "degrees of unsolvability." The very notion of computation has been generalized to various infinite contexts. In the theory of formal languages, developed as part of computer science, various limitations on Turing–Post programs turn out to correspond in a natural way to different kinds of "grammars" for these languages. There has been much work on what happens to the number of steps and amount of tape needed when the programs are allowed to operate on several tapes simultaneously instead of on just one. "Nondeterministic" programs in which a given step may be followed by several alternative steps have been studied, and a great deal of work has been done attempting to show that such programs are intrinsically capable of performing much faster than ordinary Turing–Post programs. These problems have turned out to be unexpectedly difficult, and much remains to be done.

Suggestions for Further Reading

General

Chaitin, Gregory. Randomness and mathematical proof. *Scientific American* **232** (May 1975) 47–52.

Davis, Martin and Hersh, Reuben. Hilbert's 10th problem. *Scientific American* **229** (November 1973) 84–91.

Knuth, Donald E. Algorithms. *Scientific American* **236** (April 1977) 63–80, 148.

Knuth, Donald E. Mathematics and computer science: coping with finiteness. *Science* **194** (December 17, 1976) 1235–1242.

Turing, Sara. *Alan M. Turing.* W. Heffer, Cambridge, 1959.

Wang, Hao. Games, logic and computers. *Scientific American* **213** (November 1965) 98–106.

Technical

Davis, Martin. *Computability and Unsolvability.* McGraw-Hill, Manchester, 1958.

Davis, Martin. Hilbert's tenth problem is unsolvable. *American Mathematical Monthly* **80** (March 1973) 233–269.

Davis, Martin. *The Undecidable: Basic Papers on Undecidable Propositions, Unsolvable Problems and Computable Functions.* Raven Pr, New York, 1965.

Davis, Martin. Unsolvable problems. In *Handbook of Mathematical Logic*, by Jon Barwise (Ed.). North-Holland, Leyden, 1977.

Minsky, Marvin. *Computation: Finite and Infinite Machines.* Prentice-Hall, Englewood Cliffs, 1967.

Rabin, Michael O. Complexity of computations. *Comm. Assoc. Comp. Mach.* **20** (1977) 625–633.

Trakhtenbrot, B.A. *Algorithms and Automatic Computing Machines.* D.C. Heath, Lexington, 1963.

Mathematics as a Tool for Economic Understanding

Jacob T. Schwartz

Though its influence is a hidden one, mathematics has shaped our world in fundamental ways. What is the practical value of mathematics? What would be lost if we had access only to common sense reasoning? A three-part answer can be attempted:

1. Because of mathematics' precise, formal character, mathematical arguments *remain sound even if they are long and complex.* In contrast, common sense arguments can generally be trusted only if they remain short; even moderately long nonmathematical arguments rapidly become far-fetched and dubious.

2. The precisely defined formalisms of mathematics discipline mathematical reasoning, and thus *stake out an area within which patterns of reasoning have a reproducible, objective character.* Since mathematics defines the notion of formal proof precisely enough for the correctness of a fully detailed proof to be verified mechanically (for example, by a computing mechanism), doubt as to whether a given theorem has or has not been proved (from a given set of assumptions) can never persist for long. This enables mathematicians to work as a unified international community, whose reasoning transcends national boundaries and survives the rise and fall of religions and of empires.

3. Though founded on a very small set of fundamental formal principles, mathematics *allows an immense variety of intellectual structures to be elaborated.* Mathematics can be regarded as a grammar defining a language in which a very wide range of themes can be expressed. Thus mathematics provides a disciplined mechanism for devising frameworks into which the facts provided by empirical science will fit and within which originally disparate facts will buttress one another and rise above the primitively empirical. Rather than hampering the reach

of imagination, the formal discipline inherent in mathematics often allows one to exploit connections which the untutored thinker would have feared to use. Common sense is imprisoned in a relatively narrow range of intuitions. Mathematics, because it can guarantee the validity of lengthier and more complex reasonings than common sense admits, is able to break out of this confinement, transcending and correcting intuition.

By fitting originally disparate theories to the common mold defined by the grammar which is mathematics, one gives these theories a uniform base, which among other things makes it possible for theories which initially stand in irreducibly hostile confrontation to be related and compared, perhaps by means of some comprehensive mathematical construction which includes all these theories as special cases.

On the other hand, the precise character of the axiomatic formalisms of mathematics is attained at the cost of a radical divorce between these formalisms and the real world. Therefore, in regard to real-world applications, the mathematical mode of reasoning will only retain its advantage if it is based upon some very compelling model of an aspect of the real world. Only empirical work whose direct confrontation with reality lies outside mathematics itself can validate such a model, though part of the inspiration for such models may come from within mathematics.

It is sobering to note that we possess only two techniques for ensuring the objective validity of our thought. The approach of empirical science—to anchor oneself directly to verifiable statements concerning real-world observations—is one of these techniques; and mathematics embodies the other. Science disciplines thought through contact with reality, mathematics by adhering to a set of formal rules which, though fixed, allow an endless variety of intellectual structures to be elaborated.

The external disciplines which science and mathematics impose upon their practitioners play particularly essential roles in those cases in which the truth is surprising or in which some captivating established doctrine needs to be broken down. Common sense, acting by an undisciplined principle of plausibility and intuitive appeal, cannot cope either with situations where truth is even partly counterintuitive, or with situations in which ingrained familiarity has lent overwhelming plausibility to some established error. Thus, for example, common sense can cope adequately with situations where, by shoving an object in a given direction, one causes it to move in that direction; but in those more complicated situations in which objects move in directions contrary to the dictates of intuition, the more delicate techniques of formal mathematical reasoning become indispensable.

However, application of the cautious procedures of science and of mathematics proceeds slowly, whereas the daily affairs of hurrying mankind demand constant resolution. Thus, where adequate knowledge and objective theory do not exist, fragmentary heaps of fact and primitive common sense reasoning will be drawn upon to fill the gap. For this reason, both personal decisions and social commitments of vast consequence often rest on doubtful and inadequate understanding. Moreover, just as a person committed to a given course may resist and resent efforts to point out irrational aspects of his behavior, so societies can resent and even repress criticism of intellectual structures, correct or not, which are regarded as essential to a social consensus and around which significant interests may have collected. In such situations, mathematics has often been used as a tool for defusing controversy and for allowing ideas running counter to some dominant doctrine to be developed in a neutral form.

In the present article we will try to illustrate the way in which mathematics can contribute to our understanding of social issues. In order to do this, and specifically in order to make the point that mathematical reasoning need by no means remain remote and abstract, we will venture to reason about hard-fought areas of current controversy. This makes it all the more necessary for the reader to remember that our reasoning, like all applications of mathematics to the real world, is not at all incontrovertible, but merely represents a set of conclusions to which one can be led by assuming one particular model of reality. Any such model is but one of many possible models of differing adequacy which can lead to different, and sometimes diametrically conflicting deductions. Our reasoning can therefore not tell us that our conclusions concerning reality are correct, but only that they are logically possible.

Some Classical Economic Issues

Economics, the body of theory concerned with the mechanisms by which society pushes its members to productive activity and allocates the fruits of that activity, illustrates the points made in the preceding paragraphs. Since economic institutions must link men into extensive patterns of joint action, and since these patterns must be forged in an area in which conflicting individual and group claims are never very far from the surface, no society is able to do without some unifying and generally accepted economic ideology. The economic ideology of our own American society generally emphasizes the importance of unhampered individual and corporate enterprise in production, and of individual choice in consumption. However, there also exist centralized legal mechanisms which forbid certain types of production (e.g., of unpasteurized milk except under special circumstances, use of child labor in factories) and consumption (e.g., narcotic drugs except under special circumstances). In fact, the range of coercive governmental interventions which now modify the workings of the decentralized economic machine is quite broad. For example, the government regulates allocation of the economic product among those who have cooperated in its production (e.g., by establishing minimum wage standards) and among major social subgroups (through taxing patterns and welfare programs); directly influences vital economic parameters (e.g., availability of credit, through bank regulation and the activity of the central banking system); prevents certain types of economic cooperation and encourages others (through anti-trust and labor relations laws); and prohibits certain types of misrepresentation and irresponsibility (e.g., through laws which punish commercial fraud, demand certain forms of record-keeping and information disclosure and compel individuals to contribute to a central pension fund).

To make any one of these numerous economic interventions attractive to the public and acceptable to the legislator, there is required some fragment of theory, which, whether it be primitive or sophisticated, correct or mistaken, must at least be persuasive. To commit society as a whole to an overall economic approach, a more systematic body of theory recommending this approach on moral and pragmatic grounds is required. For this reason, the stock of available economic theories has exerted a fundamental influence on American and world society. The classical case for a resolutely *laissez-faire* approach is that made in Adam Smith's *Wealth of Nations:*

> . . .As every individual, therefore, endeavours as much as he can both to employ his capital in support of domestic industry, and so to direct that industry that its produce may be of greatest value, every individual necessarily labours to render the annual revenue of the society as great as he can. He generally, indeed, neither intends to promote the public interest, nor knows how much he is promoting it. By preferring the support of domestic to that of foreign industry, he intends only his own security; and by directing that industry in such a manner as its produce may be of the greatest value, he intends only his own gain, and he is in this, as in many other cases, led by an invisible hand to promote an end which was no part of his intention.

Since their publication in 1776, these words of Adam Smith have been immensely influential, and have been repeated and refined by subsequent generations of market-oriented economic thinkers. Of course, arguments supporting entirely different conclusions would not be hard to cite.

How can we begin to find safe ground from which to judge such arguments? To begin with, it is well to note that these arguments are only justified to the extent that the pragmatic judgment implicit in them is in fact correct. If Adam Smith is correct in asserting that his "invisible hand" acts to harmonize interactions between the participants in a free economy, then his recommendations are wise; but if these interactions cause the participants in such an economy to act at crosspurposes and to frustrate each other's intentions, then he is mistaken. This implies that the free-market arguments of Smith and his disciples, like the arguments for particular measures of centralized control advanced by various of his ideological opponents, have substance only insofar as more penetrating analysis is able to establish the extent to which an economy organized on the basis of particular freedoms or controls will actually deliver to its members the goods that they want.

Thus what one needs to analyze these arguments is a neutral theoretical framework within which the interaction of individuals in an economic system can be modeled, and from which some account of the results of their interaction can be developed. In principle, the input to such a model will be a mass of rules and statistical data describing the personal, technological, and legal facts which constrain the participants, and the approaches which they typically employ in their effort to adapt to shifting economic circumstances. The model must be broad enough to allow incorporation of any element of motivation or economic response deemed significant, and should also be general and "neutral," i.e., free of preconceptions of the result that will emerge from the multiparticipant interactions being modeled.

Mathematical Models of Social and Economic Interaction

Mathematical models within which a national economy could be described by a closed self-determining system of equations began to appear in the 1870's. The limited class of models initially proposed (notably by Leon Walras in Switzerland and Vilfredo Pareto in Italy) was substantially and dramatically broadened in the late 1920's and early 1930's in a series of papers by the brilliant and universal twentieth century mathematician John von Neumann (1903–1957), whose work on these matters was rounded out and made generally available in the treatise *Theory of Games and Economic Behavior* published in 1944 by von Neumann and the late Oscar Morgenstern. The essential accomplishment of von Neumann and Morgenstern was to imbed the limited supply/demand models which had previously been considered into a much larger class of models within which substantially more varied patterns of interpersonal relationship could be brought under mathematical scrutiny.

In a model of the von Neumann–Morgenstern type, one assumes a finite number of participants, whose interactions are constrained (but not uniquely determined) by a set of precisely defined, prespecified rules. In the economic application of models of this sort, the participants represent producers and consumers, and the rules describe the legal and technical facts which constrain their economic behavior. In a sociological or political-science application of a von Neumann–Morgenstern model, participants might either be individuals or nations, and the rules of the model would then describe the institutional, traditional, or military factors defining the options open to them and the consequences which will follow from the exercise of a given pattern of options. It is useful to note that in addition to these very real sources of situations to be modeled, a stimulating variety of miniature situations readily describable by von Neumann models is furnished by the large collection of familiar board, card, and nonathletic children's games (such as Choosing or Stone-Paper-Scissors) with which everybody is familiar. It is because these

John von Neumann

John von Neumann, one of the greatest mathematicians of the twentieth century, was born in Budapest in 1903. From youth he was an obvious scientific prodigy, exhibiting remarkable talent in mathematics, physics, chemistry, and engineering. After taking simultaneous degrees in mathematics and chemical engineering in 1923, he spent the first part of his career in Germany, but came to the United States in the early 1930's, as part of the general emigration of scientists from troubled Europe. In 1933 he was appointed professor of mathematics at the Institute for Advanced Study in Princeton.

During his life von Neumann contributed massively to many branches of pure and applied mathematics, including logic, analysis, and game theory. He is also one of the founders of the computer age, having played a central role in the design of some of the first U.S. electronic computers and in the early development of programming techniques. His first papers on the theory of games were published in 1928, when he was 25 years old.

games are so suggestive of directions of analysis that von Neumann and Morgenstern used the word "Games" in the title of their book and fell into the habit of referring to their models as games and to the participants whose interaction was being modeled as players.

The formal structure of a von Neumann–Morgenstern model makes available, to each of the participants or players which it assumes, some finite number of "moves" (which might also be called "policies" or "strategies") among which the participants are allowed to choose freely. It may be that one single set of moves is available to all the participants assumed by a model; on the other hand, the model may allow different sets of moves to different participants, in this way modeling real-world situations in which participants interact in an inherently unsymmetric way. After each player has made his move, a set of rewards (and penalties, which are regarded simply as negative rewards) are distributed to the players. In a game situation, we can think of these rewards or penalties as being distributed by an umpire who administers the rules of the game. In a more realistic situation we can think of these rewards and penalties as describing the actual outcome, for each participant, of the policies followed by him and by all other participants. These rewards or penalties may, for example, be business gains or losses, given the pattern in which other participants choose to invest, buy, and sell; or they may be military victories or defeats, given the pattern in which other participants choose to fight, remain neutral, or flee. In the terminology introduced by von Neumann and Morgenstern, the reward/penalty assessed against each participant in a game is that participant's "payoff."

Suppose that players $1, 2, \ldots, n$ respectively choose moves m_1, $m_2, \ldots, m_n$ out of the collection of moves available to them. It is the collective pattern of all these moves that determines the individual payoff to each player; thus the payoff to the jth player is some mathematical function of the whole list of variables $m_1, \ldots, m_n$. Therefore if we use the standard notation for mathematical functions we can write the payoff to the jth player as $p_j(m_1, m_2, \ldots, m_n)$. The payoff functions p_j summarize all those aspects of a game or other interactive situation which a particular model is able to capture, so that we may say that each particular von Neumann model is fully defined by such a set of functions.

The participants in a von Neumann Morgenstern model are driven to interact simply because the payoff to the jth player is determined not only by his own move m_j, but also by the moves of all other participants. In selecting his move as a game is repeatedly played, each participant must therefore take account of the observed behavior of the other participants, and, if his own activity is a large enough factor in the total situation to affect these other participants significantly, must analyze the effect of his move on these other participants, and take account of the countermoves which they might make to repair their position if it is significantly changed by his choice of move. (The boxes on pages 276 and 277 illustrate the rich variety of models encompassed by this theory.)

Bluff

Bluff is a two-player game with matching gains and losses. Player 1 writes a number from 1 to 10 on a slip of paper. Without showing this slip to Player 2, he tells Player 2 what he has written, but he is allowed to lie, announcing either the same number that he wrote down, or a different number. Player 2 must then guess whether Player 1 is telling the truth or not; he does so by saying true or lie. If caught in a lie, Player 1 must pay Player 2 a dollar; if falsely accused of lying, Player 1 collects half a dollar from Player 2. If Player 1 does not lie and Player 2 correctly guesses that he has not lied, then Player 1 must pay 10 cents to Player 2. If Player 1 lies without Player 2 guessing that he has done so, then he collects fifty cents from Player 2. These payoffs are summarized by the following payoff table for Player 1:

m_1 \ m_2	true	lie
same	−10	+50
different	+50	−100

Player 2's payoffs are the exact opposite: what Player 1 wins, Player 2 loses, and vice versa. This game is, therefore, an example of a two-person zero-sum game.

Cooperation

This is a two-player game in which interplayer cooperation is possible. Both players write either a 1 or a 2 on a slip of paper. If one of the players has written 2 on his slip while the other has written 1, then the umpire gives two dollars to the player who has written 2 and collects one dollar from the other player. If both players have written 1, then the umpire collects one dollar from each of the players. If both players have written 2, then the umpire collects ninety cents from both players. (The reader may wish to work out the best way for the players to cooperate.)

Litter

This game is a many-player model of the dropping of litter in streets. There are a large number of participants, perhaps 1000. Each participant is given a slip of paper which he is allowed to either dispose of, or drop. If a player drops his piece of paper, he must then roll a pair of dice; if the dice come up snake eyes (both ones) he must pay one dollar to the umpire. (This rule is intended to model the effect of an antilittering law which prescribes a fine of $1 for littering; we are in effect assuming that the probability of being caught by a policeman in the act of littering is $1/36$, since $1/36$ is the probability that two six-sided dice both come up 1.) If a player disposes of his piece of paper, he must pay 5¢ to the umpire. (This rule, which suggests a kind of garbage disposal service charge, can also be thought of as assigning a nominal cash value to the labor needed to dispose of an item of litter in a publicly provided receptacle.) After all n participants have chosen their move, the number of slips which have been dropped rather than disposed of is counted, and each participant is required (irrespective of whether he personally has disposed of or dropped his own slip of paper), to pay $1/10$¢ to the umpire for each dropped slip. (This last rule can be thought of as assigning a nominal negative cash value to the degradation of the aesthetic environment caused by the accumulation of litter.)

Stagnation

This game is a many-participant majority game which models an everyday situation of stagnation. The model can have any number of participants; but let us suppose, to fix our thoughts, that there are ten. The participants desire to go in a group to receive some reward (e.g., to dine together). Each player can elect one of two moves, which we shall call go and stay. If more than half the players choose the move go, then those who choose the move go receive a payoff of one dollar from the umpire; in this case, those who choose the move stay receive nothing. If less than half the players choose the move go, then those who choose this move must pay a fine of 10 cents, while those who choose the move stay pay nothing. (This last rule may be thought of as assigning a nominal cash value to the embarrassment which participants breaking away from a social majority may feel, and to the trouble which they are put to in rejoining the majority.)

Strategies and Coalitions

After introducing the broad class of models described above, von Neumann and Morgenstern entered into an analysis of the behavior of participants in these models, an analysis which they were able to carry surprisingly far. The simplest cases to analyze, and those in which the most conclusive analysis is possible, are those known as "two person zero-sum games." In these games one participant's gains are exactly the losses of the other participant, so that the two participants stand in a pure relationship of opposition, no sensible grounds for cooperation existing. (Mathematically, these are the games in which the two payoff functions p_1, p_2 satisfy $p_1 + p_2 = 0$, or, equivalently, $p_1 = -p_2$. This allows the whole game to be described by a single "payoff matrix" $p(m_1, m_2) = p_1(m_1, m_2)$.) For games of this type, von Neumann and Morgenstern were able to identify a clear and compelling notion of "best defensive strategy" which they showed would always exist, provided that the notion of move was broadened to allow for an appropriate element of randomization.

A simple yet instructive example is furnished by the familiar Stone-Paper-Scissors choosing game of children, in which each of the two participants independently chooses one of three possible moves, designated as "stone," "paper," "scissors," respectively. If both make the same choice, the play ends in a tie with no winner or loser. Otherwise the winner is determined by the rule "paper covers stone, stone breaks scissors, scissors cut paper." If we take the value of a win to be $+1$ and the cost of a loss to be -1 (while ties are of course represented by 0), then the payoff matrix for this game is as given in Figure 1.

A *perfect defensive move* in any game is a move which a player will find it advantageous to repeat, even assuming that this move comes to be expected by his opponent, who then replies with the most effective riposte available to him. So if either player makes a best defensive move and the other makes his

Player 2	Player 1		
	Stone	Paper	Scissors
Stone	0	+1	−1
Paper	−1	0	+1
Scissors	+1	−1	0

Payoff Matrix

Move m	Best Response $R(m)$
Stone	Paper
Paper	Scissors
Scissors	Stone

Best Response Function

Figure 1. Stone-Paper-Scissors best response function.

best reply, then both players will simultaneously be making best defensive moves and best replies to the other's move. Suppose that we designate Player 2's best response to each possible move m_1 of Player 1 by $R_2(m_1)$, and symmetrically use $R_1(m_2)$ to designate Player 1's best response to Player 2's move m_2. Then if m_1^* and m_2^* denote the corresponding perfect defensive moves, it follows that

$$m_1^* = R_1(m_2^*) \quad \text{and} \quad m_2^* = R_2(m_1^*).$$

Hence to determine Player 1's best defensive move we have only to solve the equation $m_1^* = R_1(R_2(m_1^*))$.

In Stone-Paper-Scissors the payoff matrix is symmetrical, so the best response is the same for both players. In other words, $R_1 = R_2$, so we shall simply call it R; the best responses in this game are tabulated in Figure 1. Clearly, $R(R(\text{Stone}))$, the best response to the most effective riposte to Stone, is Scissors. Similarly, $R(R(\text{Paper}))$ is Stone, and $R(R(\text{Scissors}))$ is Paper. Thus in this game we find to our disappointment that there is no best defensive strategy (since there is no solution to the equation $R(R(m_1^*)) = m_1^*$).

However, as von Neumann was first to realize, this problem will always disappear if we allow players to *randomize* their moves, i.e., to assign fixed probabilities, totalling to 1, to the elementary (or pure) moves open to them according to the initially stated rules of a game, and then to make specific moves randomly but with these fixed probabilities. The necessary role of randomization is particularly obvious in games like Choosing, Penny-Matching, or Stone-Paper-Scissors, since it is perfectly clear that a player who always plays the same move is playing badly (no matter what single move he fixes on); to play well, a player must clearly randomize his moves in some appropriate fashion. Because of the perfect symmetry of the three options in Stone-Paper-Scissors, it is easily seen that the best randomization strategy in this game is to employ each move with equal probability.

The symmetric Stone-Paper-Scissors game offers no special advantage to either player. Other two-person zero-sum games are not so precisely balanced. The game Bluff described in the box on p. 276, for example, yields an average gain of $50/7 = 7.1$ cents per game to the first player, providing both participants play their best (randomized) strategy. (Details of these best defensive strategies are outlined in the box on p. 280.) Of course, in a zero-sum game, one person's gain is another person's loss, so the second player loses, on average, an amount equal to the first person's gain; this amount is known as the "value" of the game. In general, the value of a two-person game is the amount won by one of the players if both players use their best defensive strategies. Thus a two-person game's value measures the extent to which the game inherently favors one or the other side when two players (or perhaps two coalitions of players) face each other in a situation of pure opposition.

Having established these fundamental facts concerning the simplest two-person games, von Neumann and Morgenstern proceeded to an analysis of

Best Bluffing Strategy

We analyze the game of Bluff (p. 276) by calculating best response functions. If Player 1 plays the move "same" with probability s and "different" with probability d, then Player 2 will gain an average of $10s - 50d$ if he plays "true", and $-50s + 100d$ if he plays "lie"; the relative advantage of "lie" is therefore $150d - 60s$. Hence player 2's best response to strategy is "lie" if $150d \geq 60s$, and "true" if $150d \leq 60s$. It follows that the best defensive strategy for Player 1 is to take $150d = 60s$, i.e., $d = \frac{2}{7}$, $s = \frac{5}{7}$; and similarly the best defensive strategy for Player 2 is to randomize by selecting "true" $\frac{5}{7}$ of the time and "lie" the remaining $\frac{2}{7}$. If both players of this game play their best strategies, the average gain of Player 1 is $-10(\frac{5}{7}) + 50(\frac{2}{7}) = \frac{50}{7} = 7.1$ cents per game (and of course the average loss of Player 2 is the same). We therefore say that the value of this game to Player 1 is $\frac{50}{7}$ (and to Player 2 is $-\frac{50}{7}$).

n-person games. Their main idea was to consider all the coalitions that could form among subsets of the participants in such games, and to analyze the "strength" which each such coalition would be able to exert, and the extent to which each coalition could stabilize itself by distributing rewards amongst its participants, thus countering appeals for cooperation and promises of reward that might be made by other competing coalitions.

If we consider any subset C of the players of a game as a potential coalition, then it follows from the preceding analysis of two-person games that by agreeing amongst themselves on what their moves will be the members of C can secure for themselves a total amount $v(C)$ equal to the value of the two-person (or two-coalition) game in which all the members of C play together as a single corporate individual against all the remaining players taken as a single corporate opponent. This amount $v(C)$ is then available collectively to the members of C for distribution in some pattern chosen to stabilize the coalition, and measures the strength of the coalition C.

The coalitional analysis of von Neumann and Morgenstern begins with a study of the prototypical case where the number of players is three. In this case they were able to show that every "strategically interesting" game is equivalent to the game Choose-an-Ally in which each of three participants selects one of the two others as a partner. Then any two players who have designated each other collect a fixed sum from the third player and divide their gain. Clearly, in this game no strategy is possible other than the formation of two-person coalitions (who agree on their moves, though of course these agreements lie outside the strict rules of the game itself). With four participants, a three-parameter family of essentially different coalition strength patterns exists, so the analysis of coalitions is already intricate; the complexity of this analysis increases rapidly as the number of participants rises above four.

The coalitional analysis of von Neumann and Morgenstern uncovers a rich variety of phenomena. Nevertheless, we must assert that the direction in which this analysis proceeds is not the direction which is most appropriate for applications to economics. The reason is this: a coalitional theory of games rests on the assumption that the participants in the situation being modeled have the time and ability to work out agreements which lie entirely outside the defining rules of the game, and that these outside agreements are endorceable at least *de facto*, i.e., that once made they will be adhered to. Quite the opposite assumption, namely that the players are forbidden to or unable to communicate with each other except through the moves they make, is just as reasonable in many situations and is often much to be preferred. In particular, as von Neumann and Morgenstern point out, significant applications of their theory to economics depend on the treatment of n-person games for large n. But it is obvious that the maintenance and internal administration of coalitions of large numbers of members cannot be simple, and hence we may expect that, as the number of participants in a game increases, the tendency of the participants to be guided solely by the inherent constraints of the situation in which they find themselves and by observation of the moves made by other participants can become decisive.

Coalition-Free Equilibria

To understand what will happen in situations in which this is the case, what we need to do is to abandon coalition analysis as a false start, and to return to something very close to the original von Neumann notion of perfect defensive move, which has a direct and felicitous generalization to the n-participant situation. Specifically, we define a pattern $m_1^*, \ldots, m_n^*$ of moves in an n-player game to be an "equilibrium" if for each j, the move m_j^* of this jth player is the move which he finds it advantageous to make, given that the other players make the moves $m_1^*, \ldots, m_{j-1}^*, m_{j+1}^*, \ldots, m_n^*$. Such a (mutually confirming) pattern of moves will therefore give the jth player maximum payoff, assuming that he can count on the other players to make their equilibrium moves $m_1^*, \ldots, m_n^*$. Just as in the two-person game, the equilibrium pattern of moves is one in which every player is responding to the observed moves of all the other players in a manner which gives him maximum payoff.

This coalition-free viewpoint in game theory was first advanced seven years after the work of von Neumann and Morgenstern by the young American mathematician John Nash, and published by him under the title *Non-Cooperative Games*. The Nash approach opens the way to an effective treatment of n-person games for large n, and has the very appealing property of being based on an assumption whose validity should grow as n becomes large; by contrast, the von Neumann–Morgenstern coalitional approach founders for large n amongst rapidly growing combinatorial particularities.

From the economic point of view, the Nash equilibrium notion is particularly appealing since it can be regarded as a description of the equilibrium to which a game will move if the game's players are confined *de jure* or *de facto* by the normal rules of a "free market," namely by an inability to communicate outside the framework of the game, or by a legal prohibition against the formation of coalitions or cartels. From this angle we may say that whereas the equilibrium-point concept assumes that each player takes the context in which he finds himself as given and makes a personally optimal adjustment to this context, the von Neumann–Morgenstern coalitional analysis tends always to call the social constitution into question by asking whether any group within the economy is strong enough to force a drastic change in the established pattern. Thus, if one takes a view of economics rigorously consistent with the von Neumann–Morgenstern theory, economic analysis would necessarily include a theory of such things as Congressional negotiations on fiscal, tax, and tariff policy. However interesting such questions may be, their answers do not belong to economic theory in its most classical sense.

Having found all this to say in favor of coalition-free analysis, it is now appropriate to apply it. This will be easiest if we confine our attention to the rather large class of symmetric games, i.e., to games in which all players have essentially the same payoff function. Even though equilibria in which different players make different moves are perfectly possible even for such games, we shall for simplicity confine our attention to "symmetric equilibria" in which all players make the same move. These equilibria are determined by the equation $m^* = R(m^*)$, where $R(m)$ defines the best response of any player to a situation in which the move made by all other players is the same and is m.

Here, every participant finds the move m^* advantageous both because this move secures him a maximum payoff (given that all other players are making this same move, a fact which lies outside his control) and also because he finds the wisdom of his move confirmed by the fact that every other participant is observed to be doing the same thing. In such a situation, the response m^* can easily become so firmly established as to appear inescapable.

Optimal vs. Suboptimal Equilibria

In the symmetric case which we are considering, the noncooperative equilibrium point m^* is defined, formally, by the equation

$$p_1(m^*, m^*, \ldots, m^*) = \max_m p_1(m, m^*, \ldots, m^*), \qquad (*)$$

where the maximum on the right-hand side extends over all possible moves of the first player. This equation describes the equilibrium which will develop as each participant in a symmetric game *privately* adjusts his move to yield a personal optimum, given the moves which others are observed to make. In general this equation selects quite a different point than does the equation

$$p_1 (m^+, \ldots, m^+) = \max_m p_1(m, m, \ldots, m) , \qquad (+)$$

which describes the *collective* decision that would be taken if all players *first* agree to make the same move and then collectively choose the move which will secure greatest general benefit.

The example games given in the boxes on pp. 276 and 277 serve very well to show that although the m^* and m^+ selected by these two quite different equations can accidentally be the same, they can also differ radically. For example, consider the game Stagnation. Here there are three and only three symmetric equilibria: that in which all participants make the move "stay"; that in which all participants make the move "go"; and that in which every participant plays "stay" with probability x and "go" with probability $1 - x$, where x is chosen so as to imply a zero advantage of "stay" over "go" for a player who chooses to deviate from the average. If all players move "go," then each receives $1, the maximum possible payoff in the game, so it is quite unsurprising that this should be an equilibrium. If all players move "stay", then each receives $0, making it somewhat more surprising that this pattern of play should be an equilibrium. Nonetheless it is, since once this pattern of play is reached no player can deviate from it alone without suffering a penalty; in fact hc will suffer the penalty unless he can persuade a *majority* of the other players simultaneously to change their moves. Equilibria of this sort, in which it is possible for the situation of *every* player to be improved if the players cooperate suitably, even though it is impossible for an player by himself (and conceivably even small coalitions of players by themselves) to deviate from the equilibrium without suffering, are called *suboptimal equilibria*. We may therefore say that our stagnation game has several equilibria, one optimal and providing the maximum possible payoff $1 to every player, the others suboptimal. Of course, everyone who has fumed through those periods of immovability which affect groups of people as large as half a dozen trying to go together to lunch or to the theater will realize that the suboptimal equilibrium of this model describes a very real phenomenon.

It is even possible for a game to have exclusively suboptimal equilibria. This is clearly shown by our example game Litter (p. 277) which models the dropping of litter in streets. Suppose that all players but one play "drop" with probability x, and "dispose" with probability $1 - x$. Then the payoff to the last player is $-(^{100}/_{36}) - (^{1000}/_{10})x$ if he drops, but is $-5 - (^{999}/_{10})x$ if he plays "dispose." Hence every player will "drop," and the equilibrium value of the game for every player will be $-(^{100}/_{36}) - 100 = -103$. On the other hand, if the pattern of moves were different, i.e., if every player played "dispose," then the payoff to each player would be -5. This shows that the unique equilibrium of Litter is very distinctly suboptimal.

This result appears paradoxical at first sight, but it has a very simple explanation. The payoff to each player consists of two terms. One, which is either $-(^{100}/_{36}) - (^1/_{10}) \cong -3$ or is -5, is directly controlled by his own move; the other term, which in the preceding calculation appeared as $-^{999}/_{10}$, and which in the situation that our model is intended to represent corresponds to environmental degradation through the accumulation of heaps of litter, is controlled by the moves of others. Since each player considers the bulk damage represented by the term $^{999}/_{1000}$ to have been wreaked irrespective of his own tiny contribution to it, his own most advantageous move is to "drop", which of course subtracts a tiny amount from the payoff of all other players.

It is instructive to consider what happens if we *decrease* the payoff function in this model by increasing the fine which must be paid by persons choosing to "drop" (which they pay in the case, whose probability is $^1/_{36}$, that they are caught). Let the amount of this fine be F; then "drop" remains preferable to "dispose" as long as $(^F/_{36}) + (^1/_{10}) < 5$. When F rises above $5 \times 36 = \$1.80$, "dispose" becomes preferable. For F this large, the equilibrium payoff to every player is -5. This shows that the equilibrium value of a game with suboptimal equilibria can *increase* (even dramatically) if the game's payoff function is *decreased* in some appropriate fashion. Our calculation also shows that for minor offenses of this type to be suppressed, the penalty for the offense, multiplied by the probability of apprehension, must exceed the gain obtained by committing the offense. Among other things, this explains why the penalties for littering along highways (where the probability of apprehension is very low) must be as large as they typically are.

These game-theoretic arguments suggest a very compelling objection against classical free-market arguments. The market equilibria m^* selected by Adam Smith's invisible hand are generally not the same as the collective maximum m^+. To identify m^* with m^+, simply because their defining equations both involve maximization, is a logical fallacy comparable to identification of the assertions "every man has a woman for wife" with "a certain woman is the wife of every man." Our examples show that the equilibrium return $p_1 (m^*, \ldots, m^*)$ to every player in a symmetric game can be much inferior to the symmetric maximum return; indeed, $p_1 (m^*, \ldots, m^*)$ can lie very close to the absolute *minimum* possible value of the payoff function. If a collection of players find themselves trapped in so decidedly suboptimal an equilibrium, they may desperately require some high degree of centralized decision-making as the only way to escape catastrophe. Practical instances of this general fact are known to everyone.

Equilibria in the Economic Sphere

How then can it be argued that natural equilibrium in the economic sphere is generally optimal (whereas, as we have seen, equilibria in more general models can often be suboptimal)? To focus on this question, we must not continue to confine ourselves to the consideration of logically informative but possibly misleading thumbnail examples, as we have done till now, but must go on to give a game-theoretic account of at least the principal forces affecting the overall behavior of a real economy. However before doing so it is irresistibly tempting to use the intellectual tools we have developed to dispose of two persistent catchphrases of popular economic discussion. The first holds that government interference in the economy is necessarily undesirable simply because it is economically unproductive, i.e., "All that a government can do is take money away from some people and give it to others, and charge both for doing so." A second closely related pessimistic phrase is: "There is no such thing as a free lunch."

To contest the first slogan, let us begin by admitting its major premise: that in direct terms the economic activities of the government are purely unprofitable, since in direct terms government can only penalize and not create, so that its activities always wipe out opportunities which free individuals might have been able to exploit; and since to give something to A the government must take it from B; and since to do either some administrative cost must be incurred. To go on from this premise to the conclusion that the economic interventions of government cannot be generally and substantially beneficial is to assume that the equilibrium value of a game cannot be increased by reducing its payoff function, an assumption which the preceding examples have shown to be entirely invalid. As to the Free Lunch, our examples also show that by coercively transferring payoffs from some of the players of a game to others, one can (provided that a pattern of transfers is suitably chosen) shift the game's equilibrium in such a way as to leave *every* player with a larger actual payoff than before. In any such situation there is, indeed, such a thing as a free lunch. (It would be wrong, however, for us to conclude merely from the logical existence of such phenomena that they actually occur in the economy as it really is.)

Relevant Economic Facts

To determine if the natural equilibrium of the economy is optimal or suboptimal we must introduce additional empirical material, painting a picture of the dominant forces in the real economy detailed enough for us to derive at least a coarse idea of its motion. Unfortunately, to do so we will have to step onto considerably thinner ice than that which has borne us till now, since the economy is an enormously varied and complex thing, valid generalizations concerning which are hard to come by. In forming the picture which follows we are in effect selecting one particular model out of many possible mathematical models of reality; to judge the adequacy of this model fairly requires empirical information going far beyond the boundaries of the present article. It is entirely possible that by selecting another (perhaps better) model we might be led to conclusions diametrically different from those which we will draw. These hazards notwithstanding, we shall identify the following as describing the principal forces at work in the economy:

Material profitability of production. Let the various goods produced within an economy be numbered from 1 to k. To produce one unit of the jth commodity some (nonnegative, but possibly zero) amount π_{ji} of the ith commodity will have to be consumed as "raw material," and in addition some quantity ϕ_{ji} of the ith commodity will have to be used as "capital equipment" and this will be tied up temporarily during the actual production of the jth commodity. If the economy is not doomed to progressive exhaustion, there must exist some overall arrangement of production which is physically profitable, i.e., an arrangement which produces an amount v_i of every commodity which exceeds the amount that is consumed in production.

Pattern of distribution. Suppose that we designate the amount of labor (measured, e.g., in man-hours) required for the production of one unit of the jth commodity as π_{j0}. Then one part of the net physical economic product (this physical net being $v_i - \Sigma_{j=1}^{k} v_j \pi_{ji}$) is distributed as *wages*; the portion that then remains may be called national physical net *profit*, and is available either for consumption or for addition to the national stock of capital or consumable supplies. Wages are distributed in proportion to hours worked (i.e., to $\Sigma_{j=1}^{k} v_j \pi_{j0}$); each individual has the right to claim part of the national physical net profit (as a dividend) in proportion to the portion of capital which he owns.

Pattern of consumption. Of the portion of the net national product available to them, individuals (or families) will elect to consume a part, but will also wish to set a part aside for future consumption. The part set aside may be intended as a retirement fund, an accumulation for planned major expenditure, a reserve for contingencies, or a bequest to family or society. Generally speaking, the proportion of income reserved from immediate consumption will rise with rising income (for various reasons: e.g., people will not want their retirement income to fall drastically short of the income to which they have grown accustomed). At sufficiently high income levels the desire to consume will begin to saturate, and the bulk of income above such levels will be reserved for investment. On the other hand, individuals not constrained to do so will not react to a fall in income by immediately reducing their level of consumption; instead they will at first maintain their level of consumption by consuming a portion of the wealth to which they have claim, cutting back their habitual level of consumption only gradually.

Income reserved from consumption takes the form of claims to a portion of the inventory, capital plant, and income of productive enterprises; also claims on the future income of enterprises or of other individuals; and also exemptions from anticipated future tax payments (this latter being "government debt").

The profit maximization strategy of productive enterprises. A firm manufacturing some given commodity maintains an inventory, selling off portions of this inventory as orders are received, and initiating new production to rebuild inventory as appropriate. In attempting to maximize a firm's profitability, its managers manipulate the three control variables available to them: product price, level of current production, level of investment expenditures. If the production of one commodity consistently remains more profitable than that of another, investment will be directed towards production of the more profitable commodity. Thus over the long term we must expect the rates of profit on all types of production to converge to a common value ρ. When this "investment equilibrium" is reached, the price of each commodity will be the price which yields the normal rate of profit ρ on its production. (Thus, if the wage rate per man-hour is w and p_j is the price of the jth commodity, we expect the managers of all firms to be constrained in the long run by the following equation in their choice of prices:

$$p_j = \sum_{i=1}^{n} \pi_{ji} p_i + \pi_{j0} w + \rho \sum_{i=1}^{n} \phi_{ji} p_i.)$$

A different and shorter-term set of considerations enters into the determination of current production levels. The commodities that we have been considering, which may for example be clothing, automobiles, children's toys, or residential dwellings, are not as homogeneous or time-invariant as our argument till now may have suggested. Indeed, as time goes along the sale value of such items is progressively undermined by changing factors of fashion and technological advance, by seasonal considerations, and by storage and upkeep costs applying even to unused commodities. For example, clothing not sold promptly will miss its season and have to be held for a year, by which time it may well have grown unfashionable and have to be sold at a drastic markdown; similarly, computers built and not sold will rapidly become valueless as technology advances. To model the effect of all the important technological and deterioration effects of this kind, we can oversimplify and assume that all commodities are slightly perishable, i.e., that any part of a given commodity stock which is neither consumed nor used as raw material during a given nominal cycle of production decays at some specified rate. This decay rate can be quite large, and if it is assumed to be larger than the physical profitability of production it is obvious that all firms will have to tailor planned inventories rather closely to expected sales. This basic fact traps all firms in a situation resembling the Stagnation game discussed above: all firms collectively will profit as the level of economic activity rises, but no firm individually can afford to let its own level of production outrun that of all the others. In this relationship, which as we have seen can lead very directly to a stable suboptimal equilibrium, we see a root cause both of recessions and of protracted periods of economic stagnation.

Factors related to the housing and mortgage market. In purchasing a home, few people pay cash out of pocket. Instead, they require mortgage financing. The appeal, and even the possibility, of purchasing a house therefore depends critically on the current state of the mortgage market. Two related but significantly different factors, namely mortgage rates and mortgage terms, must be considered. High interest rates are seen by the prospective home purchaser as high monthly payments, in direct linear proportion. Thus rising interest rates choke off real estate sales. Unfavorable mortgage terms can erect an even more impenetrable barrier against home purchases. In periods of high economic activity, commercial demand for short-term loans will bring banks close to the limit of their lending ability. A bank finding itself in this situation can afford to be choosier about the mortgage loans it makes; in particular, it can afford to insure the quality of its mortgages by increasing the proportion of total purchase price that the prospective purchaser is required to pay out immediately. The prospective home purchaser will see this as an intimidating increase in required down payment, e.g., the amount of cash that he must come up with to purchase a $60,000 home may rise from 15%, i.e., $9000, to 25%, i.e., $15,000. Those unable to meet the rise in terms must quietly retreat from their would-be purchase.

Stable spending patterns of retired persons and government. If we ignore the small minority of persons who derive a substantial portion of their income from either dividends or royalties, we can say that the income of nonretired persons is proportional to the number of hours that they work, and hence is controlled in close to linear fashion by the general level of economic activity. (However, unemployment insurance payments by the government serve to lessen the impact on income of a diminished level of economic activity.) But the income of retired persons, and hence also their spending, is constant (in cash terms, if they are living on annuities or other fixed-dollar pensions, or in real terms, if they support themselves by selling previously accumulated stocks or real estate), and hence their spending levels (in cash or real terms, as appropriate) is also constant. Much the same remark applies to the activities of government, which continue at legislated levels in substantial independence from the oscillating level of economic activity which surrounds them. (Of course, the federal government has a considerably greater ability to legislate a given level of government demand independent of the level of activity in the remainder of the economy than do the state and local governments, since any state or municipality whose tax rates remain above average for long will soon find its share of economic activity, and hence its tax base, shrinking as economic institutions able to do so flee. We may say here that the overall taxing power of the separate states is held in check by the fact that they are independent players in a fifty-sided game of "offer better tax terms.")

Econometric Models

We have now drawn a complete enough sketch of the principal determinants of economic activity to explain the structure of econometric models of the sort actually used to track the economy and to guide national economic authorities. Of course, since the quantitative material, motivational, and planning factors which determine the motion of the real economy are enormously varied, the development of a quantitatively reliable model of the economy is not easy and the best currently available models are far from perfect. Nevertheless, carefully constructed linear econometric models generally give good forecasts within a forecasting range of about a year. (For example, the curves in Figure 2 show how well the influential Wharton econometric model was able to track the overall level of manufacturing production and nonfarm housing construction in the years 1948–1964.) Beyond a one-year range, the economic behavior predicted by such models will differ progressively from real events owing to the accumulating effects of inaccuracies in the model and nonlinearities in the real economy. However, a year's advance prediction gives ample time for policies to be adjusted, if

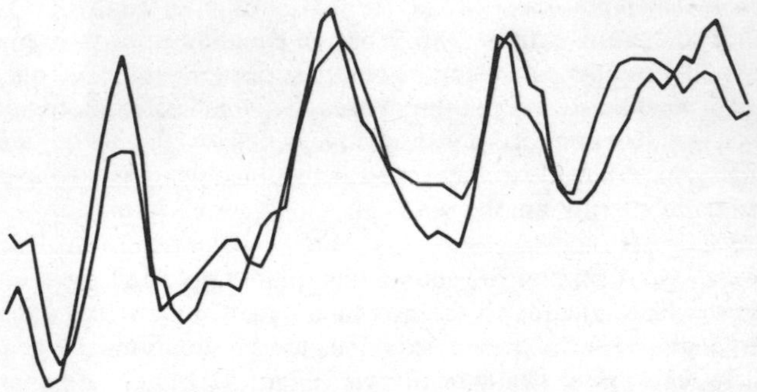

(a)

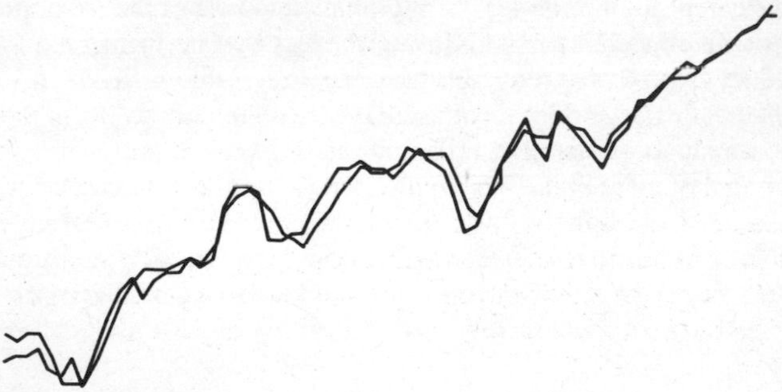

(b)

Figure 2. (a) Prediction of nonfarm housing construction and (b) Prediction of total Manufacturing Production by the Wharton Econometric Model for the Years 1948–1964. Black curves are predicted values; colored curves are actual values.

only the will to do so is present and is buttressed by a general qualitative understanding of what is happening.

In constructing such models, one begins with a list of major influences on economic activity like that which we have presented above, but expands the crude account which we have given into a more detailed but still informal picture of the manner in which economically critical consumer and corporate planning variables (e.g., purchasing plans, production, inventory, and investment levels) are likely to depend on objective factors significant to individual planners (e.g., income, security of income, sales levels, rate of inventory turnover, plant capacity in use, long- and short-term interest rates). This informal reasoning is used to set up a collection of equations, generally of the linear form

$$n = c + d_1 x_1 + \cdots + d_n x_n,$$

in which the terms which can occur are prespecified, but in which the numerical coefficients $c, d_1, \ldots, d_n$ are initially undetermined. (A specific example of such an equation—the Wharton equation used to predict levels of residential housing construction—is described in the box on p. 291.) However, it must be stated that the process by which econometric equations are formed is still uncomfortably subjective, since existing techniques generally do not give us any clear way of focusing objective data strongly enough to discriminate sharply among alternative choices of terms to include in a model.

Nevertheless, in spite of the substantial quantitative and even qualitative uncertainty which adheres to them, econometric models like the Wharton model do embody, and their success does tend to confirm, the general picture of economic forces sketched above. In particular, detailed examination of these models does tend to support our earlier suggestion that the totally unregulated economy can behave suboptimally, and hence suggest that recessions can be understood as suboptimal equilibria of the stagnation type, whose operative mechanism is found in the fact that no firm can afford to let its own level of production outrun that of all the others. From this point of view, the most noteworthy mechanism of the familiar cycle of boom and recession seems to be the gradual accumulation of inventories during boom periods, and the manner in which the rising interest rates associated with such periods act to choke off housing sales. Once these forces have caused the economy to fluctuate downward, automobile sales will ordinarily respond sharply, and much of the dynamic oscillation associated with a recession will tend to be concentrated in the automobile sector and related industries.

To remove this suboptimality, economic policy-makers have attempted since the time of Keynes to shift the economic equilibrium toward higher levels of production by raising the total of personal, investment, and government demand that determines the level of production which can be sustained without inventories beginning to accumulate to an unreasonable degree.

Residential Housing Construction in the Wharton Econometric Model

The 1967 Wharton econometric model, which served as a prototype for many subsequent econometric models, contains an equation used to fit the observed behavior of investment in residential housing. The assumed form of this equation reflects the general factors (such as total disposable income, mortgage availability), thought to influence residential housing purchases. The precise form of this equation, which we present as an example in order to make the structure of such models vivid, is:

$$I_h = 58.26 + 0.0249Y - 45.52\left(\frac{p_h}{p_r}\right)_{-3}$$
$$+ 1.433(i_L - i_s)_{-3} + 0.0851(I_h^s)_{-1}.$$

Here I_h is the rate of investment in residential housing in billions of dollars, measured in a given quarterly (3month) period; Y is total disposable personal income (in billions of dollars) for the same period; p_h is an indicator of the average price of housing and p_r an indicator of average rent levels. (The subscript -3 attached to the fraction p_h/p_r indicates that the ratio to be used is that calculated for a period three quarters (i.e., 9 months) prior to the period in which I_h is measured.) The terms i_L and i_s are the long- and short-term interest rates (again taken from a period 9 months previous to the measurement of I_h), and I_h^s is the rate of housing starts (measured three months, i.e., one quarter, prior to the measurement of I_h). The results of using this model over the sixteen year period 1948–1964 are illustrated in Figure 2.

Policies of this general kind are administered in various ways, for instance, by lowering personal income taxes, by increasing pension payments, by increasing the number of pensioners or the size of the armed forces, or by encouraging private and public investment through investment-related rebates and guarantees to private firms or through major direct public investments in, e.g., the road system, municipal improvements (including housing), or the national armament. It deserves to be noted in connection with this comment on anti-recession policy that the policy-maker's eye must focus primarily on the possibility of continuing economic underperformance rather than on the ups and downs of the business cycle, just as the prudent car owner ought to be more concerned with poor engine performance (whose cost will mount mile after mile) than with occasional, even if alarming, engine knocking. We note in this connection that the most destructive aspect of the business cycle may be the way in which it confuses macroeconomic policy, since the cyclic oscillation of the economy insures that if one applies almost any measure, sufficient or insufficient, and then waits long enough, an encouraging boomlet of economic improvement will allow one to argue for the correctness of one's policy.

The above appreciation of the mechanism of recessions, together with the general game-theoretic reasoning developed in the preceding pages, suggests an attempt to analyze other significant economic problems in the same way as suboptimal equilibria of game-like models, and to see what cures recommend themselves from this viewpoint. To follow this thought, we begin by noting that the anti-recession measures sketched above act to dissolve suboptimal equilibria by rewarding players who avoid available but globally undesirable courses of action, e.g., by subsidizing investment that would otherwise seem unwise or consumption that would otherwise be impossible. But these measures are not entirely unproblematical, since they can push the economy toward suboptimalities of other sorts. Note in particular that to implement policies like those we have listed, a government needs to dispose of some form of reward, which must be available readily and in massive quantities. The items of reward used in practice are notes of government indebtedness of various kinds; the sum of all these notes constitutes the national debt. These notes are most rationally viewed as certificates of exemption from future tax payments (since the notes themselves can be presented in payment of taxes). This fact gives the notes their basic value; and the inducement to hold them can be increased as much as desired by making them interest-bearing at a suitable rate. At first sight, it might appear that this debt-generating technique cannot safely be used over the long term, since one might fear that the accumulation of larger and larger masses of exemption certificates would eventually undermine the taxing ability which gives these certificates their value, leading ultimately to a collapse in the value of government notes and of money as well, that is, to a hyperinflationary disaster like the German hyperinflation of the 1920's or the Chinese hyperinflation of the 1940's, which would constitute another kind of suboptimal collective phenomenon. To eliminate this phenomenon, a monetary authority must ensure that whenever the accumulation of these certificates becomes troublesome some part of them will be drawn off. This can be done simply by imposing some appropriate form of property tax. In fact, such a tax can be made to act continuously, for example, by setting the estate-tax level appropriately.

Thus from our game-theoretic point of view we can recognize inflation itself as a price-related suboptimality of the economy. Analysis of theoretical models suggests (and empirical econometrics seems to confirm) that in the economy as it is the general level of prices and wages is not determined absolutely (e.g., as an equilibrium level in some macroeconomic price/wage model), but simply floats about under the control of purely frictional forces. The Wharton model equations, for example, state (see box on p. 293) that prices follow wages and wages follow prices, and hence suggest that the

Prices and Wages in the Wharton Econometric Model

In the Wharton econometric model, the general price level p_m of manufactured goods is determined by the equation

$$p_m = -0.170 + 0.514\,(W/X) + 0.2465\,(X/X_{\text{max}})$$

$$+\ 0.6064\left(\frac{(p_m)_{-1} + (p_m)_{-2} + (p_m)_{-3} + (p_m)_{-4}}{4}\right).$$

Here W/X is the labor cost per unit of product, X the overall level of manufacturing production, and X_{max} the estimated maximum capacity level of production. Translated into heuristic terms, this equation simply states that prices are proportional to labor costs, but will be raised somewhat above this as firms find themselves reaching capacity production; however, (as is shown by the last group of terms) prices do not adjust immediately to changed costs or to pressure on capacity, but adjust only gradually over a period of something like one year. The Wharton model's equation for the wage level (of manufacturing employees) is

$$W = W_{-4} + 0.050 + 4.824\,(p_{-1} - p_{-4}) - 0.1946\,(W_{-4} - W_{-8})$$

$$+\ 0.1481\left(\frac{(U - U^*)_{-1} + (U - U^*)_{-2} + (U - U^*)_{-3} + (U - U^*)_{-4}}{4}\right)$$

Here, W designates the wage level, p the price level, U the general rate of unemployment, and U^* the rate of unemployment among males 25–34 years of age, so that the difference $U - U^*$ measures the extent to which the pool of first-hired, last-fired employees is fully used. In heuristic terms, this equation shows the wage level rising five percent per year, but rising additionally in proportion to the rise in prices during the preceding year, and also rising somewhat less than otherwise expected if a substantial wage rise was registered the year before (since poor raises last year argue for better-than-average raises this year). Moreover, unemployment in the first-hired, last-fired pool acts to restrain the rate at which wages rise. (However, this rate of rise is resistant to other categories of unemployment.)

overall wage/price level does not tend to any equilibrium. Moreover, these equations indicate that the frictional force which holds back rises in the general wage level depends directly on an exceedingly undesirable econmic phenomenon, namely, unemployment. This being the case, we need to consider what would be the least unacceptable response to a coupled inflation-unemployment suboptimum.

U.S. macroeconomic policy has been considerably disoriented during the past half decade by confusion about how to deal with this problem. The significant facts reflected in the models we have discussed suggest that there may exist no satisfactory response to it other than to admit some degree of government intervention in the formation of prices and wages. At the same time, examination of the relatively mild consequences of moderate levels of inflation suggests that a mild frictional intervention, which deliberately tolerates some modest level of inflation, thereby avoiding the rigidities of a more strict wage/price control system, may be sufficient. If, for example, a central authority counts on a 2% annual rise in productivity and is prepared to tolerate a 4% inflation rate, wage/price control might consist in a law providing that any salary rise of more than 6% per annum would be taxed (and even withheld) at a close to 100% rate, and that a corporation whose prices rose by more than 5% in a given year would be liable for a surtax on the resulting excess revenue.

Of course, one must always fear that coarse, rigid centralized economic interventions will hamper the private sector's stunning ability to discern desired products, services, and productive opportunities in general. Nevertheless, the game-theoretic point of view central to the present article shows that these objections do not necessarily apply to interventions which, by grasping and manipulating some small and well-chosen set of macroeconomic parameters, move the economy away from an undesirable equilibrium. Indeed, a policy of this kind should be regarded not as contradicting but as perfecting the approach advocated by Adam Smith, specifically by adjusting the global economic environment in a way calculated to ensure that his "invisible hand" will really lead the participants in an economy to an optimal equilibrium, i.e., to a situation in which no further benefit without a compensating sacrifice is possible. It is, of course, true that even interventions which merely block off options that lead directly to a suboptimal situation will be perceived by some, and perhaps by many, of the participants in an economy as arbitrary restrictions which cut off profitable and desirable alternatives. But this view can be mistaken, since the alternatives being cut off are the gateway to an overall process, irresistible by any single firm or limited group of firms, in which everyone's most rational efforts lead all to a less desirable situation than all might otherwise have enjoyed.

Suggestions for Further Reading

General

Davis, Morton D. *Game Theory, A Non-Technical Introduction*. Basic Books, New York, 1970.
Drescher, Marvin. *Games of Strategy—Theory and Applications*. Prentice-Hall, Englewood Cliffs, 1961.

Technical

Bachrach, Michael. *Economics and the Theory of Games*. Macmillan, London, 1976.

Dusenberry, J.S., Fromm, G., Klein, L.R. and Kuh, E. (Eds). *The Brookings Quarterly Econometric Model of the United States*. Rand-McNally, Chicago, 1965.

> The theory and technique of econometric model construction.

Klein, Lawrence R. and Evans, Michael K. *The Wharton Econometric Forecasting Model*. Economics Research Unit, Department of Economics, Wharton School of Finance and Commerce, Philadelphia, 1967.

von Neumann, John and Morgenstern, Oscar. *Theory of Games and Economic Behavior*. Princeton University Pr, Princeton, 1955.

Mathematical Aspects of Population Biology

Frank C. Hoppensteadt

Why don't we see 17-year locusts every year? Are we catching too many fish? What will the age distribution be in the U.S. ten years from now? How much sewage can be treated by a bacterial population in a day? These are typical questions from population biology, and the examples presented here show how these questions and others can be studied by mathematical methods. Population biology is primarily concerned with counting, estimating, and predicting population sizes. These large complicated problems are of considerable importance to science, industry, and government, and are often solved using very simple mathematical ideas.

One common population problem is to determine the mechanisms that cause and maintain biological rhythms. Body temperature and water balance are familiar regular daily fluctuations in individuals, but there are also rhythms in populations, such as lunar and annual cycles in mating behavior. One of the most remarkable population rhythms, which we will examine in some detail, occurs in 17-year locusts (cicadas), where space limitations and hungry birds synchronize emergences so cicadas are seen only every seventeen years. Study of annual population size is frequently based on geometric devices called reproduction curves. Refined analysis of these curves reveals that in some cases a population may behave in a completely chaotic fashion even though it is governed by simple biological law. This surprising result is just beginning to be understood, and it raises many unsettling questions about whether there is any hope of modelling many real populations in detail.

A different kind of problem is posed by management of exhaustible resources like timber and fish. These difficult and complicated problems range from maximizing profits for exploiters of the resource to the protection of the resource from extinction. Mathematical methods provide ways of combining biology with economics, and give insight into how populations react to stresses placed on them by economic optimization. An example given below illustrates how some resources can completely collapse with little warning when profits from harvesting are not strictly controlled.

Mathematical techniques are used in population biology to describe many other things as well. Geographic distribution of genes, age distribution of populations, and the spread of forest disease are among the examples we draw from to illustrate the mutually reinforcing relationship between mathematics and biology. While the empirical impact of mathematics has not been as great in population biology as it has been in physics, it has provided important guidelines for determining what questions to ask, what data to collect, and what methods to use in analyzing the data. As the questions posed by population biology become more urgent, and the tools of the applied mathematician become more powerful, the relationship between the two will undoubtedly prosper.

Population Rhythms

Seventeen-year locusts (cicadas) have the remarkable property that they appear in great numbers every seventeen years, but none appear in the intervening years. Adult cicadas live for a few weeks during which time they lay fertile eggs in trees at their emergence site. These fall onto the ground and hatch as nymphs which then go underground, attach to rootlets for nourishment, and remain there for the next seventeen years. After seventeen years, they emerge from the ground as adults, fly to the trees and reproduce. The cicada life cycle is thus seventeen years.

The length of the life cycle does not, however, explain the fact that cicadas are not observed every year. There are several species of cicadas having different life spans (for example, 13, 7, 4, or 3 years) but only the 17-and 13-year cicadas have synchronized emergences. Those having shorter life spans (of seven or fewer years) appear in comparable numbers every year.

This curiosity can be explained by a careful analysis of the effect the cicada emergence has on cicada predators. These predators (mainly birds) respond to a large cicada emergence by producing more young, and by switching from other food sources. We describe this response by means of a predation threshold, that is, a population level below which extinction is certain after several generations. The predation threshold increases in response to a large emergence and decreases when there is a small emergence. We assume, in addition, that no matter how large the emergence, the cicada's offspring cannot exceed a certain fixed level known as the environmental carrying capacity. These two features—a limited carrying capacity and a responding predation threshold—can balance subtly to provide a mechanism that synchronizes emergences.

To see how this mechanism works, suppose that emergences of equal size occurred for seventeen consecutive years. (This could be rigged as an experiment by planting equal numbers of cicada nymphs for each of seventeen

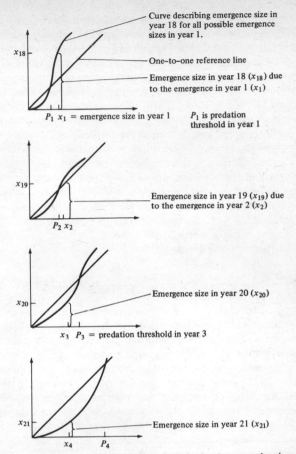

Figure 1. These diagrams show how the cicada reproduction curve changes in the first four years of the experiment, in response to emergences of previous years. The offspring of the first and second year classes grow while those of the following year classes face certain extinction.

years.) Then as the graphs in Figure 1 show, the offspring of the first two years will increase in numbers, while those of all subsequent years decline and will become extinct after several generations.

The tuning of this system is rather delicate. If the predator response to an emergence is too large, then all year classes will eventually be eliminated. If it is too small, then all year classes will persist. As mentioned earlier, cicadas with long life spans are observed to have synchronized emergences while those with life spans of seven years and less have balanced emergences. Refined analysis of the model suggested by Figure 1, using realistic values for cicada fertilities, environment carrying capacity, and predator response, shows that organisms with life spans of ten years and less approach balanced emergences while those with eleven years or more approach synchronized emergences. The results of the model for life spans of seven and thirteen years are shown in Figure 2.

Models of this kind are examples of an extremely important mathematical concept known as a nonlinear oscillator. A great many nonlinear oscillators occur in a variety of systems from electrical circuits to physiology. Many are described in terms of a model based on a differential equation that was developed by the Dutch engineer Balthasar van der Pol (1889–1959). This model, now called the van der Pol equation, has found important applications to tunnel diodes, mechanical devices and the heart and is still yielding surprising new results. In fact, the cicada model is closely connected with the van der Pol equation, as are certain other models just beginning to be understood that show how a population may behave in a totally chaotic fashion even though it is governed by simple, nonrandom laws. We will return to this intriguing phenomenon later.

Cobwebs and Fisheries

A graphical device called cobwebbing is frequently used to describe how populations change from year to year. One of the simplest examples of it occurs in fishery management. Every year the adult stock gives birth to a population of young, some of which will survive to form the adult stock in the next year. These survivors are called recruits. To simplify things, let's suppose that the fish live for only one year as adults, after which they reproduce and die. The number of recruits in year n due to a standing stock

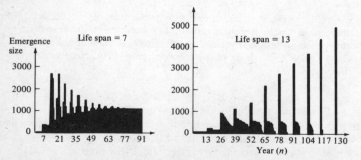

Figure 2. The results of numerical solution of the cicada model for two life spans ($L = 7$ and $L = 13$) are presented here. All other parameters in the model are identical in these calculations. The first shows that balanced emergences result, while the second results in synchronized emergences.

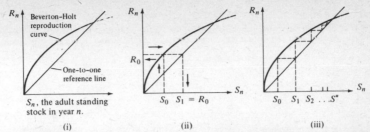

Figure 3. Beverton–Holt Model. This diagram shows how the dynamics of a fish population can be determined by the cobwebbing method once the reproduction curve is known. R_n represents the recruits to the adult stock in year $n + 1$ due to standing stock S_n. The standing stock approaches the equilibrium level S^* in successive generations.

of size S_n is denoted by R_n; then the stock in the subsequent year is $S_{n+1} = R_n$. By plotting R_n against S_n we can illustrate how these numbers are related. One such model (the Beverton–Holt model) is depicted in Figure 3, and it is typical of many real fisheries.

In principle, a stock-recruit relation can be constructed from observations of the fish population. The curve in Figure 3(i) indicates that there is a diminishing return to the adult stock: an increase in the adult stock from a large stock size results in a smaller increment of recruits than the same increase in a smaller stock. Unfortunately, precise data are difficult to get because of the inaccessibility of fish populations. No scale is given on our graphs since we seek only a qualitative description of the fishery.

The population's dynamics over several years can be determined by locating on the stock-recruit relation the successive values of S_n and R_n. An initial standing stock of size S_0 will result in a recruit stock of size R_0; so the standing stock for the next year will be $S_1 = R_0$. All this can be determined easily on the graph, as indicated in Figure 3(ii): S_1 can be found by reflecting the number R_0 through the one-to-one reference line onto the S-axis. This process can be continued easily, producing the "cobweb" pattern illustrated in Figure 3(iii). We see that the adult stock will approach a stable stock size S^* in succeeding years.

Chaotic Populations

As simple as the cobwebbing method might appear, it is closely connected with some quite difficult problems that mathematicians have been studying for the past one hundred years. One of these difficulties can be illustrated with a reproduction model (Figure 4) that appears to be only slightly different from the Beverton–Holt model. In this case, the recruit size decreases with large increasing adult stock size (e.g., because of overcrowding) until the carrying capacity K is reached. Thereafter, some calamity, such as the

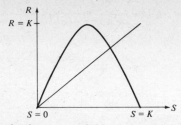

Figure 4. A reproduction curve of this kind can describe quite complicated dynamics. A typical solution is described as a function of time in Figure 5 and it has a chaotic appearance.

adults using up all of a nutrient vital to the young or eating them, leads to no young surviving.

The surprising thing about this model is that for most values of S_0, the sequence of adult stocks in later years, $S_1, S_2, S_3, \ldots$, is chaotic! (Try cobwebbing this curve.) A typical solution is described in Figure 5. While this is a somewhat contrived model, it is typical of a wide class of more reasonable ones. Thus, many populations that appear to be evolving in a random fashion may have a quite simple nonrandom reproduction curve underlying them.

Reproduction curves that produce chaotic population levels are closely related to random processes. Coin tossing is one of the most common random processes. A coin is tossed and lands showing either a head or a tail. If the coin is fair, then in a large number of tosses, heads should come up about one-half the time. We say then that there is probability $1/2$ of getting heads. Tossing a fair coin can be described by the diagram in Figure 6. If we start with heads, then on the next toss there is probability $1/2$ of getting heads and $1/2$ of getting tails. Similarly, if we start with tails. This is one description of coin tossing; we can make up another (more complicated) one based on Figure 7.

On the horizontal axis, we split the interval $0 \leq p \leq 1$ into two half-intervals, one corresponding to the probability of heads (H), the other to the

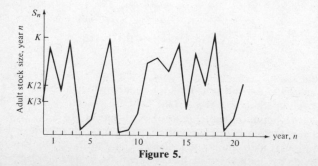

Figure 5.

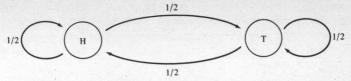

Figure 6. The coin tossing experiment described in the text can be schematized as shown here. The arrows indicate the possible occurrences and their probabilities.

probability of tails (T). The same is done on the vertical axis. Next, line segments are drawn that relate the horizontal H interval to the vertical H interval (A) and to the vertical T interval (B), equally. Similarly, half of the horizontal T interval is related to the vertical T interval (C), and the other half to the vertical H interval (D). The resulting tent-like curve is interesting to cobweb: it turns out that the sequence constructed by cobwebbing will, after a great many steps, land in the H interval about one half the time and in the T interval the other half of the time. This tent-like curve clearly resembles the reproduction curve of Figure 4 and can be changed into it by a simple (trigonometric) change of variables.

A computer simulation of this model gives an even better feeling of how solutions behave. Let's divide the horizontal interval into 1000 small intervals (cells) of equal length. Next, we take a point and iterate (cobweb) it 500 times, keeping track of the number of times this sequence lands in each of the 1000 cells. This iteration is repeated for 100 randomly chosen initial points. (This process might sound intimidating, but it is quite easy to do on a computer.) The results of this experiment can be summarized by plotting the number of times each cell is hit (Figure 8). This figure shows clearly that the solutions move all over the horizontal interval.

It is somewhat depressing to find that such a simple model can govern

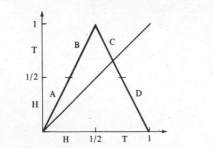

Figure 7. Another visualization of the coin tossing experiment is given by this reproduction-like curve.

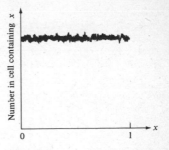

Figure 8. The result of the computer simulation described in the text shows that most solutions do migrate all over the unit interval.

such complicated behavior. In fact, such behavior is quite typical in many systems, for example, fluid turbulence or chaotic behavior of nonlinear oscillators. From this point of view, it is surprising that there is any order to things.

Economics and Ecology

Reproduction curves are used in the management of exhaustible resources to evaluate possible management programs, usually with the goal of optimizing some economic performance index. Fishery management provides a good illustration of how a mathematical model can be used to study the effects of optimization on populations. The inacessibility of fish populations for direct measurement makes mathematical methods attractive in the study and control of fisheries.

Various kinds of optimization can be illustrated for a fish population that is modelled by a Beverton–Holt model. Two quantities are observable in a fishery: the effort expended (measured in standard fishing boat days on the fishing grounds) and the harvest that is produced, called the yield. Harvesting from the population can be described graphically by a simple device. If fishing effort E is exerted, then we replace the one-to-one reference line by a steeper one whose steepness increases with increasing E (Figure 9). This automatically reduces the next year's standing stock in the proper way. For fixed effort E, the system will be maintained at a standing stock size S_E^* which is smaller than the stock size S_0^* without fishing. Figure 9 shows that the population creates R_E^* recruits for next year's population and that $R_E^* - S_E^*$ of them are harvested, to produce the (biological) yield Y of the fishery: $Y = R_E^* - S_E^*$.

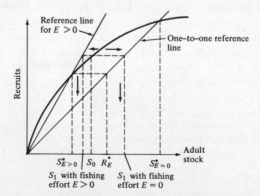

Figure 9. Given a stock-recruitment curve, the effects of increased fishing effort can be accounted for by replacing the one-to-one curve by a steeper one. The difference $R_E^* - S_E^*$ gives the yield when the effort E is exerted and the fishery is in equilibrium.

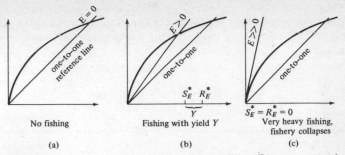

Figure 10. Three examples of yield when various efforts are exerted.

The various cases illustrated in Figure 10 reveal the dependence of Y on E; these results can be summarized by plotting effort against yield, in an Effort-Yield diagram (Figure 11). Such diagrams are used to control fisheries to maximize the sustained yield. The manager allows effort to increase until yields begin to decrease. Next, the effort is cut back until yields increase, etc. In this very natural way, one attempts to steer the fishery to its maximum sustainable yield.

Unfortunately, there are several defects in this management scheme. First, if the fish population has a reproduction curve that is just slightly different from the Beverton–Holt curve, then this management scheme can be catastrophic. For example, it can be shown that the reproduction curve in Figure 12(i), which is probably more realistic for most fish populations than the Beverton–Holt model, gives an effort-yield curve of the form illustrated in Figure 12(ii). In this case, if effort is allowed to increase beyond the critical level E_c, the fish population will die out, and so the fishery will collapse. The difference $E_c - E_{max}$ between the critical effort (which could cause collapse) and that effort which produces maximum yield measures the safety margin. If this margin is small, a manager who sees yields begin to decline for increasing effort may not be able to introduce controls on effort fast enough to prevent it from exceeding E_c.

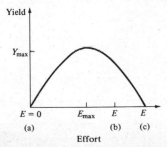

Figure 11. The Effort-Yield Curve. The three cases from Figure 10 are indicated by (a), (b), and (c).

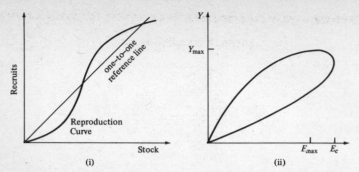

Figure 12. When a reproduction curve like this occurs, the Effort-Yield diagram takes on a bizarre appearance. If effort is allowed to exceed E_c, the fish population collapses. The difference $E_c - E_{max}$ measures the margin of safety in using the Effort-Yield curve in management of the fishery.

Another defect is that no economic factors are accounted for in this model; only the biological yield is optimized. If instead management chooses to optimize current or discounted future revenues, then quite different harvesting schedules will result, even with the Beverton–Holt reproduction model. For example, if the revenue from future harvests is discounted, it may be in the economic interest of the fisherman to completely eliminate the entire fish population as soon as possible. This and other models dealing with the interaction of economic and biological factors have been developed by Colin W. Clark of the University of British Columbia and others, and a new field termed Mathematical Bioeconomics has appeared.

Whaling provides a vivid illustration of this interaction. It is estimated that the maximum sustained biological yield of the current Antarctic fin whale population is about 1875 whales per year. If each harvested whale brings in $10,000, then the income is $18.75 million per year. However, the value of the standing stock is estimated to be $750 million. If this were collected now and invested at 5 percent interest, the proceeds would be $37.5 million per year, twice the maximum sustainable income that would be realized by allowing the whales to survive! While this calculation neglects many facts (for example, could $750 million be collected now?) it does clearly demonstrate an economic threat to natural resources.

Biology and Geography

The spruce budworm is a pest that defoliates balsam and spruce trees leading to the death of the trees. When these trees are gone, they are replaced by beech trees which do not support budworms. After the budworms are gone,

Infestation Patterns

the spruce and balsams eventually displace the beeches through competition for nutrients and sunlight, and the cycle can repeat. Spruce budworms presently threaten forests in Eastern Canada and the Northeastern U.S. How they spread from location to location is an important but complicated aspect of budworm behavior. Although this dispersal process is difficult to analyze, we can gain an intuitive idea of its behavior by simulating (on paper, or in a computer) the wave-like spreading that is observed in real situations. Although the dispersal model presented here has not yet been directly related to the spruce budworms, it illustrates well both the complexity and the potential of the model.

Our simulated forest is drawn on a rectangular grid where each space on the grid represents one tree or a clump of trees. Each space is called a site, and is designated as either infested (denoted by (+)), defoliated (−) or green (O). The pest infestation (in our paper forest) spreads according to the following rules:

1. A defoliated site becomes green next season.
2. An infested site becomes defoliated next season.
3. An infested site will next season infest those of its four nearest neighbors that are green.

These rules and their effect on certain initial infestation patterns are illustrated in the box above.

In most cases a wave of infestation propagates away from the initiation site in succeeding years. But in some cases (Figure 13) an advancing front of infested sites sheds curious spiral waves off its ends. Such phenomena are known to occur in physical systems (for example, in certain biochemical reactions) but have not yet been established in any ecological system. The initial configuration in Figure 14 also creates spiral-like waves; perhaps the reader may wish to work out the evolution of this system. (To do this, plot the results of several, say seven, years.) In this case the six square site configuration returns to the initial configuration every four years, but the interlaced spiral waves propagating out are more complicated than before.

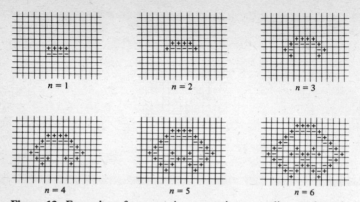

Figure 13. Examples of propagation occurring according to the rules 1, 2, 3.

While the procedure for generating infestation patterns is quite easy to state, it turns out to be very difficult to analyze. Finding formulas for the various possibilities is especially desirable, but especially difficult. Yet it is an important problem with unexpected and possibly quite useful implications. Figure 15, for example, has a well-developed wave structure in which the six sites (enclosed by heavy lines) act as a focus for infestation. If the green sites neighboring the infested sites in this focus were defoliated, then the focus would be destroyed (Figure 16). With this treatment, the infestation will remove itself from the forest after several more years.

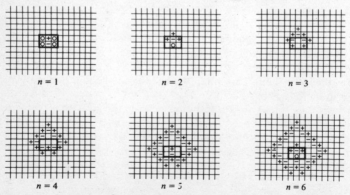

Figure 14. Examples of propagation occurring according to the Rules 1, 2, 3. Note that the boxed-in area in $n = 5$ is the same as the initial configuration. Therefore, the initial configuration acts as a pacemaker, generating waves which propagate outwards.

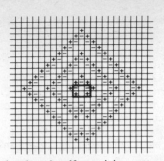

Figure 15. A well-developed, self-sustaining pattern maintained by the focus of infestation outlined.

Therefore, knowledge of foci of infestation can lead to a reasonable eradication program. The difficulty is in finding the foci. Mathematically, the problem is to locate the smallest self-reproducing patterns in an established pattern. Unfortunately, this is a surprisingly difficult problem, and no one has yet found an easy solution.

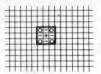

Figure 16. Defoliation as a mechanism for breaking a focus of infestation.

Models similar to the infestation simulation have been used in a wide variety of problems. For example, the spread of rumors and fads, of epidemic diseases and of an advantageous gene can all be studied by similar methods. They have also been used in cardiac physiology to describe the way that waves of contraction spread over the surface of the heart. Another version called the Ising model involves only two states (+ and ◯), and is used in physics to describe phase changes in materials. There is also a popular mathematical game called the Game of Life based on a similar two-state model (see box on p. 310). These models all provide examples of the kinds of problems being studied in one of the most active current areas of mathematical research: the study of nonlinear wave propagation.

Renewal of Populations

Before discussing recent work on human populations, it is interesting and instructive to look at one curious innovation made by the Italian mathematician Leonardo Fibonacci in 1202. Fibonacci proposed as an idealized model

The Game of Life

Consider a two-dimensional grid of sites that are either occupied (+) or vacant (○ or blank). At each time step, the configuration of occupied sites changes according to the following rules:
  a. Each occupied site having either two or three neighboring sites occupied will continue to be occupied. (There are eight sites neighboring any one.)
  b. Each occupied site having no, one, four, or more occupied neighbors becomes vacant.
  c. Each vacant site will be occupied if it has exactly three occupied neighbors.

These changes occur simultaneously at each tick of some clock.

This model was invented by J.H. Conway of Cambridge University. A great deal of work has gone into discovering which forms of excitation persist as stable configuration, which recur periodically, and which die out. Here are some examples:

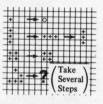

of a rabbit population that each pair of rabbits reproduce twice—at ages 1 and 2—and that at each reproduction they produce a new pair of rabbits, one female and one male. (Ages are measured in time units equal to the length of the reproductive cycle.) These two pairs will in turn go on to reproduce twice, etc. This population's growth can be visualized by plotting age against time. This is done in Figure 17, where the number of rabbit pairs having age a at time t is indicated at each age-time coordinate (a, t), and the column labelled 0 gives the number of births.

This figure shows that the number of births in successive years is given by the sequence

$$1, 1, 2, 3, 5, 8, 13, 21, 34, 55, \ldots,$$

where each term (after the second) is the sum of the two preceding terms. This is the famous Fibonacci sequence, and it arises in a surprising variety of applications, from the arrangement of leaves on plant stems to results in the theory of equations.

Let us turn now to the more serious problem of human population. In the past two hundred years the growth rate of the human population has gone

TIME	Age 0 (Number of Births)	1 (Reproductive ages)	2 (Reproductive ages)	3	4	5	6	7	8	9
0	1									
1	1	1								
2	2	1	1							
3	3	2	1	1						
4	5	3	2	1	1					
5	8	5	3	2	1	1				
6	13	8	5	3	2	1	1			
7	21	13	8	5	3	2	1	1		
8	34	21	13	8	5	3	2	1	1	
9	55	34	21	13	8	5	3	2	1	1

Figure 17. Fibonacci's model of a rabbit-pair population.

from about 0.1 per cent per year, where it had been up to that time, to about 2 percent per year at present. Now the world's population doubles itself every 35 years. Eventually, the growth rate must return to near zero, whether it is through decreased fertility or increased mortality. At one extreme the average fertility might be near eight children with a life expectancy of 15 years; at the other extreme the average fertility might be near two children with life expectancy of 75 years.

In order to understand how populations change in response to given birth and death schedules and to provide methods to compare various populations, such as those of underdeveloped countries with those of developed countries, a great deal of effort has been invested in population models. Descriptions of how a population renews itself date back to Leonhard Euler (1707–1783) and Thomas Malthus (1766–1834), and work has continued steadily since.

Although Euler's work attracted little attention at the time, Malthus's results had an astounding impact. He proposed that populations grow geometrically, in such a way that the ratio of populations in any two succeeding time periods is a constant. Much of the reaction, especially among his contemporaries, was aimed at disproving Malthus's model. Instances were found in which the model is not reliable, but surprisingly many populations are adequately described in terms of this model; see for instance, the box on p. 312 where Swedish census data is compared to the Malthusian model. In the past two hundred years it has been clearly demonstrated that Malthus's model applies well to populations in certain phases of their growth. Other models have subsequently been developed to describe the special effects of population density that cause population growth to vary from purely geometric growth.

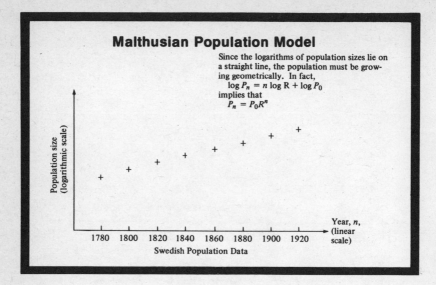

Malthusian Population Model

Since the logarithms of population sizes lie on a straight line, the population must be growing geometrically. In fact,
$$\log P_n = n \log R + \log P_0$$
implies that
$$P_n = P_0 R^n$$

Population size (logarithmic scale)

Year, n, (linear scale)

1780 1800 1820 1840 1860 1880 1900 1920

Swedish Population Data

The next major development in human population models was due to Alfred J. Lotka (1880–1949) and A.G. McKendrick who derived a way to predict a population's age structure. Their work, close to the spirit of Fibonacci's rabbit model, is the basis of what is now refered to as renewal theory. Data is grouped in time periods (usually five or ten years in length) chosen to correspond to the intervals at which censuses are taken. The number of births in the Nth time period is viewed as the sum of contributions to current births by those surviving from earlier years. As in the Fibonacci model, the entire age structure can be reconstructed from the record of births.

Suppose, for example, that a census is taken every five years; we will split the population up into five-year age groups. Moreover, for simplicity, we will consider only the female population. The birth rate will then be described by a sequence of numbers B_0, B_1, B_2, . . . ,where B_N is the number of females born in the Nth census period. The data which relate the age structure to the birth rate are the survival probabilities and the fertilities. The former, denoted by l_N, is the proportion of those who will survive from the $(N - 1)$st census period to the Nth period. For example, l_1 is the probability of a newborn reaching age 5, while l_{13} is the probability of an individual surviving from age 60 to 65. (The survival probabilities l_N are usually given in a life table produced and used by life insurance companies. Figure 18 gives the survival probabilities for United States females for 1966.) Combining these survival probabilities with the birth rate yields, for instance, the formula $l_1 B_{N-1}$ for the number of females having ages between 5 and 10 in the Nth census.

Age class N	Age	Proportion of female births surviving to age class N l_N	Number of births to those in age class N b_N
2	10–14	0.97493	8,128
3	15–19	0.97345	620,426
4	20–24	0.97058	1,297,990
5	25–29	0.96711	827,786
6	30–34	0.96285	474,542
7	35–39	0.95667	252,526
8	40–44	0.94805	74,440
9	45–49	0.93486	4,436
10	50–54	0.91519	0

Mortality and fertility data for females in the United States in 1966.

Figure 18. Life table for United States, 1966

To predict the number of births, and hence to be able to predict the age distribution of the population in future years, we need fertility data for the various age groups. Let b_N be the number of births (per 100,000) to females having ages between $5(N-1)$ and $5N$. For humans, these numbers are zero except for ages between 15 and 55. The actual fertilities for U.S. females in 1966 are given in Figure 18. The renewal equation, which completes our model, is constructed by observing that

Number born = number born to + ... + Number born
in Nth census   survivors of first   to survivors
census   of $(N-1)$st
census.

We write this more concisely as

$$B_N = (b_N l_1 l_2 \cdots l_N B_0) + (b_{N-1} l_1 l_2 \cdots l_{N-1} B_1) + \cdots$$
$$+ (b_2 l_1 l_2 \, B_{N-2}) + (b_1 l_1 B_{N-1}).$$

This equation, which forms the basis of renewal theory, can be solved easily on a computing machine.

Renewal theory has many uses. It provides fundamental models for economic and community planning, and for insurance through actuarial sciences. On the other hand, renewal theory continues to be a lively area of mathematical interest. Models similar to the renewal equation derived here for human populations are used to describe the age structure of nonhuman populations, such as bacterial populations used in breweries or in sewage treatment. A typical problem involved with bacterial populations is to control their nutrient in such a way as to optimize the amount of material metabolized. To do this, the renewal equation is used to describe the population, and the birth and death rates (the control parameters) are manipulated to guide the system to optimum production. Also renewal theory is used in industry to predict when machinery will fail and have to be replaced. These problems often require optimization methods, such as dynamic programming, coupled with renewal theory.

Genetic Structure

It was work on garden peas carried out by Abbot Gregor Mendel in the last century that eventually led to our current understanding of how various genetic traits are passed from one generation to the next through reproduction in plants, animals and humans. Mendel's work was not noticed for nearly half a century, but when it was rediscovered in 1900 it led to a science of genetics that has grown into a huge research effort today.

It is now known that human cells contain large molecules called chromosomes that carry the chemical code for construction of the entire organism. Chromosomes, long string-like arrangements of deoxyribonucleic acid (DNA) molecules, normally appear in humans in twenty-three matched pairs with each pair being joined at one position called the centromere. The chemical codes at various locations on the chromosomes are known to manifest themselves in dramatic ways such as eye, hair, and skin color, as well as in more subtle ways as determination of various types of proteins that are the basic building blocks of the body. Each code segment is called a gene, and its location (locus) can be reliably identified by its distance from the centromere. Each joined chromosome pair carries two genes at the same location, one on each strand. The number of genes and how they are expressed is, for the most part, unknown. These and a great many other important questions about genetics remain to be answered.

On the other hand, a great deal is known about the way many genetic traits are inherited, and in particular about the frequency and transmission of various genes in a population. Mendel observed that gene frequencies behave in certain predictable ways. His observations, which have come to be known as Mendel's Laws, describe the distribution of a parent's genes among its offspring. Other properties have since been determined both through experimental observations of various organisms such as bacteria (E. Coli), insects (fruit flies and mosquitoes) and rodents (mice and rats), and through the formulation of a mathematical theory based on Mendel's Laws. In particular, several scientists, among them the British mathematician Godfrey H. Hardy (1877–1947), showed that in the absence of outside influences, a population's gene frequencies will not change from one generation to the next (see box on p. 315). The most general formulation of this genetic law has come to be known as the Hardy–Weinberg equilibrium, after Hardy and the German physician Wilhelm Weinberg. This work, although derived from a simple idealized model, provided powerful guidelines for

Hardy – Weinberg Law

Parent Population			Parent Population Gene Pool	
genetic type at the specified locus	number having this type	proportion having this type	gene type	proportion having this type
AA	N_1	$P_1 = N_1/N$		
			A	$p_A = \dfrac{2N_1 + N_2}{2N} = P_1 + (P_2/2)$
AB	N_2	$P_2 = N_2/N$		
			B	$p_B = \dfrac{2N_3 + N_2}{2N} = P_3 + (P_2/2)$
BB	N_3	$P_3 = N_3/N$		
	N	1		1

Offsping Population			Offspring Population Gene Pool	
genetic type at the specified locus	number having this type	proportion having this type	gene type	proportion having this type
AA	$p_A{}^2 N$	$p_A{}^2$		
			A	$p'_A = p_A{}^2 + p_A p_B$
AB	$2p_A p_B N$	$2p_A p_B$		
			B	$p'_B = p_B{}^2 + p_A p_B$
BB	$p_B{}^2 N$	$p_B{}^2$		
	N	1		1

A simplified geometric model assuming random mating (but no natural selection or mutation) in a population whose size is constant over generations illustrates how gene frequencies change. Random mating means that an offspring is formed by selecting two genes from the parent gene pool at random; we denote the probability of selecting an A gene by p_A, and the probability of selecting a B gene by p_B. The critical observation is that $p'_A = p_A{}^2 + p_A p_B = p_A(p_A + p_B) = p_A$. Thus, the gene pool proportions do not change over generations. The distribution of genetic types $p_A{}^2 : 2p_A p_B : p_B{}^2$ are the Hardy–Weinberg proportions, sometimes referred to as the Hardy–Weinberg Law.

later development. The Hardy–Weinberg equilibrium is observed in most natural populations, and it has been used as a basis for many experiments.

The genetic structure of human populations is frequently described in terms of the gene pool carried by the population. Among the simplest genetic traits to study are those involving only gene traits at a single locus. If the possible genes appearing at this locus are denoted by A and B, then every individual in the population falls into one of the three categories:

AA—those having both genes at this locus of type A;
AB—those having one gene of type A and the other of type B;
BB —those having both genes of type B.

If the total population size is N, then the population carries $2N$ genes at this locus, some of type A and some of type B. The gene pool is usually described in terms of the proportions of the genes that it comprises. These proportions may change in time, for example, from 25% type A to 75% type A.

Keeping track of the gene pool proportions gets a little complicated. However there are cases that illustrate some important results without becoming bogged down in technicalities. Suppose, for example, that a population remains relatively constant and is synchronized so that all reproductions take place at the same time and the parents don't participate in more than one reproduction. Although this sounds like an exceedingly artificial reproduction scheme, the results agree qualitatively with those derived from more realistic population models. Now we allow natural selection to act. The various genetic types have various abilities to reproduce and leave offspring. In this more general case, the Hardy–Weinberg proportions will not be maintained over generations. The model proceeds like this: Let g_n denote the proportion of the gene pool that is of type A after the nth reproduction time. Mendel's Laws can be used to derive a formula relating g_{n+1} to g_n. By plotting these relations, we get curves, much like reproduction curves; by cobwebbing these, we can determine how the gene pool changes in succeeding generations. There turn out to be only four essentially different cases, each described in Figure 19.

Within twenty years of the rediscovery of Mendel's work, three outstanding scientists—Ronald A. Fisher, Sewell Wright, and J.B.S. Haldane—firmly established population genetics on a foundation of extensive mathematical theories. Fisher, Wright, and Haldane attacked diverse problems, developed and used effectively many mathematical models, and introduced statistical methods for evaluating experiments suggested by their theories. A long list of scientists subsequently extended and embellished their work, benefitting as well a number of mathematical areas, notably, probability, statistics, and partial differential equations.

The Geography of Genes

The ABO blood group system in humans is widely known. It is caused by three possible genes that can appear at a certain location on a specific

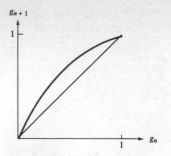

A Dominant. Cobwebbing shows that no matter where we start ($g_n > 0$), then $g_n \to 1$ in succeeding generations, and so the A gene eventually dominates the gene pool.

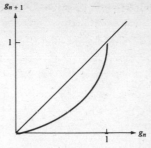

B Dominant. In this case, if $g_o < 1$, then $g_n \to 0$. Therefore, B eventually dominates the gene pool.

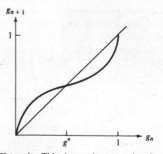

Heterosis. This interesting case is observed frequently. The best established case is that of sickle cell trait in tropical regions where AA types are susceptible to malaria, but otherwise healthy, the BB types suffer from severe anemia, but the AB types enjoy some immunity to malaria and do not suffer from anemia. These natural forces of disease thus act to maintain both A and B genes in the population. Cobwebbing shows that $g_n \to g^*$ which indicates that both genes are maintained.

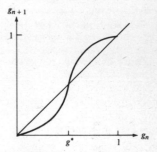

Disruptive Selection. Here if $g_o < g^*$, then $g_n \to 0$ so B eventually dominates the gene pool, but if $g_n > g^*$, then A wins. This bizarre behavior is believed to occur in some blood groups, but it is difficult to document.

Figure 19

chromosome pair which determine the blood type. (Actually, there are fourteen other major blood groups, although they are less well known.) Blood groups are important in clinical work since they enable physicians to avoid blood incompatitilities in blood transfusions and to correct mother-fetus incompatibilities. They also have legal uses in paternity cases.

DISTRIBUTION OF BLOOD GROUP GENE A IN THE ABORIGINAL POPULATIONS OF THE WORLD

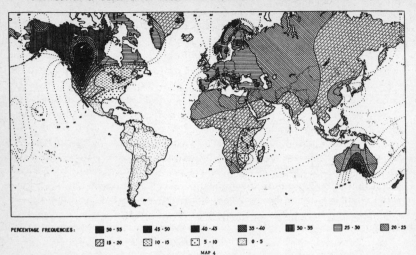

Figure 20. Distribution of blood group gene A in the aboriginal populations of the world.

Blood groups also provide a powerful tool in anthropology. An impressive example of this was given by Arthur E. Mourant who plotted the global frequencies of blood types (see Figure 20). The gradual variation of blood groups with geography is clearly illustrated by this map. This phenomenon is described by the term *genetic cline*.

Mathematical studies of clines have been directed at understanding the mechanisms that can maintain clines. This work proceeds in the following way. A gene pool is visualized as being distributed over a geographic region by lumping local populations together to form units, called demes. Neighboring demes interact because of movement of individuals. At the same time, natural selection favors some genes in some demes but other genes in other demes. That is, certain genes make individuals more fit in some demes than in others. A mathematical model of genetic clines must, therefore, take account of both dispersal and natural selection. Although this is quite similar to the dispersal models used to trace the spread of infestation in a forest, the rates and number of states quickly become more complicated. We will not elaborate on these models.

Genetic Engineering

One of the most amazing recent developments in biology has been in artificial genetic recombination in bacteria. A bacterium is a cell enclosing, typically, a single circular chromosome. Bacteria contain pieces of DNA in addition to that which comprise their chromosomes. These extra-chromosomal DNA elements are called plasmids, and they play important roles in the bacterium. For example, some plasmids make the bacterium resistant to antibiotics such as penicillin or streptomycin. In addition, they enable the bacterium to bind toxins, such as gold, platinum, and hydrocarbons, in a harmless way. These facts have motivated research into the use of bacteria in mining gold and platinum from water, in clearing up oil spills, and in cleaning oil tankers. Their genetic structure is very important for these and other industrial applications.

Mathematical methods are used extensively in studies of plasmid frequencies in bacterial colonies. These models basically require a form of renewal theory, and the methods described above are used in their analysis. However, here the age structure of the population also is significant, and this aspect of the population must be analyzed. Since this work requires a more technical development than is intended here, details of this analysis will not be attempted.

Conclusion

Mathematical methods have provided real insight into some biological phenomena. This is most clearly illustrated in population genetics where sound models such as the Hardy–Weinberg equilibrium were derived from first principles by way of Mendel's Laws. Unfortunately, such precise principles are not yet available in other population areas, and we must be satisfied for now with models that mimic specific observed phenomena. There is no doubt that this situation will eventually change, but it does for the moment cast shadows on the credibility of a great deal of current work in theoretical ecology, for example.

On the other hand, population problems have led to the initiation and development of many theories and methods that are central to mathematics. In particular, theories of probability, dynamical systems, and wave propagation have evolved in this way. Of course, these areas also received considerable stimulation from the physical sciences and from within mathematics itself. Still, several novel developments are due solely to population biology, and new questions from population biology continue to stimulate interest in some mathematical topics that had grown stale.

Suggestions for Further Reading

Cavalli-Sforza, L.L. and Bodmer, W.F. *The Genetics of Human Populations*. Freeman, San Francisco, 1971.

Clark, Colin W. *Mathematical Bioeconomics: The Optimal Management of Renewable Resources*. Wiley, New York, 1976.

Keyfitz, N.S. and Flieger, W. *Population: Facts and Methods of Demography*. Freeman, San Francisco, 1971.

Part Four

The Relevance of Mathematics

The Relevance of Mathematics

Felix E. Browder
Saunders Mac Lane

"Relevance" has been a favorite concept in the last few years. Questions about the relevance of an institution, an activity, or a subject are often asked (and less often answered). We take it that relevance must refer, at least implicitly, to a relation with some body of values or purposes. Thus a subject may be relevant in the first instance by way of its applications to another subject—which in its turn may then be tested for its further relevance, ultimately to human welfare or to an overriding conception of the *good*. In short, the relevance of mathematics involves both the various applications of mathematics and the position of mathematics in the spectrum of human values.

Let us note a significant historical precedent for the discussion of the relation of mathematics and values. A little more than 2300 years ago, a celebrated lecture was given in Athens on a closely related theme. This was the famous "Lecture on the Good" by Plato, the most influential of the world's philosophers, and it is the only lecture by Plato of which there is any objective evidence. Among his listeners was Aristotle, then a student in the Platonic Academy and later a critic of Plato's doctrines. Aristotle put down his testimony on the contents of Plato's lecture in a treatise in three books called "On the Good" of which no fragment remains, but which was paraphrased in the writings of Aristotle's disciples. "Many attended the lecture under the impression that they would obtain some of the human goods, such as riches, health, power, or above all a wonderful blissfulness. However, when the exposition began with mathematics, number, geometry, and astronomy and the thesis 'The class of the Limit taken as One is the Good', the surprise became general. A part lost interest in the subject, the others criticized him."

Plato is reported to have said: "The foundations of all things are the One and Indeterminate Magnitude, or the 'More or less'." In present day terms, what he seems to have said is that the Archai, the foundations of both the physical and moral orders (which he did not distinguish from one another) are the processes by which there is generated the sequence of natural numbers 1, 2, 3, 4, . . . and the continuum of all numbers. Thus the report by Aristotle on Plato's most fundamental Unwritten Doctrine is that Plato solved the problem of the relationship between the foundations of the Good and of mathematics by identifying the two.

Plato's solution of the problem of value by identifying the Good with mathematics is one that very few people would defend today. Nevertheless, many active branches of our present intellectual world are dominated by great visions of mathematical order which are essentially platonic. In astronomy, the Ptolemaic cosmology with sun and planets moving in epicycles within concentric spheres was such a mathematical vision. When this cosmology was overthrown in the sixteenth and seventeenth centuries by Copernicus, Kepler, and Newton, the resulting Newtonian cosmos was even more mathematical, both in its use of the new mathematical apparatus of the calculus, and in its extensive subsequent development. When the Newtonian cosmology was overthrown in its turn, the new special and general theories of relativity involved an even more perfected vision of a platonic character. The contemporary speculative cosmologies of finite universes again take mathematical form and involve new mathematical techniques developed in the geometrical study of the configurations which are called manifolds. The main course of modern physics has moved through the channel of the revolutionary development of quantum mechanics in the late 1920s, yielding a new and precise mathematical formalism whose consequence has been a fundamental mathematization of the basic principles of chemistry. In more recent decades, we have seen the development of molecular biology founding the basic machinery of biological inheritance upon the geometry of the DNA molecule and the combinatorics of sequences of amino acids. New cosmological visions of the platonic order have been formulated for the origin and possibly the vanishing of the whole physical cosmos.

The development of mathematics itself has involved many platonic visions made manifest. For example, the notions of counting and measuring both involve numbers; this notion of number has evolved continually, from whole numbers to rational, real, and complex numbers. In another direction, the ordinary finite numbers such as 2, 7, 1,000,000 and the like have been extended to encompass infinite magnitude, both "countable" infinities (the number of whole numbers) and "uncountable" infinities (the number of points on a line). These visions of the infinite have pricked, delighted, tormented, and inspired mathematicians over the past hundred years.

Platonic visions of large-scale mathematical structure continue to be attractive even in the human sciences, though they usually avoid the platonic identification. Many such speculative visions are being developed, some of them highly uncertain, like the world models of doom promoted by Jay Forrester, Dennis Meadows, and their followers. The movement called structuralism in the humanistic and social disciplines in Europe, as led by Claude

Levi-Strauss and Jean Piaget, has argued eloquently for a formal mathematical structure underlying all the varied spheres of human action and meaning. In America, the social disciplines pride themselves on being more hard-headed and quantitative—but this again is often expressed by the use of mathematical models. In fact, if not in name, the platonic vision of mathematical structure is very much alive today.

From this vantage point the customary division of mathematical research into pure mathematics and applied mathematics is not the most effective way to understand the relevance of mathematics. One and the same mathematical idea can apply to totally different disciplines. One mathematical notion can arise initially in the context of "pure" mathematics, only to find some later application. Conversely, some specific application may lead to a notion which later has a development within pure mathematics in quite different directions. In other words, there can be short-run applications of mathematics, done with forethought and intended just for one application, as compared with long-run applications where an old idea is used in a surprising way.

Conics

One of the best-known examples of a surprising application concerns the curves known as the conics. They are so called because each of these curves can be obtained by taking a plane cross-section of a (double) cone. There are four different sorts of curves in which the plane cuts the cone, each illustrated in Figure 1. One is the *circle* when the cross-section is perpendicular to the axis of the cone; another is the *ellipse* when the section is somewhat tipped. The cross-section may be tipped further till it becomes parallel to one

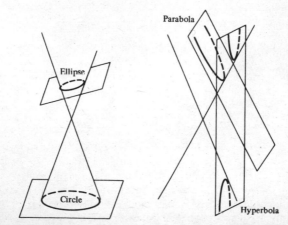

Figure 1. Sections of a cone: Ellipse, Circle, Parabola, Hyperbola

of the lines in the surface of the cone; this cross-section is an open curve, the *parabola*. Finally, some planes will cut both upper and lower halves of the cone, giving a curve in two pieces, the *hyperbola*.

In 200 B.C., the Greek geometer Apollonius of Perga wrote a celebrated treatise on conic sections, describing systematically all the properties of these curves. The circle is the path of a point P moving in a plane at a constant distance r (the radius) from a center; he observed that an ellipse could similarly be described as the path of a point P moving so that the sum of its distances PF and PF' from two fixed points F and F' remained constant (see Figure 2). These points F and F' are called the *foci* of the ellipse. The parabola has only one focus F; a point P on the parabola moves so that its distance from the focus F always equals its distance from a line (outside the parabola). All these and many other geometric properties were developed by Apollonius. This study was carried out as an exercise in pure mathematics, and aside from later applications to burning mirrors, few applications of conics were considered or made in the classical world. In particular, there were no applications to astronomy, although Apollonius was himself a major

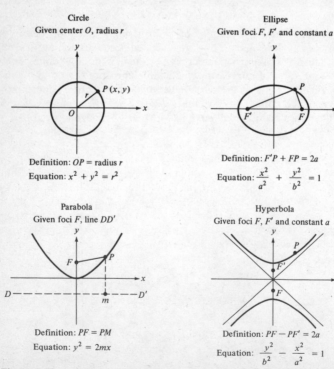

Circle
Given center O, radius r

Definition: OP = radius r
Equation: $x^2 + y^2 = r^2$

Ellipse
Given foci F, F' and constant a

Definition: $F'P + FP = 2a$
Equation: $\dfrac{x^2}{a^2} + \dfrac{y^2}{b^2} = 1$

Parabola
Given foci F, line DD'

Definition: $PF = PM$
Equation: $y^2 = 2mx$

Hyperbola
Given foci F, F' and constant a

Definition: $PF - PF' = 2a$
Equation: $\dfrac{y^2}{b^2} - \dfrac{x^2}{a^2} = 1$

Figure 2. Equations for conic sections.

contributor to the Greek study of mathematical astronomy. Instead, the Greek mathematical theory of the motions of the planets was organized in the Ptolemaic system in which the planets move around the earth by circles on circles—more exactly, on circular "epicycles" which themselves move around the earth in circles.

In 1604, 1800 years later, the German mathematician and mathematical physicist Johannes Kepler read the writings of Apollonius and their Islamic commentaries and studied their application in optics to parabolic mirrors. In 1609, he made the brilliant observation (impossible without the availability of the ancient theory) that the orbits of the planets should be described as ellipses, rather than by means of circles and epicycles. He thereby laid the principal foundation for Newton's later theory of gravitation.

This use of conics played a major role in the development of mathematics in the seventeenth century: Descartes's development of analytic geometry made it possible to describe the conics by the now standardized equations in rectangular coordinates (Figure 2) and the problem of constructing lines tangent to a conic at a point was one of the motives for the development of calculus.

This example of an application is somewhat remarkable, since most important advances in pure mathematics rarely wait 1800 years for application. Yet in this case, we can surely say that the magnitude of the application made it worth waiting 1800 years. Moreover, with the increasing momentum of mathematical and scientific activity, the speed of application has also increased (contrary to some people's impressions) with the time interval of delay growing considerably shorter. Thus, it took 60 years from the development of matrix theory as a part of pure mathematics in 1860 to its application in physics. It began when Arthur Cayley used matrices—square arrays of numbers—in order to describe linear geometric transformations, such as rotations, shears, and similarities. In 1925, Werner Heisenberg used matrices as the mathematical tool ("matrix mechanics") needed in quantum mechanics to describe atomic systems. It took 30-odd years from the development of tensor calculus by the geometers of Italy in the 1870's to its application as the basic mathematical tool of relativity theory by Albert Einstein in the 1910's, and 20 years from the development of the eigenfunction expansions of differential and integral operators by David Hilbert in 1906–1910 (following on the theory of Charles Sturm and Joseph Liouville of 1840) to its application in wave mechanics in 1927.

Constructible Functions

Another surprising application developed from the pure mathematical notion of "constructible" functions. The story begins with the trigonometric functions of an angle θ, such as the functions $\sin \theta$ and $\cos \theta$. They first appear as ratios of the sides of a right triangle. A point P moving on a circle of

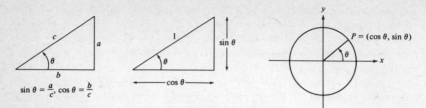

Figure 3. Trigonometric functions sine and cosine by triangles and coordinates.

radius 1 has coordinates $x = \cos\theta$, $y = \sin\theta$ (see Figure 3) when its radius vector is at angle θ with the horizontal. As the angle θ increases from 0 to 2π (that is, from 0 to 360 degrees) $\cos\theta$ runs through all its values from 1 to -1 and back, as does $\sin\theta$. From 2π to 4π these values repeat (Figure 4); one says then that $\sin\theta$ and $\cos\theta$ are periodic functions of θ. For that matter, $\sin2\theta$, $\cos3\theta$, and so on are also periodic, with a shorter period, as shown in Figure 4. This is the source of the remarkable fact that trigonometry, started from measurement of triangles, turns out to be a powerful way of handling periodic phenomena.

Early in the nineteenth century the French scientist Joseph Fourier, in studying the conduction of heat, had been led to represent more general "periodically varying" functions $f(x)$ as infinite sums of trigonometric functions $\sin nx$ or $\cos nx$ for $n = 1, 2, 3, \ldots$ (see box below). His successors, in studying these "Fourier series," were much concerned with their convergence properties, that is, with the question whether the infinite "sum" of the terms in such a series equalled the desired function. In the late nineteenth century the German mathematician George Cantor, in working on this problem, found that he had to deal with quite general sets consisting of those numbers at which the series could converge (or fail to converge). He was thus led to the explicit study of infinite sets of numbers or of other mathematical objects. Moreover, it presently appeared that the formal lan-

The Fourier Series

A function f is periodic with period 2π if $f(x + 2\pi) = f(x)$, where 2π (radians) is 360°. Its Fourier series is

$$f(x) \sim a_0 + a_1 \cos x + b_1 \sin x + \ldots + a_n \cos x + b_n \sin x + \ldots$$

where a_n and b_n are given by the formulas

$$a_n = \frac{1}{\pi} \int_0^{2\pi} f(x)\cos nx, \qquad b_n = \frac{1}{\pi} \int_0^{2\pi} f(x)\sin nx$$

guage used to describe such sets could be employed very generally in the formal description of most mathematical ideas. It was therefore a shock when Bertrand Russell and others about 1900 discovered paradoxes resulting from the unrestricted application of a set theoretic language.

It thus became important to limit the rules for the manipulation of arbitrary sets by giving an explicit system of axioms. Once this had been done, there still remained the possibility that someone would find a new paradox. David Hilbert then proposed to prove that there could be no such paradox—that such a system of axioms, properly formulated, would never lead to a contradiction. He proposed to do this by considering mathematical proofs as purely formal operations upon strings of symbols expressing the axioms—so that a step-by-step analysis of proofs could itself be treated mathematically. Despite vigorous efforts, Hilbert's proposed consistency proof was carried through only for some very simple formal systems, not strong enough to cover all of set theory and mathematics. Then in 1931 the young Austrian mathematician Kurt Gödel proved his celebrated incompleteness theorem which shows that such consistency proofs are in fact impossible for the usual systems of mathematical logic; it appeared to be entirely a result in the foundations of pure mathematics.

Later Gödel came to the Institute for Advanced Study in Princeton, and there lectured on a further analysis of the idea underlying his incompleteness theorem. He pointed out that the formal proofs being analyzed could be

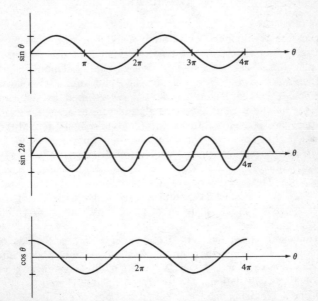

Figure 4. Sinθ and cosθ as periodic functions. (θ is measured in radians, where 180° = π radians.)

replaced by numbers, and that the successive steps of these proofs depended on constructive processes applied to numbers. These processes he called "recursive," meaning that they can be carried out effectively, step by step. From this impetus, at least four other mathematicians in 1936 pushed further the understanding of recursive (i.e., constructible) operations on numbers (see box on p. 331). And in the same year, 1936, the brilliant young English logician Alan Turing identified the results of such procedures, the general recursive functions, with the outcomes of what could be computed on a *machine in general*. It is with this analysis, and its impact on the minds of such men as John von Neumann and others, that the theoretical concept and the analysis of the digital computer in the modern sense began.

It remains true to this very day that the theoretical description of what can be computed in general and its more penetrating analysis are rooted in that soil of mathematical logic which Gödel turned over for the first time in his memoir of 1931. In current terms, the problems of the theory of computability have been put in terms of relative quantitative practicality, whether a given problem can be solved in a number of steps which grows no faster than a fixed power of its number of components. The basic achievements of this study of the complexity of *computation* are fully in the spirit of Gödel's original analysis.

Numbers

The successive development of different sorts of numbers has represented successively wider visions of mathematical reality. At first, it seemed to the Greeks that all measurements could be made by rational numbers—ratios of two integers, such as $3/4$ or $7/5$. However, the Pythagorean school discovered that the square root of 2 could never be represented by such a rational number (see box on p. 331) even though it is sure to arise, for example, as the diagonal of a square with unit sides. This discovery forced the Greek mathematicians to consider "incommensurable" quantities such as $\sqrt{2}$. It has led to our present system of rational *and* irrational numbers—the so-called "real" numbers, which represent all possible distances between the points on a line. These real numbers include not only $\sqrt{2}$, but also the square roots $\sqrt{3}$, $\sqrt{7}$, . . . and indeed the square roots of all positive whole numbers. However, there is no real square root of a negative number (see box on p. 331), and in particular, no square root of -1.

Nevertheless, it turned out to be useful to calculate as if there were such a square root of -1—an imaginary number, called $i = \sqrt{-1}$. Calculating with i meant using numbers like i, $3 + 2i$, $7 - \sqrt{2}i$ and the like; the box on p. 332 shows how these calculations lead to a full-fledged arithmetic with numbers of the form $x + iy$. Using these numbers it was possible to provide solutions for all sorts of polynomial equations beside the equation $i^2 = -1$, and in a later and even more surprising development, to represent a variety of physical phenomena, from electric currents to the shapes of airfoils.

Recursive Functions = Computable Functions

Recursion. Multiplication (of whole numbers m, n) can be computed "recursively" from addition by the equations

$$m \cdot 1 = m,$$
$$m(n + 1) = mn + m.$$

The first equation starts the definition, giving the product of m by 1. Once the product of m by some whole number n is defined, the second equation gives the product of m by the next integer $n + 1$.

The λ-calculus (Alonzo Church, 1928) provides a basic formalism to show how an expression in x determines a function. Thus the expression $x^2 + x + 3$, regarded as a function of x, is written $(\lambda x)(x^2 + x + 3)$. Similarly $(\lambda x)\sin x$ means "the function sine". Church (1936) used this calculus to produce an unsolvable problem in number theory.

General Recursive Functions (Stephen C. Kleene, 1936). A function $F(n)$ of a whole number n is defined by recursion from given functions if all its values $F(1)$, $F(2)$, $F(3)$, ..., can be calculated from a finite number of equations using the given functions. Kleene showed this definition equivalent to definability by the λ-calculus.

Finite Combinatory Processes were codified by Emil L. Post in 1936. He considered a symbol space which was an infinite sequence of boxes. The process starts at one box, and may mark that box, or may erase the mark already there, or may move on to the next box to the left or to the right, all according to a coded set of directions. The idea is parallel to Church's λ-calculus, and equivalent to a Turing machine.

Impossible Square Roots

The square root of 2 cannot be rational:

Suppose that $\sqrt{2}$ could be written in lowest terms as a rational number $\sqrt{2} = m/n$, with m and n integers and not both even. Then squaring both would give $2n^2 = m^2$. This would force m to be even, hence m^2 to be divisible by 4, and hence n to be even—so that m/n would not be in lowest terms after all! This contradiction shows $\sqrt{2}$ is irrational.

The square root of -1 cannot be real:

Suppose that $\sqrt{-1}$ were a real number b, so $b^2 = -1$. If b is positive, so is b^2, so then $b^2 \neq -1$. If b is negative, it is $b = -a$, so $b^2 = (-a)(-a) = a^2$ is again positive, so $b^2 \neq -1$ again. If b is zero, so is b^2. Hence $b^2 = -1$ is impossible for real b.

Complex Numbers

Complex numbers $x + iy$ for real numbers x and y are represented by points in the complex plane. Algebra is done in the ordinary way, subject to the rule that $i^2 = -1$. Multiplication by the complex number i means rotation by 90° counterclockwise, as in

$$(x+iy)i = xi + yi^2 = -y + ix \, .$$

Also, the standard functions of real numbers can be extended to apply to complex numbers. For example, the exponential function e^x, where $e = 2.71828\ldots$ is the base of natural logarithms, becomes

$$e^{i\theta} = \cos \theta + i \sin \theta \, .$$

This formula is useful in Fourier series and elsewhere.

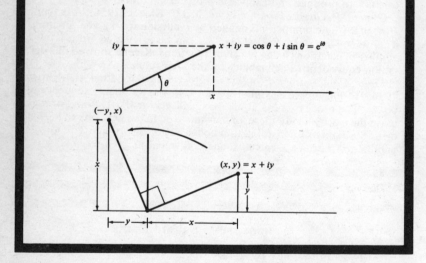

The development of numbers didn't stop there. Since complex numbers worked so well to describe the plane, it was natural to search for some similar numbers which would work in *three* dimensions. That search failed, but it then turned out that there could be four-dimensional numbers called the "quaternions" involving 1 and three more units i, j, and k where each of i, j, and k is a (different) square root of -1 (see the box on p. 333). These generalized numbers provided a good description of rotations in three dimensions, just as complex numbers did in two dimensions, as well as a formalism for vector analysis, all at the cost of having a multiplication rule $ij = -ji$ in which the result of multiplication *depends* on the order of factors. In other words, multiplication of quaternions is not commutative.

The Quaternions

The quaternions are constructed from 1 and three additional units i, j, and k, so that each quaternion $q = a + xi + yj + zk$ is a sum of a "scalar" a (a real number) and a "vector" $xi + yj + zk$, where x, y, and z are real numbers.

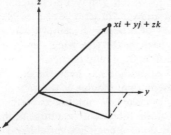

Rules

$$i^2 = j^2 = k^2 = -1$$
$$ij = k = -ji$$
$$jk = i = -kj$$
$$ki = j = -ik$$

The quaternions can be used to describe rotations in three-dimensional space. Specifically, a rotation of three-dimensional space is an operation taking a point with coordinates x, y, z into some other point. It can always be written as the operation

$$(xi + yj + zk) \rightarrow q(xi + yj + zk)q^{-1}$$

where q is a suitable quaternion. For example, if we take q to be k, we get rotation by 180° about the z axis, while rotation by 90° about the z-axis is given by $q = (1 + k)/\sqrt{2}$.

If one searches for other number systems of this sort, one can find only these three: the real numbers, the complex numbers, and the quaternions. This is because of a mathematical result (due to McLagen Wedderburn in 1905) which asserts that if a number system obeys all the laws of ordinary algebra for addition, subtraction, multiplication, and division—except for the law that $ab = ba$, noted just above—and if the number system has a finite number of units (one for the real numbers, two for the complex numbers and four for the quaternions) then that number system must be the real numbers, the complex numbers, or the quaternions.

This result seems clearly to be a negative one, cutting off further development. But it has not turned out so. By dropping other laws or by starting with just the rational numbers rather than with all the real numbers, algebraists have been able to devise many other types of number systems. As a recent example, the so-called graded Lie algebras to be discussed below have been of use in the classification of elementary particles in physics.

Meanwhile, the complex numbers $z = x + iy$ have turned out to be much more than just a system of numbers. Ordinary functions such as x^2, 2^x, or $\sin x$ can be extended to make sense for complex numbers, so that one may form $w = z^2$ or $w = 2^z$ or $w = \sin z$ for z and w both complex. This discovery led to a theory of functions of a complex variable, called analytic functions. This theory starts with the extension of calculus to complex functions of a complex argument z, defining $w = f(z)$ to be an analytic function of z when it has a suitable derivative, measuring the rate of change of w with respect to z, independent of the direction in which z may be changing (see box on p. 335). The resulting theory of analytic functions of a single complex variable was the great masterwork of nineteenth century mathematics. Its impact on physics can be measured by the following sentence which opens a recent treatise on the physics of fundamental particles: "The great discovery of theoretical physics in the last decade has been the complex plane."

One of the major pure mathematical themes of the past three decades has been the extension (begun by Karl Weierstrass, Henri Poincaré, and Friedrich Hartogs at the end of the nineteenth century) of this theory to the theory of analytic functions of several complex variables $z_1, z_2, \ldots, z_n$. Here a function is said to be analytic in these variables if it is analytic in each variable separately, when all the others are held fixed. One of the most important topics in this theory from the point of view of applications is the generalization of what is known as the theory of residues. For the case of one variable, knowledge of the "residues" of a function at its "poles" makes it possible to calculate the integral around any closed curve (see box on p. 336). The generalization of this sort of conclusion to integrals over hypersurfaces in n variables demands deep general results from the algebraic geometry of n variables. While the theory of analytic functions of several complex variables has pressed forward in recent decades, mathematicians and physicists in the past decade have used such sophisticated results to calculate certain integrals invented by the physicist Richard Feynman in quantum field theory.

Differential Equations

Another example of the interrelation of "pure" and "applied" mathematics is the study of the long-term properties (also called the asymptotic properties) of the solutions of differential equations. Since the time of Newton, following his example in the formulation of mechanics and of celestial mechanics, it has been customary to describe the basic laws for physical processes in terms of differential equations. When the state of a physical system is described by means of several state variables, then the differential equations prescribe the rates of change of these state variables. For example, Newton's second law of motion specifies for a moving body that the

Analytic Functions of a Complex Variable

Consider $w = f(z)$, where the complex number $w = u + iv$ depends on the complex number $z = x + iy$. Then u, the real part of w, depends on both x and y, so is a function $u = u(x,y)$. Similarly $v = v(x,y)$, so that w can be written as

$$w = u(x,y) + iv(x,y).$$

To apply the calculus, one wishes to find the rate of change of w with respect to x (the real part of z) or with respect to iy. These rates are given by "partial" derivatives:

$$\frac{\partial w}{\partial x} = \frac{\partial u}{\partial x} + i\,\frac{\partial v}{\partial x}$$

$$\frac{1}{i}\,\frac{\partial w}{\partial y} = \frac{1}{i}\left(\frac{\partial u}{\partial y} + i\,\frac{\partial v}{\partial y}\right) = \frac{\partial v}{\partial y} - i\,\frac{\partial u}{\partial y}.$$

If these two rates are equal, their real parts and their imaginary parts are equal, as in

$$\frac{\partial u}{\partial x} = \frac{\partial v}{\partial y}, \qquad \frac{\partial v}{\partial x} = -\frac{\partial u}{\partial y}.$$

These two equations are called the Cauchy–Riemann equations. When they hold, $w = u + iv$ is called an analytic function of z.

Such an analytic function maps the plane consisting of all complex numbers z into the plane of all complex numbers w in a way which preserves angles. This is used in many ways; for example, to analyze the profile of airplane wings, choose a function to map a circle (in the z-plane) to a wing profile (in the w-plane).

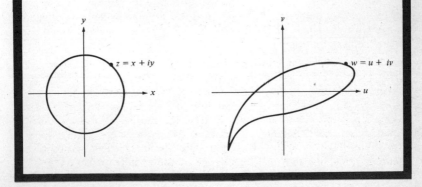

Residues of an Analytic Function

The residues of $w = f(z)$ arise at points z where the function is not defined because it becomes "infinite" (a pole). For example

$$f(z) = \frac{3}{z-1} + z^2 \qquad \text{Res}_1 f = 3$$

is infinite at $z = 1$, and has residue 3 there, while

$$f(z) = \frac{1}{z^2} - \frac{2}{z} + 1 + z \qquad \text{Res}_0 f = -2$$

is infinite at the origin ($z=0$) and has residue -2 there. The Cauchy integral theorem states that the integral of an analytic function f around a closed curve C is the sum of the residues of f at all its poles inside C:

$$\int_C f(z)dz = \sum_{z_0 \in \text{pole}(f)} \text{Res}_{z_0} f.$$

rate of change of its velocity (in any one direction) is proportional to the imposed force in that direction.

More generally, consider a physical system whose state at any time is specified by two variables x and y (position and velocity, for example). A corresponding differential equation will specify the time rates of change dx/dt and dy/dt as they depend on the instantaneous values of x and y, that is, as functions $f(x,y)$ and $g(x,y)$. This amounts to a pair of "first order" differential equations

$$\frac{dx}{dt} = f(x,y) \qquad \frac{dy}{dt} = g(x,y) .$$

One can think of the state of the system as represented by a point with coordinates x,y in the plane; these two differential equations specify at each position in the plane the velocity with which the point will move. Given an initial position of the point, a solution will express the resulting position as a function of time. The solution may be expressed in terms of known functions $x = x(t)$ and $y = y(t)$ of t, or may be determined numerically.

In the decade before the first World War, the great French mathematician and mathematical astronomer Henri Poincaré introduced a new mode of analysis of the ultimate or "asymptotic" properties of such solutions—that is, of their behavior as the time t becomes very large. He showed, under suitable hypotheses, that each such solution either converged to a *singular point* (x_0,y_0) in the plane or to a *limit cycle* $u(t)$, $v(t)$. By a singular point he meant a point (x_0,y_0) at which both the given rates of change are zero. By a limit cycle he meant a solution—a possible motion $u(t)$, $v(t)$ satisfying the given differential equations—which had a period T. (Both of these types can be illustrated in the simplest case, that of harmonic motion; see box on p. 338.) Poincaré also gave information on the relation between the number of singular points and the number of limit cycles; for example, he showed that the closed curve representing the limit cycle must contain at least one singular point. The results obtained (known as the Poincaré–Bendixson theory) were applied in subsequent decades to a variety of problems in science and engineering, especially in the theory of servo-mechanisms and control processes, and more recently as one of the principal tools in formulating models in developmental biology. The extension of these results to more than two variables was first attacked seriously in the past two decades and offers substantial difficulties of a qualitative and geometric nature which have been only partially resolved.

The asymptotic properties of such systems as those obtained from solutions of differential equations may be considered from another point of view, that of the theory of probability. Probability had started from games of chance—calculating the chances that a coin will come up heads and not tails or that a point dropped at random will land in a given region. This latter probability is proportional to the area of that region. In its modern form the theory of probability has developed as a precise and powerfully sophisticated discipline on the basis of the theory of the "measure" of areas developed in 1902 by the French mathematician, Henri Lebesgue. He was led to his measure of area and to his corresponding general notion of an integral as a way of answering the difficult pure mathematical problem of finding a simple way to compute the Fourier coefficients of very general functions—the same Fourier coefficients a_n and b_n already exhibited as integrals in the box on p. 328.

Given this start with Lebesgue measure, probability theory developed in this century using a variety of ideas and techniques introduced by pure mathematicians such as the Russian Andrei N. Kolmogoroff, the American Norbert Wiener, and the Frenchman Paul Levy. One central question concerned the long-range behavior of a mechanical system with a large number of components. Ludwig von Boltzmann in Germany and Josiah Willard Gibbs in the United States had put forward the hypothesis that a quantity depending on such a system should have identical averages when averaged over *space* or over *time*. This so-called "ergodic" hypothesis was long debated. The first two general theorems giving a sharp mathematical formulation of this hypothesis were proved in 1931 by John von Neumann and George D. Birkhoff. These ergodic theorems represented a decisive advance.

An important example of an ergodic system is given by a Bernoulli trial—in simplest form, the statistical process of repeatedly tossing a fair coin. Starting around 1960, Kolmogoroff and some of his students, es-

Harmonic Motion

Harmonic motion is the motion of a particle at distance x from a center moving with velocity y to satisfy the equations

$$\frac{dx}{dt} = y, \quad \frac{dy}{dt} = -x .$$

(The first equation states that y, the velocity, is the rate of change of x; the second equation states that the acceleration of x is opposite to x.) One solution of these equations, for any constant A, is

$$x = A \cos t, \qquad y = A \sin t .$$

This solution is periodic; after the time t runs from 0 to 2π, the solution repeats itself. The solution can be pictured as the coordinates (x,y) of a point P moving with constant velocity around a circle of radius A in the counterclockwise direction. The projection M of this point on the x-axis then represents the original particle, moving back and forth periodically:

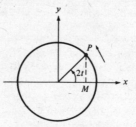

Another solution of these equations is $x = 0$, $y = 0$—the particle simply stays fixed. This is a "singular solution." So, these equations give the first example of limit cycles (periodic solutions) and singular solutions.

pecially Y. Sinai, generalized the physical concept of entropy in a subtle way to serve as a mathematical tool for the study of the transformation of probabilistic systems such as these Bernoulli trials. The development of ergodic theory in this new and sophisticated form created the tools for the much deeper analysis of the statistical behavior of mechanical systems. In work initiated by Sinai and carried to completion by the American mathematician Donald Ornstein surprising conclusions have been obtained: A few relatively crude statistical hypotheses imply that any such system is equivalent in its probability structure to one of the standard systems of Bernoulli trials for coins. Using the ergodic theorem, one therefore obtains very sharp conclusions about the asymptotic behavior in time of such systems.

The description of physical phenomena in a continuous form (of wave motion or diffusion as well as of equilibrium phenomena), has been given since the eighteenth century in the form of partial differential equations. These are equations prescribing rates of change of the state variables in the various spatial directions as well as in time. The pure mathematical theory of the solutions of partial differential equations deals especially with those solutions which satisfy further conditions called "boundary-value conditions." These usually give the values of the state variables on the boundaries of the given regions of space-time in which they are defined. This theory was begun by the celebrated German mathematician Bernhard Riemann in the 1850's. Since that time, it has been carried further in conjunction with the development of a whole framework of methods and concepts in mathematical analysis.

Since 1945, in particular, this theory has undergone a particularly intensive development through its interaction with two other mathematical theories. The first of these is what is called the theory of (linear) functional analysis, the study of (linear) functions defined not on the real or complex numbers, but on infinite-dimensional spaces of other functions. Using these basic ideas, the French mathematician Laurent Schwartz created in the 1950's a new theory of generalized functions (he called them *distributions*) through which new classes of solutions could be generated for partial differential equations. Another such theory (created in the 1930's by men like Michel Plancherel, Salomon Bochner, Norbert Wiener, Torsten Carleman, and others) was the theory of the Fourier transform—a continuous analogue of Fourier series. The Fourier transform replaces a given function $f(X)$ by a new function $f(t)$ obtained from f by an averaging process. In the simplest case the averaging process is given by an integral (see box on p. 340). The classical problems concerning all the various broad classes of partial differential equations have been solved in a simple and systematic way by the use of these tools. Many new classes have been attacked involving the foundations of relatively inaccessible theories such as the theory of analytic functions of several complex variables. Some very sophisticated tools have been invented in recent years, enabling the rapid transformation of problems and their solutions. Such tools include "pseudo-differential" operators introduced originally by Alberto Calderon and Antoni Zygmund, and the more general theory of Fourier integral operators introduced and studied by the Swedish mathematician, Lars Hörmander and the Russian mathematician, Yuri Egoroff.

Symmetry and Groups

A striking example of the development of a platonic vision is the mathematical analysis of symmetry which has led to the theory of groups. In geometry, the presence of different sorts of symmetry, rotational symmetry or symmetry by reflection, had long been recognized. It was developed classically

Fourier Coefficients and Fourier Transforms

The Fourier coefficients of a (periodic) function $f(x)$ are given by integrals

$$a_n = \frac{1}{\pi} \int f(x) \cos nx \, dx, \qquad b_n = \frac{1}{\pi} \int f(x) \sin nx \, dx.$$

In complex form $c_n = a_n + ib_n$ with $e^{inx} = \cos nx + i \sin nx$ they are

$$c_n = \frac{1}{\pi} \int f(x) \, e^{inx} \, dx.$$

The Fourier transform $\hat{f}$ of a function $f(x)$ is defined as a function of the (real) variable t given by the analogous formula

$$\hat{f}(t) = \frac{1}{\sqrt{2\pi}} \int e^{-ixt} f(x) \, dx.$$

The Fourier transform has the advantage that complicated properties of the original function f are replaced by simpler properties of its transform $\hat{f}$.

in the symmetry of ornaments, friezes, and vases. The fact that there could be a mathematical formulation of symmetry first appeared in the study of the roots of algebraic equations in the work of Joseph Lagrange and Évariste Galois in the late eighteenth and early nineteenth centuries. The symmetry involved here was that among several roots of one equation. In the simplest case of a quadratic equation such as $x^2 - 2 = 0$, the only symmetry is the process of interchanging (permuting) the two roots $x = \sqrt{2}$ and $x = -\sqrt{2}$. The same sort of symmetry is present for any quadratic equation $ax^2 + bx + c = 0$, which has two roots, as given by the quadratic formula. Similarly, equations of third degree such as $x^3 - 2 = 0$ have three (complex) roots. Any permutation of these roots (see box on p. 341) counts as a symmetry.

Lagrange and Galois were concerned initially with the longstanding problem of finding formulas for the roots of equations of higher degree. For example, a polynomial equation

$$x^5 + 7x^4 - 2x^3 + x^2 - 3x + 11 = 0$$

of fifth degree should have five (complex numbers as) roots. If these five roots are permuted or interchanged suitably, they still are roots of the same equation. Lagrange and Galois found that the study of the suitable permutations of these roots—more exactly, the "group" of all these permutations —was the key to understanding the process of solving the equation. In this

Symmetries of Polynomial Equations

The equation $x^3 - 2 = 0$ has three roots. To find them, use the auxiliary equation $y^3 = 1$, which has a root $y = 1$ and two complex roots

$$\omega = -1/2 + i\sqrt{-3}/2, \qquad \omega^2 = -1/2 - i\sqrt{-3}/2$$

Then the original equation has the following three roots

$$x = \sqrt[3]{2}. \qquad x = \omega\sqrt[3]{2}, \qquad x = \omega^2\sqrt[3]{2}.$$

As a result the left-hand side of the equation is a product of three factors

$$(x^3 - 2) = (x - \sqrt[3]{2})\,(x - \omega\sqrt[3]{2})\,(x - \omega^2\sqrt[3]{2}).$$

The three roots can be pictured in the plane of complex numbers as three points (equally spaced) on a circle of radius $\sqrt[3]{2}$:

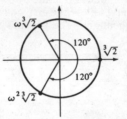

Any operation interchanging (or permuting) these roots leaves the equation fixed. There are six such operations: Here are three:

Rotate 120° counterclockwise: $\sqrt[3]{2} \quad \omega\sqrt[3]{2} \quad \omega^2\sqrt[3]{2}$

Rotate 240° counterclockwise: $\sqrt[3]{2} \quad \omega^2\sqrt[3]{2} \quad \omega\sqrt[3]{2}$

Reflect on the horizontal axis: $\sqrt[3]{2} \quad \omega\sqrt[3]{2} \quad \omega^2\sqrt[3]{2}$.

There are two more reflections (axes at 120°) and one more operation, the identity (change nothing). Any one of these six operations, followed by another, is still one of these six operations. Together, they form the group of symmetries of this equation.

way they succeeded in proving that in general there could not be any formula involving just extraction of roots which would give all the solutions of an equation of 5th degree. Thus they solved a problem which had been of active concern for two hundred years.

At the same time, they recognized that these suitable permutations of the

How to Multiply Permutations

A permutation of *n* things is an operation on these things, sending each one of them to another one (or the same one). For example, reading down, here are four different permutations of three things:

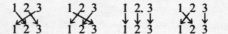

There are two more like the last; the operation interchanging 2 and 3 (and leaving 1 fixed) and the operation interchanging 1 and 3 (and leaving 2 fixed). Hence all told, there are six different permutations of three things. A permutation, like any other operation, can be labelled with a single letter, say *S* or *T*.

To multiply two permutations, simply operate with the first permutation, and then with the second one. For example, on the left below are two permutations *S* and *T*, one above the other, of the four things 1, 2, 3, 4. To multiply them, first follow the arrows of *S* and then those of *T*; the result of this multiplication is the product *ST* shown on the right

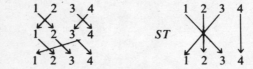

Clearly this product of two permutations *S* and *T* is another permutation; also this multiplication of permutations has all the algebraic properties of ordinary multiplication except that *TS* may not be *ST* (the product depends on the order). Because of these properties, the set of all permutations of four things is said to be a *group*. Similarly, the six different permutations of three things, as listed above, form a group.

roots described the "symmetry" of the equation. The essential aspect of this collection of permutations was that two suitable permutations, one following the other or one following the inverse of the other, was still a suitable permutation (see box above). Such sets of permutations (or other transformations) are called *groups*.

Gradually, in the course of the nineteenth century, the theory of these groups of symmetries or transformations became one of the leading themes of mathematical development. They provided an explicit example of objects which are not numbers or functions but which are still subject to algebraic manipulation. It turned out that groups appeared in many other parts of nineteenth century mathematics—in number theory, in the study of analytic functions of a complex variable, in the foundations of geometry, in the theory of matrices and in the study of ordinary and partial differential equations. Mathematicians examined the possible finite groups, and studied quite abstractly the various way in which a given finite group could be "represented" by groups of matrices. For geometry and to analyze differential equations, they also studied infinite and "continuous" groups and their representations. From this start, the concepts of group theory spread to give a new way of doing algebra and mathematics.

In the twentieth century, the notion of a group has become the fundamental conceptual and formal tool for mathematical descriptions of the physical world. In crystallography, the starting point is the description of all the possible space groups—there are exactly 230 of them. In quantum theory, the basic symmetries of the atoms and molecules lead directly to the applications of group theory; indeed, one of the famous early books on the subject, by Hermann Weyl, was entitled *Group Theory and Quantum Mechanics*. The matrix representation of groups, originally studied just as a topic in algebra, is today extensively used in physics. Some of the best descriptions of the fundamental particles of high energy physics by the "8-fold way" are descriptions in terms of group theory.

At the same time, the purely mathematical study of groups continues unabated. The problem of describing *all* possible finite groups still seems beyond reach, but there are certain natural building blocks, the finite *simple* groups. Many such simple groups are known, but the last dozen years has seen the discovery of 20 or more new such groups. Today group theorists are hot on the trail of one such group which should have

$$2^{46} \cdot 3^{20} \cdot 5^{9} \cdot 7^{6} \cdot 11^{2} \cdot 13^{3} \cdot 17 \cdot 19 \cdot 23 \cdot 29 \cdot 31 \cdot 41 \cdot 47 \cdot 59 \cdot 71 \approx 8 \times 10^{53}$$

elements—which they have called the "monster." Despite the existence of such exotic groups, there is good hope that they will shortly classify all the finite simple groups.

The infinite groups, such as groups involving an infinite number of symmetries, are a different matter. They, too, have been extensively studied. For example, the descriptions of such groups by generators and relations provide one of the most striking examples of a mathematical problem which cannot be solved in a constructible way. In these and other respects group theory provides an outstanding example of the interpenetration of pure and applied mathematics.

Formalism

Mathematics in the past century has begun to study its own methods of reasoning and its own structure as the objects of new mathematical methods and disciplines. In the first decade of the twentieth century, in particular, fundamental attitudes towards mathematics were transformed in the discussion of the foundations of mathematics.

Under the impetus of problems generated by one of the most original innovations in the mathematics of modern times, Cantor's theory of infinite sets, an intense division of opinion developed among leading mathematicians and logicians as to the legitimate basis for justifying the validity of mathematical knowledge. Three major currents of ideology arose and fought for supremacy: intuitionism, logicism, and formalism. Logicism, first formulated by Gottlieb Frege in Germany and popularized by Bertrand Russell, argued for founding mathematics upon the transparencies of logical truth. It turned out that logical truth is not transparent, and this program, which was the favorite of logicians rather than mathematicians, collapsed. Intuitionism, founded and led by L.E.J. Brouwer, took a more constructive tack. The existence of mathematical objects would not be asserted unless they could be explicitly exhibited. In effect, this view denied the validity of most mathematical reasoning about infinite processes and in particular cut out the classical "principle of the excluded middle" used in logic ever since Aristotle. Intuitionism thus tended to be a very restrictive point of view.

The third movement, formalism, endorsed and led by the most prestigious German mathematician of the time, David Hilbert, won a victory by default (though Hilbert's technical program collapsed under Gödel's analysis mentioned above). The formalist doctrine in foundations justified the correctness (not the significance) of mathematical theories by trying to show that they did not lead to contradictions. By its accession to the position of an orthodoxy, it achieved effects that its rather sophisticated originators certainly never dreamed of.

In its most vulgar form (and it is the vulgar forms of sophisticated intellectual doctrines that tend to sweep across the intellectual landscape), the formalist doctrine was taken to say that mathematics consists simply of the formal manipulation of uninterpreted symbols, or of reasoning by formal deduction (itself reduced simply to symbolic manipulation) from any assumptions whatever as long as they could be presented in an explicitly symbolic form. Taken in this form (and it is clear from his explicit statements that Hilbert would have found this form of the thesis horrifying), the vulgar formalist doctrine argued against even the possibility of any objective content for any part of mathematics. It even seemed to argue against the significance and the content of the historically conditioned mathematical fields, as well as against their intuitions and central problems. In the context of applying mathematics to the analysis of the phenomena of the natural world, it makes any significant application a miracle on principle and a triumph of will over content. It is in terms of this vulgar formalist conception of mathematics that Eugene Wigner's well-known phrase "the unreasonable effectiveness of mathematics in the physical sciences" gains its psychological force. Indeed, how could a game with meaningless symbols bear any intrinsic or significant relation to the processes of the physical world?

An important aggravation of this tendency took place as a consequence of one of the major turning points in the history of modern physics, the creation of quantum mechanics and its envelopment of most of physics during the late 1920's and early 1930's. Perhaps the most thoroughly mathematical in form of all physical theories which had developed up to that time, quantum mechanics and its mathematical machinery failed to achieve an easy *rapprochement* with conventional physical intuition and failed to obliterate the classical ideal of physics which it officially replaced. What it did succeed in doing was to destroy the strong influence of *rationalism* as a cornerstone of physical theory. Perhaps the most fruitful in detailed application of any of the major physical theories, quantum mechanics might justifiably have said about it in the words of G.K. Chesterton: "We have seen the truth; and the truth makes no sense."

In the subsequent development of theoretical physics, the disjunction between mathematical formalism and the demands of a confining intuition has become ever greater as the theories have grown more elaborate. While the vulgar formalist attitude towards mathematics has never achieved a totally dominant influence upon mathematical research even when unchallenged as an ideology, the very similar instrumentalist viewpoint of the contemporary theoretical physicist towards mathematics became dominant in physics. This is the domain of basic physical concepts in which, before the quantum mechanical revolution, the most creative interaction took place between the concepts, intuitions, and techniques of mathematics and the physical sciences.

Certainly in the past decade, if not over a longer period, the influence of formalist doctrines in any form has waned in terms of the attitudes of mathematicians carrying on research on a significant level. This has not been accompanied by any great thrust of counter-ideology. Indeed, over the same period, vulgar formalism has been spread on a much more explicit level and to a much wider public than it ever reached before through the formalist thrust of a large part of the new curricula in the elementary and secondary schools.

"Formalism" in this sense has a natural root within the common sense assumptions of our present day culture. It is a trivially natural way to do mathematics if one disbelieves in universals. We find it easy to believe in the reality of physical things and (except for the behaviorists) in the reality of thoughts in people's heads. We find it difficult in principle, except after long conditioning, to believe in the reality of relations. The fact that the most fundamental and firmly accepted parts of our general scientific knowledge of the world involve mathematical relations as complex and sophisticated as those involved in Newtonian mechanics, in the Maxwell electromagnetic theory, in special relativity, or in the operator formalism of quantum mechanics seems to elude us even when we take these doctrines for granted.

Mathematics in its own distinct way is an objective science dealing with certain basic intuitive themes of human experience and practice in constantly elaborated forms by a continually more ingenious apparatus of technical devices and simplifying concepts. Because of its origins and its nature, mathematics is not unreasonably effective in the physical sciences, simply reasonably effective. With a certain amount of ingenuity and insight (amounting perhaps to genius), someone will probably even make it reasonably effective in the biological sciences.

Mathematics is being ever more extensively applied and in a way that fulfills its basic character. Every serious contribution to mathematical thought is fundamentally applicable since it represents, if it is serious, a new insight into relations of a fundamental kind. By an extension of the principle of plentitude, one might even venture the assertion that every significant mathematical relation must have a significant role within the phenomena of the physical, biological, or social worlds. (Of course, this begs the metaphysical question by the use of the adjective *significant*, but not the practical question.)

Indeed, who at the time of the writing of Apollonius' book on conic sections in the third century B.C. would have dreamed that almost two thousand years later, Kepler would have employed Apollonius' results to describe the motion of the planets? Who would have thought in 1820 at the time when Évariste Galois introduced the concept of the group into the study of the solvability properties of algebraic equations in terms of radicals that a century later and beyond, the fundamental principles of physics would be formulated in terms of concepts from group theory? Who (except for Leibniz, of course) might have foreseen that the development of mathematical logic and of proof theory in the first thirty-odd years of the twentieth century would be the basic conceptual tool for the introduction of digital computers and the basis for their great transformation of modern technology?

Why has there been a tendency to estrangement between most fields of scientific research and some of the main currents of mathematical research during the past fifty years? In part, this is due to the influence of formalist currents in mathematics and an instrumentalist view of mathematics in the physical sciences. A role, though somewhat lesser, has been played by the pressures of specialization and the enormous complexities generated in the technical development of the individual fields of science and mathematics. The most damaging effect of this lies in the field of scientific and mathematical education and in the tendency to treat the education of students in any field as a process of indoctrination in the given field to the exclusion of any openness to other types of experience and insight.

Within mathematics itself, we cannot neglect the role in this estrangement which has been played by the emphasis upon the primacy of new fields of mathematical research which have originated from more classical domains of mathematics but not from areas of problems and insights connected with the natural sciences. The most important examples of such fields have been mathematical logic, the theory of numbers and especially algebraic number theory, algebraic topology, and algebraic geometry. Such an emphasis has been strongly buttressed by the role of influential groups of mathematicians pushing this position, particularly by mathematicians associated with the Bourbaki group in France. It is not the case, however, that in their consequences, the development of such disciplines needs to remain irrelevant on a permanent basis to applications in other scientific disciplines. As one cannot emphasize too strongly, examples to the contrary are numerous and of importance.

The Variety of Mathematics

There are many other examples of contact between sophisticated concepts arising in "pure" mathematics and various scientific applications. Let us list here just a few.

(a) In the general theory of relativity, global solutions of Einstein's field equations can be found using concepts from Riemannian geometry "in the large" (that is, globally) and ideas of differential topology. In recent years Roger Penrose and S.W. Hawking in England have employed these techniques to derive fascinating cosmological consequences.

(b) In the nineteenth century the study of certain continuous groups (especially for differential equations as noted above) at the hands of the Norwegian mathematician Sophus Lie led to a corresponding algebraic notion, that of a "Lie algebra." The *simple* Lie algebras (i.e., the basic building blocks of such algebras) were classified completely in the early twentieth century by the French mathematician, Élie Cartan. Subsequently, the development of algebraic topology indicated the importance of algebras where the elements have degrees such as degree 0, 1, 2, . . . ; these are the *graded* algebras. Henri Cartan (Élie's son) was influential in emphasizing the importance of graded objects in algebra. Today it turns out that one can classify the simple graded Lie algebras, but the results differ from those in the ungraded case. This study is used in the description of super-symmetries and super-selection rules in elementary particle physics.

(c) The properties of "differential forms" such as $f(x,y)dx\,dy$ were used extensively in geometry by the same Élie Cartan. They have been recently seen to provide an elegant reformulation of the Maxwell field equations for electrodynamics, and to have other uses in the gauge theories of physics.

(d) The large finite simple groups have been used in algebraic coding theory.

(e) Bifurcation theory is a topic in nonlinear functional analysis which arose originally in celestial mechanics and astrophysics. It has been applied in a great variety of areas, e.g., in the descriptions of flow problems, chemical reactions, and fracture and buckling problems.

(f) Combinatorial analysis—the higher reaches of the calculations of the numbers of combinations of given things under specified conditions—has recently developed in a very sophisticated way, with striking applications to various theoretical problems in crystallography. One example is the problem of finding the closest packing of spheres.

(g) Quantum field theory, which has long been beset with problems of mathematical rigor requiring "renormalization," has recently been redeveloped in one and two space dimensions in a constructive and rigorous form. The development makes extensive use of formal analogies between field theory and statistical mechanics.

(h) A *graph* is a diagram which consists of a finite number of points connected by edges in a specified way. In the early 1930's there was a study of the problem of representing graphs without singularities or crossings in the plane. For example, the graph with five vertices, each pair connected by an edge, cannot be so represented. In the early 1930's Casimir Kuratowski in Poland found exact conditions to insure that a graph could be represented in the plane; his work was further developed in the United States by Hassler Whitney and Saunders Mac Lane. Today this has been much more extensively analyzed because it deals with the practical question of when an electrical circuit can be realized as a printed circuit.

(i) Conversely, many practical problems lead to theoretical developments. For example, Ludwig Prandtl's ideas on the boundary layers which occur in calculating the lift of an airplane wing have contributed a technique which led to a general singular perturbation theory for differential equations.

What kind of conclusions should one draw from examples, like the ones we just cited? We should like to suggest some, which we believe are validated by a broad range of experience.

The potential usefulness of a mathematical concept or technique in helping to advance scientific understanding has very little to do with what one can foresee before that concept or technique has appeared.

Such usefulness has very little to do with the purity or applied character of the motivation underlying the creation of the technique or concept, or with its degree of abstraction.

Concepts or techniques are only useful if they can be eventually put in a form which is simple and relatively easy to use in a variety of contexts.

We don't know what will be useful (or even essential) until we have used it. We can't rely upon the concepts and techniques which have been applied in the past, unless we want to rule out the possibility of significant innovation.

To summarize, mathematics as the science of significant form interacts in an ever widening way with the whole framework of human thought and practice.

It was a thought of this sort that was expressed by the Anglo-American philosopher Alfred North Whitehead when he put forward his own rewriting of Plato's Lecture on the Good in an intellectual credo entitled "Mathematics and the Good" written for the volume dedicated to him in the Library of Living Philosophers. Whitehead wrote: "The notion of the importance of pattern is as old as civilization. Every art is founded on the study of pattern. The cohesion of social systems depends on the maintenance of patterns of behavior, and advances in civilization depend on the fortunate modification of such behavior patterns. Thus the infusion of patterns into natural occurrences and the stability of such patterns, and the modification of such patterns is the necessary condition for the realization of the Good. Mathematics is the most powerful technique for the understanding of pattern, and for the analysis of the relation of patterns. Here we reach the fundamental justification for the topic of Plato's lecture. Having regard to the immensity of its subject matter, mathematics, even modern mathematics, is a science in its babyhood. If civilization continues to advance in the next two thousand years, the overwhelming novelty in human thought will be the dominance of mathematical understanding."

Acknowledgments and Suggestions for Further Reading

Many of the ideas of this paper are drawn from two previous papers:

Browder, Felix E. The relevance of mathematics. *American Mathematical Monthly* **83** (1976) 249–254.

Browder, Felix E. Does pure mathematics have a relation to the sciences? *American Scientist* **64** (1976) 542–549.

The quotations we have used first appeared in the following works:

Chesterton, G. K. *The Innocence of Father Brown*. Penguin Books, New York, 1975.

Schilpp, P. A. (Ed.). *The Philosophy of Alfred North Whitehead*. Tudor Pub, New York, 1941.

Wigner, Eugene P. The unreasonable effectiveness of mathematics in the natural sciences. *Comm. Pure Appl. Math.* **13** (1960) 1–14.

We list below a number of books and papers which are relevant to the themes of this article and which develop some of the mathematical topics we have sketched above in considerably more detail. We have chosen references which we hope are as accessible and non-technical as one can find, considering the inevitable fact that precision and the fruitful use of mathematical concepts demand technical detail for their realization.

Finite Structures and Their Application

Berlekamp, Elwyn R. *Algebraic Coding Theory*. McGraw-Hill, New York, 1968.

Hopcroft, John E. and Ullman, Jeffrey D. *Formal Languages and Their Relations to Automata*. Addison-Wesley, Reading, 1969.

Mac Lane, Saunders and Birkhoff, Garrett. *Algebra*. Macmillan, New York, 1967.

MacWilliams, F. Jessie and Sloane, Neil J. A. *The Theory of Error-Correcting Codes I, II*. North-Holland, Amsterdam, 1977.

Slepian, Paul. *Mathematical Foundations of Network Analysis*. Springer-Verlag, New York, 1968.

Complex Analysis and Its Applications to Theoretical Physics

Eden, R. J., Olive D. I., Landshoff, P. V. and Polkinghorne, J. C. *The Analytic S-Matrix*. Cambridge University Pr, New York, 1966.

Levinson, Norman and Redheffer, Raymond. *Complex Variables*. Holden-Day, San Francisco, 1970.

Differential Equations

Friedman, Avner. *Partial Differential Equations*. Holt, Rinehart and Winston, New York, 1969.

Gavalas, George R. *Nonlinear Differential Equations of Chemical Systems*. Springer Tracts in Natural Philosophy, Vol. 17. Springer-Verlag, New York, 1968.

Hirsch, Morris and Smale, Stephen. *Differential Equations, Dynamical Systems and Linear Algebra*. Academic Pr, New York, 1975.

Treves, Francois. *Basic Linear Partial Differential Equations*. Academic Pr, New York, 1975.

Mathematics Applied to the Physical Sciences

Lipkin, Harry J. *Lie Groups for Pedestrians*. North-Holland, Amsterdam, 1965.

Meyer, Richard E. *Introduction to Mathematical Fluid Dynamics*. Wiley, New York, 1971.

Reed, Michael and Simon, Barry. *Methods of Mathematical Physics, Vol. I, Functional Analysis*. Academic Pr, New York, 1972.

Schwartz, Laurent. *Mathematics for the Physical Sciences*. Hermann, Paris and Addison-Wesley, Reading, 1966.

Other Applications

Hawking, Stephen W. and Ellis, G. F. R. *The Large Scale Structure of Space-Time*. Cambridge University Pr, New York, 1973.

Sinai, Ya. G. *Introduction to Ergodic Theory*. Mathematical Notes. Princeton University Pr, Princeton, 1977.

About the Authors

Jonathan L. Alperin (*Groups and Symmetry*, pp. 65-82) is Professor of Mathematics at the University of Chicago. He received his Ph.D. from Princeton in 1961, two years after graduating from Harvard College. After teaching for a year at M.I.T. he joined the faculty at the University of Chicago in 1963. He has done extensive research in the theory of finite groups, and has held a Guggenheim fellowship.

Kenneth Appel (*The Four Color Problem*, pp. 153-180) is Professor of Mathematics at the University of Illinois. He received his bachelor's degree from Queens College in 1953 and his Ph.D. from the University of Michigan in 1959. After two years on the technical staff of the Institute for Defense Analyses in Princeton, he joined the Illinois faculty in 1961. Although most of his work has involved combinatorial problems in logic and group theory, he has used computers to give insight into combinatorial problems for twenty years. Appel was a member of the city council of Urbana, Illinois from 1971 to 1975.

Felix E. Browder (*The Relevance of Mathematics*, pp. 323-350) is Louis Block Professor of Mathematics at the University of Chicago. He received his Ph.D. from Princeton in 1948, two years after completing undergraduate studies at M.I.T. He has held Guggenheim and Sloan Foundation fellowships, and is a member of the National Academy of Sciences. His research interests are in nonlinear functional analysis and partial differential equations.

Martin Davis (*What is a Computation?*, pp. 241-267) is Professor of Mathematics at the Courant Institute of Mathematical Sciences, New York University, where he teaches both mathematics and computer science.

He received his bachelor's degree from City College in New York in 1948 and his Ph.D. from Princeton University in 1950. His research interests have been in mathematical logic and theoretical computer science. In 1975 Davis was awarded three prizes by the Mathematical Association of America and the American Mathematical Society for articles on the unsolvability of Hilbert's tenth problem.·

Ronald L. Graham (*Combinatorial Scheduling Theory*, pp. 183–211) is Head of the Discrete Mathematics Department of Bell Laboratories in Murray Hill, New Jersey. He received his B.S. degree (in physics) from the University of Alaska in 1959, and his Ph.D. (in number theory) from the University of California at Berkeley in 1962. He has been at Bell Labs since 1962 except for 1975 when he was Regents Professor of Mathematics at UCLA. In 1972 he was co-recipient of the Pólya Prize in combinatorics awarded by the Society for Industrial and Applied Mathematics. Graham is also a Past-President of the International Jugglers Association.

Frank C. Hoppensteadt (*Mathematical Aspects of Population Biology*, pp. 297–320) is currently Professor of Mathematics at the Courant Institute of Mathematical Sciences, New York University, and the University of Utah. He received his Ph.D. in mathematics from the University of Wisconsin. His research and teaching are devoted to problems in mathematical biology, and in perturbation and computer methods in applied mathematics.

Wolfgang Haken (*The Four Color Problem*, pp. 153–180) is Professor of Mathematics at the University of Illinois. He received his doctor's degree in 1953 in Kiel, Germany and worked from 1954 to 1962 as a development engineer at Siemens in Munich. He then joined the Illinois faculty in 1965, following two years as a temporary member of the Institute for Advanced Study in Princeton, New Jersey. Haken's research has been in topology; he has devoted more than a decade to research on the still-unresolved Poincaré conjecture on 3-dimensional manifolds.

Allen L. Hammond (*Mathematics—Our Invisible Culture*, pp. 15–34) is Research News Editor of *Science*, the journal of the American Association for the Advancement of Science. He graduated from Stanford and in 1970 took his Ph.D. in applied mathematics at Harvard, where he worked on geophysical problems. Since then he has been a writer and editor for *Science*, where he originated and directs the Research News section. Hammond is the principal author of a 1973 book on energy technologies, *Energy and the Future*, and his writing has appeared in *Harpers*, the *New York Times*, and other publications.

Saunders Mac Lane (*The Relevance of Mathematics*, pp. 323–350) is Max Mason Distinguished Service Professor of Mathematics at the University of Chicago and Vice-President of the National Academy of Sciences.

He received his D.Phil. from the University of Göttingen, Germany, in 1934. He has taught at Harvard, Cornell and the University of Chicago, and has done research in algebra, in algebraic topology, and in the foundations and conceptual organization of mathematics. He has served as President of the Mathematical Association of America and of the American Mathematical Society.

David S. Moore (*Statistical Analysis of Experimental Data*, pp. 213–239) is Professor of Statistics at Purdue University. He graduated from Princeton University and received his Ph.D. from Cornell University in 1967, at which time he joined the faculty at Purdue. His research interests include the behavior of statistical procedures in large samples and tests of the goodness-of-fit of data to assumed models. He has a particular concern for the teaching of statistics.

Roger Penrose (*The Geometry of the Universe*, pp. 83–125) is Rouse Ball Professor of Mathematics at Oxford University and Fellow of Wadham College. He obtained his B.Sc. degree from London University in 1952 and his Ph.D. from Cambridge University in 1957. He was a Research Fellow at St. John's College, Cambridge during 1957–1960, then held various temporary posts in the U.S. and U.K. and taught at Birkbeck College, London between 1964 and 1973. He is a Fellow of the Royal Society. His research interests range from linear algebra, geometry and topology to general relativity and particle physics.

Ian Richards (*Number Theory*, pp. 37–64) is Associate Professor in the School of Mathematics at the University of Minnesota, from where he graduated in 1957. He received his Ph.D. from Harvard University in 1960 in complex analysis. Recently his research interests have shifted to number theory, largely because of his curiosity about prime numbers and the Riemann hypothesis. An expert canoeist, Richards has appeared on television as a canoe instructor, although never as a mathematician. Richards wishes to thank his friend Myran Lutter, whose criticism greatly improved the article.

Jacob T. Schwartz (*Mathematics as a Tool for Economic Understanding*, pp. 269–295) is Chairman of the Department of Computer Science at the Courant Institute of Mathematical Sciences, New York University. He has worked in a variety of mathematical fields, including functional analysis, mathematical economics, and programming language design, and is the author of books and research papers in these areas. He received a B.S. degree from City College in New York, and the Ph.D. degree from Yale. He is a member of the National Academy of Sciences.

Lynn Arthur Steen (*Mathematics Today*, pp. 1–12) is Professor of Mathematics at St. Olaf College in Northfield, Minnesota. He received his

B.A. from Luther College in Decorah, Iowa in 1961 and his Ph.D. from M.I.T. in 1965. He twice received the Lester R. Ford award for mathematical exposition from the Mathematical Association of America. His writing on mathematics has appeared in *Scientific American*, *Science*, and *Science News*, as well as in various mathematical publications. Steen is currently editor of *Mathematics Magazine* and contributing editor for mathematics for *Science News*.

Philip D. Thompson (*The Mathematics of Meteorology*, pp. 127–152), is Senior Scientist at the National Center for Atmospheric Research. A theoretical meteorologist, Thompson has worked in geophysical fluid dynamics, theory of turbulence and the application of high-speed computing techniques to the problem of weather prediction. He received a B.S. degree from the University of Chicago in 1943, and an Sc.D. degree from M.I.T. in 1953. Thompson was one of the original members of von Neumann's Meteorology Project at the Institute for Advanced Study.

Further Reading

General Surveys

Aleksandrov, A. D., Kolmogrorov, A. N., and Lavrent'ev, M. A. (Eds.). *Mathematics—Its Content, Methods and Meaning*, 3 Vols. MIT Pr, Cambridge, 1969.

Behnke, H., et al. (Eds.). *Fundamentals of Mathematics*, 3 Vols. MIT Pr, Cambridge, 1974.

Boehme, George A. W. *The New World of Mathematics*, Dial Pr, New York, 1959.

Courant, R. and Robbins, H. *What is Mathematics?* Oxford University Pr, New York, 1941.

Dedron, P. and Itard, J. *Mathematics and Mathematicians*. Transworld Pub, London, 1973.

Dieudonné, Jean A. Mathematics. *Collier's Encyclopedia* 15 (1976) 541–552.

Gaffney, Matthew P. and Steen, Lynn Arthur, *Annotated Bibliography of Expository Writing in the Mathematical Sciences*. Math. Assoc. of America, Washington, 1976.

Khurgin, Ya. *Did You Say Mathematics?* MIR, Moscow, 1974.

Kline, Morris. *Mathematics in Western Culture*. Oxford University Pr, New York, 1964.

Linn, Charles F. (Ed.). *The Ages of Mathematics*, Vol. I–IV. Doubleday, New York, 1977.

Mathematics in the Modern World. Freeman, San Francisco, 1968.

National Research Committee on Support of Research in the Mathematical Sciences (COSRIMS) (Ed.). *The Mathematical Sciences. A Collection of Essays*. MIT Pr, Cambridge, 1969.

Newman, James R. *The World of Mathematics*, 4 Vols. Simon and Schuster, New York, 1959–1960.

Pólya, George. *Mathematics and Plausible Reasoning*, Vol. I and II. Princeton University Princeton. Vol. I, 1954; Vol. II, rev. ed., 1969.

Pólya, George. *Mathematical Discovery*, 2 Vols. Wiley, New York, 1962.

Rademacher, Hans and Toeplitz, Otto. *The Enjoyment of Mathematics*. Princeton University Pr, Priceton, 1957.

Saaty, Thomas L. (Ed.). *Lectures on Modern Mathematics*, 3 Vols. Wiley, New York, 1963–1965.

Saaty, Thomas L. and Weyl, F. Joachim (Eds.). *The Uses and Spirit of the Mathematical Sciences*. McGraw-Hill, New York, 1969.

Sawyer, W. W. *Introducing Mathematics*, 4 Vols. Penguin Books, London, 1964–1970.

Steinhaus, H. *Mathematical Snapshots*, 2nd ed. Oxford University Pr, New York, 1969.

Stewart, Ian. *Concepts of Modern Mathematics*. Penguin Pr, New York, 1975.

Whitehead, Alfred North. *An Introduction to Mathematics*. Oxford University Pr, New York, 1958.

Foundations of Mathematics

Baum, Robert J. *Philosophy and Mathematics: From Plato to the Present*. Freeman, San Francisco, 1974.

Lakatos, Imre. *Proofs and Refutations: The Logic of Mathematical Discovery*. Cambridge University Pr, New York, 1976.

Nagel, E. and Newman, J. R. *Gödel's Proof*. New York University Pr, New York, 1958.

Wang, Hao. *From Mathematics to Philosophy*. Humanities Pr, New York, 1973.

Wilder, R. L. *Introduction to the Foundations of Mathematics*. Wiley, New York, 1965.

Philosophy of Mathematics and Its Relation to Science

Benacerraf, P. and Putnam, H. (Eds.). *Philosophy of Mathematics: Selected Readings*. Prentice-Hall, Englewood Cliffs, 1964.

Poincaré, H. *The Foundations of Science*. Science Pr, Lancaster, 1946.

Weyl, H. *Philosophy of Mathematics and Natural Science*. Princeton University Pr, Princeton, 1949.

Whitehead, A. N. *Science and the Modern World*. Macmillan, New York, 1925.

History of Mathematics

Bourbaki, N. *Éléments d'Histoire des Mathématiques*, 2nd ed. Hermann, Paris, 1969.

Boyer, Carl B. *A History of Mathematics*. Wiley, New York, 1968.

Dantzig, Tobias. *Number, the Language of Science*, 4th ed. Free Pr, New York, 1967.

Eves, Howard. *An Introduction to the History of Mathematics*, 4th ed. Holt, Rinehart and Winston, New York, 1976.

Kline, Morris. *Mathematical Thought from Ancient to Modern Times.* Oxford University Pr, New York, 1972.

Kramer, Edna E. *The Nature and Growth of Modern Mathematics*, 2 Vols. Fawcett World Library, New York, 1974.

LeVeque, William J., et al. History of mathematics. *Encyclopaedia Britannica*, 15th Ed. 11 (1974) 639–670.

Struik, D. J. *A Concise History of Mathematics.* Dover, New York, 1967.

van der Waerden, Bartel L. *Science Awakening.* Oxford University Pr, New York, 1961.

Zaslavsky, Claudia. *Africa Counts: Number and Pattern in African Culture.* Prindle, Weber & Schmidt, Boston, 1973.

Biography

Bell, Eric T. *Men of Mathematics.* Simon and Schuster, New York, 1937.

Grattan-Guinness, Ivor. *Joseph Fourier*, 1768–1830. MIT Pr, Cambridge, 1972.

Hardy, Godfrey H. *A Mathematician's Apology*, rev. ed. Cambridge University Pr, New York, 1969.

Hardy, Godfrey H. *Ramanujan.* Chelsea, New York, 1968

Hoffman, Banesh. *Albert Einstein, Creator and Rebel.* Viking Pr, New York, 1972.

Infeld, Leopold. *Whom the Gods Love.* Whittlesey House, 1948.

Ore, Oystein. *Niels Henrik Abel: Mathematician Extraordinary.* Chelsea, New York, 1974.

Reid, Constance. *Courant.* Springer-Verlag, New York, 1976.

Reid, Constance. *Hilbert.* Springer-Verlag, New York, 1970.

Wierner, Norbert. *I Am A Mathematician.* MIT Pr, Cambridge, 1964.

Index

VINTAGE WORKS OF SCIENCE AND PSYCHOLOGY

V-635	**HEILBRONER, ROBERT L.** / Between Capitalism and Socialism
V-283	**HENRY, JULES** / Culture Against Man
V-882	**HENRY, JULES** / Pathways to Madness
V-465	**HINTON, WILLIAM** / Fanshen
V-95	**HOFSTADTER, RICHARD** / The Age of Reform: From Bryan to F. D. R.
V-795	**HOFSTADTER, RICHARD** / America at 1750: A Social Portrait
V-9	**HOFSTADTER, RICHARD** / The American Political Tradition
V-317	**HOFSTADTER, RICHARD** / Anti-Intellectualism in American Life
V-686	**HOFSTADTER, RICHARD AND MICHAEL WALLACE (eds.)** / American Violence
V-540	**HOFSTADTER, RICHARD AND CLARENCE VER STEEG** / Great Issues in American History, From Settlement to Revolution, 1584-1776
V-541	**HOFSTADTER, RICHARD (ed.)** / Great Issues in American History, From the Revolution to the Civil War, 1765-1865
V-542	**HOFSTADTER, RICHARD (ed.)** / Great Issues in American History, From Reconstruction to the Present Day, 1864-1964
V-385	**HOFSTADTER, RICHARD** / The Paranoid Style in American Politics and Other Essays
V-591	**HOFSTADTER, RICHARD** / The Progressive Historians
V-201	**HUGHES, H. STUART** / Consciousness and Society
V-241	**JACOBS, JANE** / Death & Life of Great American Cities
V-584	**JACOBS, JANE** / The Economy of Cities
V-433	**JACOBS, PAUL** / Prelude to Riot
V-459	**JACOBS, RAUL AND SAUL LANDAU WITH EVE PELL** / To Serve the Devil: Natives & Slaves Vol. I
V-460	**JACOBS, RAUL AND SAUL LANDAU WITH EVE PELL** / To Serve the Devil: Colonial & Sojourner Vol. II
V-2017	**JUDSON, HORACE FREELAND** / Heroin Addiction: What Americans Can Learn from the English Experience
V-790	**KAPLAN, CAROL AND LAWRENCE (eds.)** / Revolutions: A Comparative Study
V-361	**KOMAROVSKY, MIRRA** / Blue-Collar Marriage
V-675	**KOVEL, JOEL** / White Racism
V-367	**LASCH, CHRISTOPHER** / The New Radicalism in America
V-560	**LASCH, CHRISTOPHER** / The Agony of the American Left
V-280	**LEWIS, OSCAR** / The Children of Sanchez
V-421	**LEWIS, OSCAR** / La Vida
V-634	**LEWIS, OSCAR** / A Death in the Sanchez Family
V-533	**LOCKWOOK, LEE** / Castro's Cuba, Cuba's Fidel
V-406	**MARCUS, STEVEN** / Engels, Manchester, and the Working Class
V-480	**MARCUSE, HERBERT** / Soviet Marxism
V-2001	**MARX, KARL AND NICOLOUS, MARTIN (trans.)** / The Grundrisse: Foundations of the Critque of Political Economy
V-2002	**MARX, KARL AND FERNBACH, DAVID (ed.)** / Political Writings, Vol. I: The Revolution of 1848
V-2003	**MARX, KARL AND FERNBACH, DAVID (ed.)** / Political Writings, Vol. II: Surveys from Exile
V-2004	**MARX, KARL AND FERNBACH, DAVID (ed.)** / Political Writings, Vol. III: The First International and After
V-619	**McCONNELL, GRANT** / Private Power and American Democracy
V-386	**McPHERSON, JAMES** / The Negro's Civil War
V-928	**MEDVEDEV, ROY A.** / Let History Judge: The Origins and Consequences of Stalinism
V-816	**MEDVEDEV, ZHORES AND ROY** / A Question of Madness
V-112	**MEDVEDEV, ZHORES** / Ten Years After Ivan Denisovich
V-614	**MERMELSTEIN, DAVID (ed.)** / The Economic Crisis Reader
V-971	**MILTON, DAVID AND NANCY, SCHURMANN, FRANZ (eds.)** / The China Reader IV: Peoples China
V-93	**MITFORD, JESSICA** / Kind and Usual Punishment: The Prison Business
V-539	**MORGAN, ROBIN (ed.)** / Sisterhood is Powerful